LINEAR SECOND ORDER
ELLIPTIC OPERATORS

LINEAR SECOND ORDER
ELLIPTIC OPERATORS

Julián López-Gómez

Complutense University of Madrid, Spain

NEW JERSEY • LONDON • SINGAPORE • BEIJING • SHANGHAI • HONG KONG • TAIPEI • CHENNAI

Published by

World Scientific Publishing Co. Pte. Ltd.
5 Toh Tuck Link, Singapore 596224
USA office: 27 Warren Street, Suite 401-402, Hackensack, NJ 07601
UK office: 57 Shelton Street, Covent Garden, London WC2H 9HE

Library of Congress Cataloging-in-Publication Data
López-Gómez, Julián
    Linear second order elliptic operators / by Julián López-Gómez, Complutense University of
Madrid, Spain.
    pages cm
    Includes bibliographical references and index.
    ISBN 978-981-4440-24-0 (hardcover : alk. paper)
    1. Elliptic operators. I. Title.
    QA329.L67 2013
    515'.7242--dc23

                                                                    2013001024

British Library Cataloguing-in-Publication Data
A catalogue record for this book is available from the British Library.

Printed in Singapore

Julián, you know...

no hard work, no glory...

and trust me...

if you want to be modern, you should be classical...

with all my love.

Preface

The main goal of this book is to provide a reasonably self-contained proof of a scalar version of the *characterization of the strong maximum principle* established by Theorem 2.1 of J. López-Gómez and M. Molina-Meyer [148], which was extended by H. Amann and J. López-Gómez [13] to a generalized class of *non-classical* mixed boundary conditions. Precisely, it will be shown that, for any second order linear elliptic operator

$$\mathfrak{L} := -\mathrm{div}\,(A\nabla\cdot) + \langle b, \cdot \rangle + c$$

in a regular bounded domain Ω, and any boundary operator $\mathfrak{B}$ of non-classical mixed type on $\partial\Omega$, as discussed in the beginning of Chapter 4, the following five conditions are equivalent:

i) $(\mathfrak{L}, \mathfrak{B}, \Omega)$ satisfies the *strong maximum principle*, in the sense that

$$\begin{cases} \mathfrak{L}u = f & \text{in } \Omega \\ \mathfrak{B}u = g & \text{on } \partial\Omega \end{cases} \quad \text{and} \quad (f, g) > (0, 0) \quad \Longrightarrow \quad u \gg 0.$$

By $h > 0$ it is meant that $h \geq 0$ with $h \neq 0$, and we write $u \gg 0$ if $u(x) > 0$ for all $x \in \Omega$ and

$$\frac{\partial u}{\partial n}(x) < 0 \quad \text{for all } x \in u^{-1}(0) \cap \partial\Omega,$$

where n stands for the outward unit normal to Ω at $x \in \partial\Omega$.

ii) $(\mathfrak{L}, \mathfrak{B}, \Omega)$ satisfies the *maximum principle*, in the sense that

$$\begin{cases} \mathfrak{L}u = f & \text{in } \Omega \\ \mathfrak{B}u = g & \text{on } \partial\Omega \end{cases} \quad \text{and} \quad (f, g) \geq (0, 0) \quad \Longrightarrow \quad u \geq 0.$$

iii) $(\mathfrak{L}, \mathfrak{B}, \Omega)$ admits a *positive strict supersolution* $h \in W^{2,p}(\Omega)$, $p > N$, i.e.,

$$(\mathfrak{L}h, \mathfrak{B}h) > (0, 0) \quad \text{in } \Omega \times \partial\Omega.$$

iv) The *principal eigenvalue* of the linear eigenvalue problem

$$\begin{cases} \mathfrak{L}\varphi = \tau\varphi & \text{in } \Omega \\ \mathfrak{B}\varphi = 0 & \text{on } \partial\Omega \end{cases} \tag{0.1}$$

is positive. It will be throughout denoted by σ_0.

v) The *resolvent operator* of the linear problem

$$\begin{cases} \mathfrak{L}u = f \in L^p(\Omega) & \text{in } \Omega \\ \mathfrak{B}u = 0 & \text{on } \partial\Omega \end{cases}$$

is *strongly positive*.

This characterization theorem, which is Theorem 7.10 of Chapter 7, has tremendously facilitated the development of the modern theory of *spatially heterogeneous* reaction diffusion systems. In particular, it has enhanced the mathematical analysis of a great variety of nonlinear parabolic equations whose dynamics are regulated by *metasolutions* (see R. Gómez-Reñasco [85], R. Gómez-Reñasco and J. López-Gómez [86], J. López-Gómez [145], J. López-Gómez and M. Molina-Meyer [151], and the list of references in each of these monographs).

As a consequence from Theorem 1.1 of H. Berestycki, L. Nirenberg and S. R. S. Varadhan [27], the equivalence of ii) and iv) holds true under no regularity constraints on Ω when $\mathfrak{B}u = u$ for all u. Also, by Corollary 2.4 of [27], iii) implies ii) under Dirichlet boundary conditions. But, seemingly, the lack of regularity of Ω did not allow H. Berestycki, L. Nirenberg and S. R. S. Varadhan [27] to obtain the complete equivalence of J. López-Gómez and M. Molina-Meyer [148]. Both papers, [27] and [148], appeared simultaneously, early 1994, and both were submitted for publication in 1992.

More recently, H. Amann [11] completed the previous list of five items by proving that each of them is also equivalent to each of the next two:

vi) $(\mathfrak{L}, \mathfrak{B}, \Omega)$ satisfies the *weak maximum principle*.
vii) $(\mathfrak{L}, \mathfrak{B}, \Omega)$ satisfies the *very weak maximum principle*.

But these natural extensions and refinements are left outside the general scope of this book, because, instead of encyclopedic generality, we are paying special attention to simplicity and transparency in its exposition so that it can be comfortably read by graduate and even undergraduate students.

Our proof of the previous characterization theorem is based on the classical minimum principles of E. Hopf [103, 104], M. H. Protter and H. F. Weinberger [183] and J. M. Bony [28]. Consequently, though Theorem 7.10

is relatively recent, this is a rather classical textbook on linear second order elliptic operators.

As this book focuses on linear operators, some important *nonlinear* applications of the classical minimum principles, like the *moving plane method*, have not been included here. The moving plane method has shown to be a milestone in analyzing the symmetry properties of the positive solutions in a variety of *spatially homogeneous* nonlinear elliptic problems. Though it was introduced by A. D. Alexandroff [5] and later refined by J. Serrin [201], it became famous when B. Gidas, W. M. Ni and L. Nirenberg [78] used it to prove the radial symmetry of the positive solutions of a special class of spatially homogeneous semilinear elliptic problems in balls. Naturally, the moving plane method has been covered by most of the existing literature on nonlinear elliptic equations (see, e.g., T. Suzuki [216], P. Pucci and J. Serrin [184], and the references therein).

Although this book might have shared title with the recent monograph of P. Pucci and J. Serrin [184], both texts are really complementary, as there is almost no overlapping between them. Essentially, [184] generalizes the minimum principle of E. Hopf to a variety of quasilinear elliptic problems, whereas this monograph characterizes the range of applicability of the generalized minimum principle of M. H. Protter and H. F. Weinberger [183], and uses it to characterize whether or not the resolvent of a linear elliptic operator is positive. As a consequence from this characterization, it becomes apparent that the generalized minimum principle of [183] holds if, and only if, $\sigma_0 > 0$. As this result seems to be new, some of the contents of P. Pucci and J. Serrin [184] might be substantially generalized in its light.

Anyway, in order to illustrate the degree of independence between the book of P. Pucci and J. Serrin [184] and the present one, the reader should be aware that some of the most fundamental references pioneering the present monograph, like J. López-Gómez and M. Molina-Meyer [148], H. Amann and J. López-Gómez [13], and the elegant paper of H. Berestycki, L. Nirenberg and S. R. S. Varadhan [27], were not incorporated to the bibliography of [184].

This book consists of nine chapters. The first seven provide us with a self-contained proof of the characterization theorem that we have just established above, which can be delivered at an undergraduate level, whereas the last two chapters apply it to obtain some of the most relevant properties of σ_0 from the point of view of the applications.

Essentially, Chapter 1 studies the classical minimum principles of E. Hopf [103, 104], and M. H. Protter and H. F. Weinberger [183], and Chapter 2 analyzes the more recent theorems of classification of supersolutions of W. Walter [224] and J. López-Gómez [144]. As a result from these classification theorems, the generalized minimum principle of M. H. Protter and H. F. Weinberger [183] has been completed and substantially polished. We decided to begin this book by studying these classical topics, as they can be easily delivered at an undergraduate level with no previous knowledge of generalized solutions, Hilbert space techniques, or elliptic regularity theory. Indeed, the first two chapters can be taught as part of an elementary course on classical minimum principles within the fruitful spirit of the paradigmatic monograph of M. H. Protter and H. F. Weinberger [183].

Essentially, the next three chapters constitute an informal advanced course on Hilbert space techniques for linear second order elliptic operators, which may be taught independently of the more advanced materials treated in the last four chapters. Actually, we have included these chapters to make this book easily accessible for undergraduate students. Though most of the contents of Chapters 3–5 are classical and can be found in a number of well known textbooks, the reader should appreciate the generality of our treatment, as we are not focusing our attention exclusively on the Dirichlet problem, as it occurs in most of the existing textbooks (see, e.g., L. C. Evans [60]), but on a rather general class of boundary value problems.

More precisely, Chapter 3 studies the theorem of G. Stampacchia [212] and derives from it the celebrated theorem of P. D. Lax and A. N. Milgram [122], much within the spirit of Chapter 6 of H. Brézis [29]. Besides polishing some of the contents of Chapter 6 of the textbook of H. Brézis [29], this book adds the construction of the projections on any closed convex set of a uniformly convex Banach space, which might be a new result published here for the first time.

Chapter 4 discusses the concept of weak solution for the problem

$$\begin{cases} (\mathfrak{L} + \omega)u = f & \text{in } \Omega \\ \mathfrak{B}u = 0 & \text{on } \partial\Omega \end{cases} \tag{0.2}$$

and shows the existence of $\omega_0 \in \mathbb{R}$ such that, for every $\omega > \omega_0$ and $f \in L^2(\Omega)$, (0.2) has a unique weak solution. The proof of this result is based on the theorem of P. D. Lax and A. N. Milgram [122] for classic mixed boundary conditions, but it goes back to J. López-Gómez [147] under the general non-classical boundary conditions of this book. Astonishingly, the existence of weak solutions in the general context of [147] is based on a device introduced by M. H. Protter and H. F. Weinberger [183] to derive their generalized minimum principle from the minimum principle of E. Hopf [103].

Naturally, in order to establish the existence and the uniqueness of the weak solution of (0.2) one must introduce some fundamentals on Sobolev spaces and trace theorems. This is accomplished in the first part of Chapter 4. However, the main theorems about Sobolev spaces are not proven in this textbook, because here we are not focusing our attention on the generalized Lebesgue spaces used in this book, but on the pivotal problem of characterizing the positivity of the resolvent operator of (0.2). Actually, most of these proofs are well documented in some recent undergraduate textbooks, as, e.g., L. C. Evans [60], where the interested readers are sent to study the proofs, if necessary.

Chapter 5 gives a first glance on the problem of the regularity of the weak solutions. Essentially, it is a short introduction to this topic through the L^p-theory of A. P. Calderón and A. Zygmund [36, 37], the ulterior refinements of S. Agmon, A. Douglis and L. Nirenberg [4], and the *method of continuity*. Obviously, the contents of this chapter might have been expanded to generate an extremely specialized monograph on L^p-*regularity theory*. But this is far from being our attempt in an introductory chapter like Chapter 5, where we restrict ourselves to collect, without proofs, all the regularity results used in the last four chapters of this book.

The materials of the last four chapters, the core of this book together with Chapters 1 and 2, have been incorporated to a textbook for the first time here with the generality adopted in this monograph.

Chapter 6 gives a self-contained proof of the theorem of M. G. Krein and M. A. Rutman [120]. Basically, it polishes and completes the short elementary proof of P. Takác [218].

Chapter 7 begins by establishing the generalized minimum principle of J. M. Bony [28], as well as the weak counterparts of the classical minimum principles already revisited in Chapters 1 and 2. The materials of Chapters 3–5 allow us to generalize all the classical results of Chapters 1 and 2 to the weak context of the Sobolev spaces, providing them with its most powerful versatility.

Alternatively, if we had wished to diffuse our own contributions to the theory through a more specialized monograph, instead of writing down a textbook appropriate for undergraduate students, we should have not included in this book most of the materials of Chapters 3–5, but simply completed from the very beginning the first two chapters with the minimum principle of J. M. Bony [28] and invoked to [147] for the existence and uniqueness of weak solutions.

Then, Chapter 7 combines the weak minimum principles with the theorem of M. G. Krein and M. A. Rutman [120] to establish the existence, the uniqueness and the dominance of the principal eigenvalue σ_0, as well as the scalar version of Theorem 2.1 of J. López-Gómez and M. Molina-Meyer [148] and Theorem 2.4 of H. Amann and J. López-Gómez [13].

The last two chapters give a series of fundamental applications of the main theorem of Chapter 7, which are imperative to study the dynamics of wide classes of *spatially heterogeneous* reaction-diffusion equations and systems. As a first application, Chapter 7 concludes by sharpening the classical minimum principles of E. Hopf and M. H. Protter and H. F. Weinberger through the characterization of its precise range of applicability. Naturally, these refinements might enjoy a huge number of linear and nonlinear applications.

Precisely, Chapter 8 studies some of the most fundamental properties of σ_0. Among them, it establishes the monotonicity and continuity properties of σ_0 with respect to c, Ω, $|\Omega|$, and the boundary operator $\mathfrak{B}$. Finally, Chapter 9 characterizes the existence of principal eigenvalues for the weighted linear boundary value problem

$$\begin{cases} \mathfrak{L}\varphi = \tau W\varphi & \text{in } \Omega \\ \mathfrak{B}\varphi = 0 & \text{on } \partial\Omega \end{cases} \tag{0.3}$$

where $W \neq 0$ is a bounded and measurable function. Our results are extremely sharp refinements of some pioneering results by R. Courant and D. Hilbert [44], when W is bounded away from zero, and P. Hess and T. Kato [97], when W changes sign. Most of the contents of the last two chapters go back to J. López-Gómez [137] and S. Cano-Casanova and J. López-Gómez [39].

This book is indebted to the fruitful scientific collaboration of the author with Professors S. Cano-Casanova and M. Molina-Meyer, members of the author's research teams since the beginning of their scientific carriers, as it is triggered by many of their own mathematical findings. Actually, the Master thesis of Professor S. Cano-Casanova [38] really was a first attempt to detail a self-contained proof of Theorem 2.1 of J. López-Gómez and M. Molina-Meyer [148] under general boundary conditions.

Overall, the whole process of elaboration of this book has taken over ten years. In the mean time, the author's research teams have been supported by a number of projects from the Spanish Government. Among them, we should mention the following:

- BFM2000-0797, of the Ministry of Science and Technology.

- REN2003-00707, of the *National Plan of Natural Resources* (Ministry of Science and Technology).
- CGL2006–00524/BOS, of the *National Plan of Global Change and Biodiversity* (Ministry of Science and Innovation).
- MAT2009-08259, of the *National Plan of Mathematics* (Ministry of Science and Innovation).
- MTM2012-30669, of the *National Plan of Mathematics* (Ministry of Economy and Competitiveness).

Part of the research necessary to prepare this monograph has been tremendously facilitated by them, though, rather astonishingly, only MAT2009-08259 deserved the assignation of students in formation, which might be explained by the high degree of corruption of the Spanish Administration.

Madrid, December 31st, 2012

J. López-Gómez

Contents

Chapter 1

The minimum principle

This chapter considers second order differential operators of the form

$$\mathfrak{L} := -\sum_{i,j=1}^{N} a_{ij}(x)\frac{\partial^2}{\partial x_i \partial x_j} + \sum_{j=1}^{N} b_j(x)\frac{\partial}{\partial x_j} + c(x), \quad x \in \Omega, \qquad (1.1)$$

where $N \in \mathbb{N}$, $N \geq 1$, Ω is a domain (open and connected by arcs) of $\mathbb{R}^N$, and

$$a_{ij}, b_j, c \in L^{\infty}_{\text{loc}}(\Omega), \qquad i, j \in \{1, ..., N\}, \qquad (1.2)$$

i.e., all coefficients of $\mathfrak{L}$ are measurable functions, as discussed by Lebesgue, bounded on compact subsets of Ω. Without loss of generality, we can assume that

$$a_{ji} = a_{ij}, \qquad i, j \in \{1, ..., N\}.$$

Then, the matrix of the coefficients

$$A_x := \big(a_{ij}(x)\big)_{1 \leq i,j \leq N}, \qquad x \in \Omega, \qquad (1.3)$$

is symmetric. Subsequently, the operator

$$\mathfrak{L}_p := -\sum_{i,j=1}^{N} a_{ij}(x)\frac{\partial^2}{\partial x_i \partial x_j}$$

is referred to as the *principal part of* $\mathfrak{L}$, and a_{ij}, $1 \leq i, j \leq N$, are said to be the *principal coefficients* of $\mathfrak{L}$.

Throughout this book, := means equality by definition and I stands for the identity map of the underlying linear space (in principle, anyone). Also, for a given measurable function $f : \Omega \to \mathbb{R}$, it is said that $f \geq 0$ if $f(x) \geq 0$ almost everywhere in Ω, and, given another measurable function $g : \Omega \to \mathbb{R}$, it is said that $f \geq g$ if $f - g \geq 0$. Most precisely, it is said

1

that $f > 0$ when $f \geq 0$ and $f > 0$ on a set of positive measure, and we
write $f > g$ if $f - g > 0$. Similarly, for every $f, g \in \mathcal{C}(\partial\Omega)$, it is said that
$f \geq 0$ (resp. $f > 0$) when $f(x) \geq 0$ for all $x \in \partial\Omega$ (resp. $f \geq 0$ and there
exists $x \in \partial\Omega$ such that $f(x) > 0$), whereas we write $f \geq g$ (resp. $f > g$) if
$f - g \geq 0$ (resp. $f - g > 0$).

Moreover, given two arbitrary Banach spaces U and V, we denote by
$\mathcal{L}(U, V)$ the Banach space of the linear and continuous operators between
U and V, by $\text{Iso}(U, V)$ the set of isomorphisms between U and V, and by
U' the topological dual space of U, i.e.,

$$U' := \mathcal{L}(U, \mathbb{R}).$$

Notations are shortened by setting

$$\mathcal{L}(U) := \mathcal{L}(U, U), \qquad \text{Iso}(U) := \text{Iso}(U, U).$$

Further, $\mathcal{M}_N(\mathbb{K})$ will stand for the linear space of matrices of order $N \geq 1$
with entries in the numerical field $\mathbb{K} \in \{\mathbb{R}, \mathbb{C}\}$, i.e.,

$$\mathcal{M}_N(\mathbb{K}) = \mathcal{L}(\mathbb{K}^N),$$

and, for every $L \in \mathcal{L}(U, V)$, $N[L]$ and $R[L]$ will stand for the kernel (null
space) and the image (range) of L, respectively.

This chapter introduces the concept of ellipticity for $\mathfrak{L}$ and gives a series
of sufficient conditions so that it satisfies the classical *minimum principle*
in the domain Ω.

Besides $L^\infty_{\text{loc}}(\Omega)$, in this book we consider the Banach space $L^\infty(\Omega)$ of all
measurable and bounded functions in Ω, and, for every integer $k \geq 0$, the
Banach space $\mathcal{C}^k(\bar{\Omega})$ of all real functions in $\bar{\Omega}$ with continuous derivatives
of order k in $\bar{\Omega}$. By simplicity, we write $\mathcal{C}(\bar{\Omega}) := \mathcal{C}^0(\bar{\Omega})$. Similarly, we also
consider

$$\mathcal{C}^k(\Omega) := \bigcap_{\substack{D \text{ open} \\ \bar{D} \subset \Omega}} \mathcal{C}^k(\bar{D}).$$

1.1   Concept of ellipticity. First consequences

Throughout this book, $|\cdot|$ stands for the Euclidean norm of $\mathbb{R}^N$, $N \geq 1$,

$$|\xi| := \sqrt{\sum_{j=1}^N \xi_j^2}, \qquad \xi = (\xi_1, ..., \xi_N)^T \in \mathbb{R}^N,$$

where T means *transposition*, as coordinate vectors of $\mathbb{R}^N$ will always be written vertically. Also, for every $R > 0$ and $x_0 \in \mathbb{R}^N$, we denote by $B_R(x_0)$ the open ball of radius R centered at x_0,

$$B_R(x_0) := \{x \in \mathbb{R}^N \ : \ |x - x_0| < R\}.$$

The following definition introduces the concept of ellipticity.

Definition 1.1 (Elliptic operator). *Given $x \in \Omega$, it is said that $\mathfrak{L}$ is elliptic at the single point x if there exists a constant $\mu_x > 0$ such that*

$$\sum_{i,j=1}^{N} a_{ij}(x)\xi_i\xi_j \geq \mu_x |\xi|^2 \qquad \text{for every} \quad \xi \in \mathbb{R}^N,$$

i.e., when the bilinear form associated to the matrix A_x introduced in (1.3),

$$\mathfrak{B}(\xi, \eta) := \xi^T A_x \eta, \qquad (\xi, \eta) \in \mathbb{R}^N \times \mathbb{R}^N,$$

is positive definite; in other words, when all the eigenvalues of A_x are positive. In such case, μ_x is said to be an ellipticity constant of $\mathfrak{L}$ at x.

The operator $\mathfrak{L}$ is said to be elliptic in Ω when it is elliptic at every point $x \in \Omega$. In such case, $\mathfrak{L}$ is said to be uniformly elliptic in Ω when the ellipticity constant μ_x can be chosen independently of $x \in \Omega$, i.e., when there exists $\mu > 0$ such that

$$\sum_{i,j=1}^{N} a_{ij}(x)\xi_i\xi_j \geq \mu |\xi|^2 \qquad \text{for all} \quad (x, \xi) \in \Omega \times \mathbb{R}^N.$$

The largest μ for which this condition holds is called the **ellipticity constant of $\mathfrak{L}$ in Ω**.

The paradigm of elliptic operator is the 'minus Laplacian',

$$\mathfrak{L} = -\Delta := -\sum_{j=1}^{N} \frac{\partial^2}{\partial x_j^2}.$$

Although this operator seems to be of a very special nature, the next result shows that the principal part of any elliptic operator can be transformed, through a local change of coordinates, into $-\Delta$.

Proposition 1.1. *Suppose $\mathfrak{L}$ is elliptic at $x_0 \in \Omega$. Then, there exists an invertible matrix $M \in \mathcal{M}_N(\mathbb{R})$ such that, for every $u \in \mathcal{C}^2(\Omega)$, the transformed function*

$$v(y) := u(x), \qquad y := Mx, \tag{1.4}$$

satisfies

$$\mathfrak{L}_p u|_{x=x_0} = -\Delta v|_{y=y_0} \left(= -\sum_{j=1}^{N} \frac{\partial^2 v}{\partial y_j^2}(y_0) \right) \qquad (1.5)$$

where $\mathfrak{L}_p$ is the principal part of $\mathfrak{L}$ and $y_0 := Mx_0$.

Proof. Let $x_0 \in \Omega$, $u \in \mathcal{C}^2(\Omega)$, and an invertible matrix
$$M = (m_{ij})_{1 \le i,j \le N} \in \mathcal{M}_N(\mathbb{R}).$$
Then,

$$\mathfrak{L}_p u\Big|_{x=x_0} = -\sum_{i,j=1}^{N} a_{ij}(x_0) \frac{\partial^2 u}{\partial x_i \partial x_j}\bigg|_{x=x_0} = -\nabla_x^T A_{x_0} \nabla_x u\Big|_{x=x_0} \qquad (1.6)$$

where ∇_x stands for the column gradient vector in the x-coordinates. By (1.4), we find that

$$\frac{\partial u}{\partial x_j} = \sum_{i=1}^{N} \frac{\partial v}{\partial y_i}\frac{\partial y_i}{\partial x_j} = \sum_{i=1}^{N} m_{ij}\frac{\partial v}{\partial y_i} = (m_{1j}, ..., m_{Nj})\nabla_y v$$

for all $1 \le j \le N$, and hence,

$$\nabla_x = M^T \nabla_y.$$

Consequently, (1.6) can be expressed in the form

$$\mathfrak{L}_p u\Big|_{x=x_0} = -\nabla_y^T M A_{x_0} M^T \nabla_y v\Big|_{y=y_0}$$

and, therefore, to prove (1.5) it suffices to construct a matrix M satisfying

$$M A_{x_0} M^T = I. \qquad (1.7)$$

To carry out this construction, let $\lambda_j(x)$, $1 \le j \le N$, denote, for every $x \in \Omega$, the eigenvalues of the matrix A_x counted according to their algebraic multiplicities. As A_x is symmetric, $\mathbb{R}^N$ possesses a basis consisting of orthogonal eigenvectors of A_x. Let C_x be the matrix whose columns are the coordinates of these eigenvectors. Then, C_x is orthogonal, i.e.,

$$C_x^{-1} = C_x^T,$$

and, by construction,

$$J_x := \mathrm{diag}\left(\lambda_j(x)\right)_{1 \le j \le N} = C_x^T A_x C_x \qquad (1.8)$$

is the Jordan form of A_x. As $\mathfrak{L}$ has been assumed to be elliptic at x_0, $\lambda_j(x_0) > 0$ for all $1 \le j \le N$, and hence, by making the choice

$$M := D C_{x_0}^T, \qquad D := \mathrm{diag}\left(\frac{1}{\sqrt{\lambda_j(x_0)}}\right)_{1 \le j \le N},$$

condition (1.7) holds. Indeed, owing to (1.8), it becomes apparent that

$$M A_{x_0} M^T = D C_{x_0}^T A_{x_0} C_{x_0} D = D J_{x_0} D = I.$$

This concludes the proof. $\qquad\qquad\qquad\qquad\qquad\qquad\qquad\qquad\square$

As a consequence from Proposition 1.1, the next result holds.

Theorem 1.1 (Classical minimum principle). *Suppose the differential operator $\mathfrak{L}$ is elliptic in Ω with $c \geq 0$, and $u \in C^2(\Omega)$ satisfies*

$$\mathfrak{L}u(x) > 0 \qquad \text{for every} \quad x \in \Omega.$$

Then, u cannot attain a non-positive local minimum in Ω.

Proof. On the contrary, assume that there exist $x_0 \in \Omega$ and $R > 0$ such that $B_R(x_0) \subset \Omega$,

$$m := u(x_0) \leq 0,$$

and

$$u(x_0) \leq u(x) \quad \text{for all} \quad x \in B_R(x_0).$$

By Proposition 1.1, there exists a linear change of coordinates $y = Mx$ such that

$$\mathfrak{L}_p u\big|_{x=x_0} = -\Delta v(y_0),$$

where

$$v(y) := u(x), \qquad y_0 := Mx_0.$$

As x_0 is a critical point of u, necessarily $\nabla_x u(x_0) = 0$ and, hence,

$$0 < \mathfrak{L}u(x_0) = \mathfrak{L}_p u(x_0) + c(x_0)u(x_0) = -\Delta v(y_0) + c(x_0)u(x_0).$$

Therefore,

$$\Delta v(y_0) < c(x_0)u(x_0) \leq 0$$

because $c(x_0) \geq 0$ and $u(x_0) \leq 0$. This is impossible, for as y_0 is a local minimum of v and, hence, $\Delta v(x_0) \geq 0$. $\qquad\square$

1.2 Minimum principle of E. Hopf

The following result extends Theorem 1.1 to cover the more general case when $\mathfrak{L}u \geq 0$ vanishes somewhere in Ω.

Theorem 1.2 (Minimum principle of E. Hopf). *Suppose $\mathfrak{L}$ is uniformly elliptic in Ω with $c \geq 0$, and $u \in C^2(\Omega)$ satisfies*

$$\mathfrak{L}u \geq 0 \quad \text{in} \quad \Omega, \qquad \text{and} \qquad m := \inf_\Omega u \in (-\infty, 0].$$

Then, either $u = m$ in Ω, or $u(x) > m$ for all $x \in \Omega$. In other words, u cannot attain m in Ω, unless $u = m$ in Ω.

In particular, when Ω is bounded and $u \in C^2(\Omega) \cap C(\bar{\Omega})$, then

$$\inf_\Omega u = \inf_{\partial\Omega} u = m.$$

Throughout this book, a function $u \in \mathcal{C}^2(\Omega)$ is said to be a *harmonic* function of $\mathfrak{L}$ in Ω if

$$\mathfrak{L}u = 0 \quad \text{in} \quad \Omega,$$

while it is said to be *superharmonic* if

$$\mathfrak{L}u \geq 0 \quad \text{in} \quad \Omega,$$

and *subharmonic* when $-u$ is superharmonic.

According to this terminology, Theorem 1.2 establishes that no non-constant superharmonic function u can reach a non-positive absolute minimum in Ω. Consequently, interchanging u by $-u$, no non-constant subharmonic function can attain a non-negative absolute maximum in Ω. Therefore, no non-constant harmonic function can attain its absolute maximum or its absolute minimum in Ω.

Proof.   It proceeds by contradiction. Suppose there exist $x_0, x_1 \in \Omega$ such that

$$m = u(x_0) = \inf_{\Omega} u \leq 0, \qquad u(x_1) > m. \tag{1.9}$$

Establishing a contradiction concludes the proof. Let $\gamma \in \mathcal{C}([0,1];\Omega)$ be a continuous curve connecting x_0 with x_1 in Ω, i.e., such that $\gamma(t) \in \Omega$ for each $t \in [0,1]$ and

$$\gamma(0) = x_0, \qquad \gamma(1) = x_1.$$

According to (1.9), by the continuity of $t \mapsto u(\gamma(t))$, there exists $t_m \in [0,1)$ such that

$$u(\gamma(t_m)) = m \quad \text{and} \quad u(\gamma(t)) > m \quad \text{for all} \quad t \in (t_m, 1],$$

i.e., $y_0 := \gamma(t_m)$ is the first point along the arc of curve γ from x_1 to x_0 where u attains m. Note that $t_m = 0$ if $u(\gamma(t)) > m$ for every $t \in (0,1]$. Though in such case $y_0 = x_0$, in general, $y_0 \neq x_0$. This is the situation illustrated by Figure 1.1. Now, set

$$\text{Tray}\,\gamma := \{\gamma(t) \,:\, 0 \leq t \leq 1\}, \qquad d := \text{dist}\,(\text{Tray}\,\gamma, \partial\Omega) > 0,$$

and pick

$$y_1 \in \{\gamma(t) \,:\, t_m < t < 1\}$$

such that

$$|y_0 - y_1| < d/2.$$

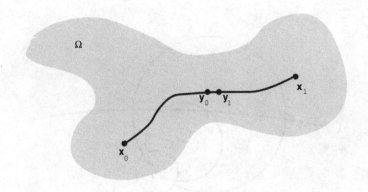

Fig. 1.1   The construction of y_0 and y_1

By construction, $u(y_1) > m$. Thus, by the continuity of u, there exists $r > 0$ such that

$$u(x) > m \quad \text{for each} \quad x \in B_r(y_1). \tag{1.10}$$

Moreover, since $u(y_0) = m$, necessarily

$$r \le |y_0 - y_1| < d/2. \tag{1.11}$$

Let ρ denote the maximal $r > 0$ satisfying (1.10). By (1.11), $\rho > 0$ is well defined and it satisfies

$$\rho \le |y_0 - y_1| < d/2. \tag{1.12}$$

In particular, $\bar{B}_\rho(y_1) \subset \Omega$ (see Figure 1.2). By the continuity of u and the maximality of ρ, there exists

$$y_2 \in \partial B_\rho(y_1)$$

such that

$$u(y_2) = m; \tag{1.13}$$

we might take $y_2 = y_0$ if $\rho = |y_0 - y_1|$. Subsequently, we consider the middle point of the segment linking y_1 to y_2,

$$z := \frac{y_1 + y_2}{2},$$

and the ball $B_{\frac{\rho}{2}}(z)$. This ball is tangent to $B_\rho(y_1)$ at y_2 and it satisfies

$$\bar{B}_{\frac{\rho}{2}}(z) \setminus \{y_2\} \subset B_\rho(y_1) \tag{1.14}$$

(see Figure 1.2). As $u(x) > m$ for all $x \in B_\rho(y_1)$, (1.14) implies

$$u(x) > m \quad \text{for all} \quad x \in \bar{B}_{\frac{\rho}{2}}(z) \setminus \{y_2\}. \tag{1.15}$$

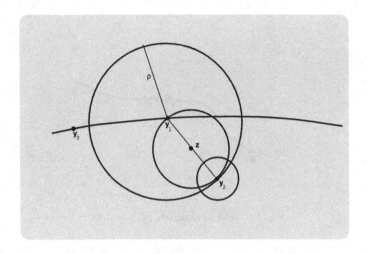

Fig. 1.2   The construction of $B_{\frac{\rho}{4}}(y_2)$

Finally, consider the ball of radius $\frac{\rho}{4}$ centered at y_2, $B_{\frac{\rho}{4}}(y_2)$. Figure 1.2 sketches the previous construction. Note that (1.12) implies

$$\bar{B}_{\frac{\rho}{4}}(y_2) \subset \Omega$$

and, hence, by (1.2), the coefficients of $\mathfrak{L}$ are bounded in $\bar{B}_{\frac{\rho}{4}}(y_2)$. The rest of the proof consists in constructing a non-constant function

$$w \in \mathcal{C}^2 \left(\bar{B}_{\frac{\rho}{4}}(y_2) \right)$$

such that

$$w(y_2) = m, \tag{1.16}$$

$$w(x) > m \quad \text{for every} \quad x \in \partial B_{\frac{\rho}{4}}(y_2), \tag{1.17}$$

and

$$\mathfrak{L}w(x) > 0 \quad \text{for every} \quad x \in B_{\frac{\rho}{4}}(y_2). \tag{1.18}$$

Thanks to (1.16) and (1.17), w attains its absolute minimum (necessarily below $m = w(y_2) \leq 0$) in $B_{\frac{\rho}{4}}(y_2)$. According to Theorem 1.1, such a function w cannot satisfy (1.18). This contradiction concludes the proof of the theorem.

Subsequently, we consider the functions

$$v(x) := e^{-\alpha |x-z|^2} - e^{-\alpha \frac{\rho^2}{4}}, \qquad x \in \mathbb{R}^N, \tag{1.19}$$

and

$$w(x) := u(x) - \epsilon v(x), \qquad x \in \Omega,$$

for some appropriate constants $\alpha > 0$ and $\epsilon > 0$ to be chosen later. By construction, $w \in \mathcal{C}^2(\bar{B}_{\frac{\rho}{4}}(y_2))$. Thus, to complete the proof of the theorem it suffices to show that there exist $\epsilon > 0$ and $\alpha > 0$ for which w satisfies (1.16), (1.17) and (1.18).

Since $|y_2 - z| = \rho/2$, we have that $v(y_2) = 0$ and, hence, (1.13) implies

$$w(y_2) = u(y_2) - \epsilon v(y_2) = u(y_2) = m.$$

Thus, (1.16) holds.

Subsequently, we will prove that, for sufficiently large $\alpha > 0$,

$$\mathcal{L}v(x) < 0 \quad \text{for every} \quad x \in \bar{B}_{\frac{\rho}{4}}(y_2). \tag{1.20}$$

Since $\mathcal{L}u \geq 0$ in Ω, (1.20) implies that

$$\mathcal{L}w(x) = \mathcal{L}u(x) - \epsilon\mathcal{L}v(x) > 0$$

for all $x \in \bar{B}_{\frac{\rho}{4}}(y_2)$ and $\epsilon > 0$, and, consequently, (1.18) holds. Indeed, for each $j \in \{1, ..., N\}$ and $x \in \mathbb{R}^N$, we find from (1.19) that

$$\frac{\partial v}{\partial x_j}(x) = -2\alpha\,(x_j - z_j)\,e^{-\alpha|x-z|^2},$$

$$\frac{\partial^2 v}{\partial x_j^2}(x) = \left[4\alpha^2(x_j - z_j)^2 - 2\alpha\right]e^{-\alpha|x-z|^2},$$

where x_i and z_i, $i \in \{1, ..., N\}$, stand for the i-th coordinates of x and z, respectively. Moreover, for every $i, j \in \{1, ..., N\}$, with $i \neq j$, and $x \in \mathbb{R}^N$,

$$\frac{\partial^2 v}{\partial x_i \partial x_j}(x) = 4\alpha^2\,(x_i - z_i)\,(x_j - z_j)\,e^{-\alpha|x-z|^2}$$

and hence,

$$\mathcal{L}v(x) = -\sum_{i,j=1}^{N} a_{ij}(x)\frac{\partial^2 v}{\partial x_i \partial x_j}(x) + \sum_{j=1}^{N} b_j(x)\frac{\partial v}{\partial x_j}(x) + c(x)v(x)$$

$$= -c(x)e^{-\alpha\frac{\rho^2}{4}} + \left\{-4\alpha^2\sum_{i,j=1}^{N} a_{ij}(x)(x_i - z_i)(x_j - z_j)\right.$$

$$\left. +2\alpha\sum_{j=1}^{N}[a_{jj}(x) - b_j(x)(x_j - z_j)] + c(x)\right\}e^{-\alpha|x-z|^2}.$$

As $\mathfrak{L}$ is uniformly elliptic in Ω, there exists a constant $\mu > 0$ such that

$$\sum_{i,j=1}^{N} a_{ij}(x)(x_i - z_i)(x_j - z_j) \geq \mu \, |x - z|^2 \quad \text{for all} \quad x \in \Omega,$$

and hence,

$$\sum_{i,j=1}^{N} a_{ij}(x)(x_i - z_i)(x_j - z_j) \geq \mu \frac{\rho^2}{16} \quad \text{for all} \quad x \in B_{\frac{\rho}{4}}(y_2),$$

because $x \in B_{\frac{\rho}{4}}(y_2)$ implies $|x - z| \geq \frac{\rho}{4}$ (see Figure 1.2). Moreover, since $\bar{B}_{\frac{\rho}{4}}(y_2)$ is a compact subset of Ω, it follows from (1.2) that there exist two constants $C_1 > 0$ and $C_2 > 0$ such that

$$\sum_{j=1}^{N} |a_{jj}(x) - b_j(x)(x_j - z_j)| \leq C_1, \qquad |c(x)| \leq C_2,$$

for all $x \in \bar{B}_{\frac{\rho}{4}}(y_2)$. Thus, for every $\alpha > 0$ and $x \in \bar{B}_{\frac{\rho}{4}}(y_2)$,

$$\mathfrak{L}v(x) \leq \left(-\mu \frac{\rho^2}{4} \alpha^2 + 2\alpha C_1 + C_2 \right) e^{-\alpha |x - z|^2},$$

since $c \geq 0$. Therefore, (1.20) holds for sufficiently large $\alpha > 0$. Throughout the rest of this proof, we assume that $\alpha > 0$ has been chosen in this way.

To complete the proof of the theorem it remains to show that (1.17) holds for sufficiently small $\epsilon > 0$. Indeed, setting

$$S_1 := \partial B_{\frac{\rho}{4}}(y_2) \cap \bar{B}_{\frac{\rho}{2}}(z), \qquad S_2 := \partial B_{\frac{\rho}{4}}(y_2) \setminus S_1,$$

it is apparent that S_1 is a compact subset of $\bar{B}_{\frac{\rho}{2}}(z) \setminus \{y_2\}$ (see Figure 1.3) and, in particular, according to (1.15), we have that

$$u(x) > m \quad \text{for all} \quad x \in S_1.$$

Thus, since u is continuous, there exists $\xi > 0$ such that

$$u(x) \geq m + \xi \quad \text{for all} \quad x \in S_1. \tag{1.21}$$

Subsequently, we consider any $\epsilon > 0$ satisfying

$$0 < \epsilon < \frac{\xi}{1 - e^{-\alpha \frac{\rho^2}{4}}}.$$

According to (1.19), it is apparent that

$$\begin{cases} v(x) > 0 & \text{if and only if} \quad x \in B_{\frac{\rho}{2}}(z), \\[2mm] v(x) = 0 & \text{if and only if} \quad x \in \partial B_{\frac{\rho}{2}}(z), \\[2mm] v(x) < 0 & \text{if and only if} \quad x \notin \bar{B}_{\frac{\rho}{2}}(z). \end{cases} \tag{1.22}$$

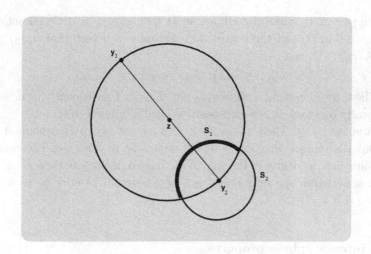

Fig. 1.3 S_1 is the thicker portion of $\partial B_{\frac{\rho}{4}}(y_2)$

By (1.22), for every $x \in S_1$, we have that

$$0 \leq v(x) = e^{-\alpha|x-z|^2} - e^{-\alpha\frac{\rho^2}{4}} < 1 - e^{-\alpha\frac{\rho^2}{4}}$$

and hence, by the choice of ϵ,

$$0 \leq \epsilon v(x) < \epsilon \left(1 - e^{-\alpha\frac{\rho^2}{4}}\right) < \xi.$$

Consequently, (1.21) implies that

$$w(x) = u(x) - \epsilon v(x) > m + \xi - \xi = m \qquad \text{for all} \quad x \in S_1.$$

Finally, by (1.22), for every $x \in S_2 := \partial B_{\frac{\rho}{4}}(y_2) \setminus S_1$ we have that $v(x) < 0$ and, consequently,

$$w(x) = u(x) - \epsilon v(x) > u(x) \geq \inf_\Omega u = m.$$

Thus, since

$$S_1 \cup S_2 = \partial B_{\frac{\rho}{4}}(y_2),$$

condition (1.17) is fulfilled, and the proof is complete. $\qquad\square$

Remark 1.1.

(a) Although condition $c \geq 0$ is not strictly necessary for the validity of Theorem 1.2, when c is negative somewhere in Ω the theorem might not be true.

(b) Suppose $\mathfrak{L}$ is uniformly elliptic in Ω with $c \geq 0$, $u \in \mathcal{C}^2(\Omega)$ satisfies $\mathfrak{L}u \geq 0$ in Ω, and there exist $x_0 \in \Omega$ and $r > 0$ such that $u(x_0) \leq 0$, $\bar{B}_r(x_0) \subset \Omega$, and

$$u(x_0) \leq u(x) \quad \text{for every} \quad x \in B_r(x_0).$$

Then, by Theorem 1.2, $u = u(x_0)$ in $B_r(x_0)$. Consequently, u must be locally constant at any non-positive local minimum in Ω.

(c) Suppose $c = 0$. Then, for every $m \in \mathbb{R}$, $u - m$ is superharmonic if u is superharmonic, since $\mathfrak{L}m = 0$. Consequently, in this case, Theorem 1.2 holds independently of the sign of m. Indeed, if $\mathfrak{L}u \geq 0$, then $u - m \geq 0$ is superharmonic and, therefore, by Theorem 1.2, $u(x) > m$ for all $x \in \Omega$ if $u \neq m$.

1.3   Interior sphere properties

The next definition introduces some important geometrical properties of $\partial\Omega$ that will be used throughout this book.

Definition 1.2. *Ω is said to satisfy the* **interior sphere property** *at a single point $x \in \partial\Omega$ if there are $z_x \in \Omega$ and $r > 0$ such that*

$$|x - z_x| = r, \qquad B_r(z_x) \subset \Omega.$$

Now, let Γ_0 be a closed and open subset of $\partial\Omega$. Then:

(a) *Ω is said to satisfy the* **uniform interior sphere property** *on Γ_0 if there exists $r > 0$ such that for every $x \in \Gamma_0$ there is a point $z_x \in \Omega$ for which*

$$|x - z_x| = r, \qquad B_r(z_x) \subset \Omega.$$

In such case, it is said that Ω satisfies the uniform interior sphere property on Γ_0 with parameter r. When $\Gamma_0 = \partial\Omega$, it is simply said that Ω satisfies the uniform interior sphere property.

(b) *Ω is said to satisfy the* **uniform interior sphere property in the strong sense** *on Γ_0 if there exists $r > 0$ such that for every $z \in \Omega$ with $\mathrm{dist}\,(z, \Gamma_0) \leq r$ there is a point $x_z \in \Gamma_0$ for which*

$$\mathrm{dist}\,(z, \partial\Omega) = |z - x_z|, \qquad B_r\left(x_z + r\frac{z - x_z}{|z - x_z|}\right) \subset \Omega.$$

In such case, it is said that Ω satisfies the uniform interior sphere property in the strong sense on Γ_0 with parameter r. When $\Gamma_0 = \partial\Omega$, it is simply said that Ω satisfies the uniform interior sphere property in the strong sense.

Note that if Ω satisfies the interior sphere property at $x \in \partial\Omega$, then there exist $z_x \in \Omega$ and $r > 0$ such that

$$B_r(z_x) \subset \Omega, \qquad \bar{B}_r(z_x) \cap \partial\Omega = \{x\}. \tag{1.23}$$

Indeed, by definition, there are $\tilde{z}_x \in \Omega$ and $\tilde{r} > 0$ such that

$$|x - \tilde{z}_x| = \tilde{r}, \qquad B_{\tilde{r}}(\tilde{z}_x) \subset \Omega.$$

Obviously, (1.23) holds by making the choice

$$r := \tilde{r}/2, \qquad z_x := (x + \tilde{z}_x)/2.$$

If Ω satisfies the uniform interior sphere property on Γ_0, then it satisfies the interior sphere property at every $x \in \Gamma_0$, of course. Moreover, the following result holds.

Lemma 1.1. *Let $\Omega \subset \mathbb{R}^N$, $N \geq 1$, be an open set with boundary $\partial\Omega$, and Γ_0 a closed and open subset of $\partial\Omega$. Then, the following assertions are true:*

(a) *If Ω satisfies the uniform interior sphere property in the strong sense on Γ_0 with parameter $r > 0$, then it also satisfies the uniform interior sphere property on Γ_0 with the same parameter r.*

(b) *If Γ_0 is of class $\mathcal{C}^1$ (see Definition 1.5 and Theorem 1.8, if necessary) and it satisfies the uniform interior sphere property on Γ_0 with parameter r, then it also satisfies the uniform interior sphere property in the strong sense on Γ_0 with the same parameter r.*

Therefore, Definitions 1.2(a) and (b) are equivalent when Γ_0 is of class $\mathcal{C}^1$.

Proof. Suppose Ω satisfies the uniform interior sphere property in the strong sense on Γ_0 with parameter $r > 0$. According to Definition 1.2(b), for every $z \in \Omega$ with $\text{dist}(z, \Gamma_0) = r$ there exists $x_z \in \Gamma_0$ such that

$$\text{dist}(z, \partial\Omega) = |z - x_z|, \qquad B_r\left(x_z + r\frac{z - x_z}{|z - x_z|}\right) \subset \Omega.$$

As $\Gamma_0 \subset \partial\Omega$ and $x_z \in \Gamma_0$, this implies

$$r = \text{dist}(z, \Gamma_0) \geq \text{dist}(z, \partial\Omega) = |z - x_z| \geq \text{dist}(z, \Gamma_0) = r$$

and hence,

$$r = \text{dist}(z, \Gamma_0) = \text{dist}(z, \partial\Omega) = |z - x_z|.$$

Thus,

$$x_z + r\frac{z - x_z}{|z - x_z|} = x_z + r\frac{z - x_z}{r} = z$$

and, therefore,

$$B_r(z) \subset \Omega.$$

Let $x \in \Gamma_0$ and $n_0 \in \mathbb{N}$ such that $n_0 \geq 1/r$, and for any integer $n \geq n_0$ pick $z_n \in \Omega$ such that

$$|x - z_n| \leq \frac{1}{n} \leq \frac{1}{n_0} \leq r.$$

As

$$\mathrm{dist}(z_n, \Gamma_0) \leq |x - z_n| \leq r,$$

it follows from Definition 1.2(b) that, for every $n \geq n_0$, there is a point

$$x_n := x_{z_n} \in \Gamma_0$$

for which

$$\mathrm{dist}\,(z_n, \partial\Omega) = |z_n - x_n|, \qquad B_r\left(x_n + r\frac{z_n - x_n}{|z_n - x_n|}\right) \subset \Omega.$$

Setting

$$\tilde{z}_n := x_n + r\frac{z_n - x_n}{|z_n - x_n|}, \qquad n \geq n_0,$$

we have that $B_r(\tilde{z}_n) \subset \Omega$, $|\tilde{z}_n - x_n| = r$, and $x_n \in \Gamma_0$. Hence,

$$\mathrm{dist}(\tilde{z}_n, \Gamma_0) = \mathrm{dist}(\tilde{z}_n, \partial\Omega) = r, \qquad n \geq n_0.$$

Thus, as the set

$$\Gamma_r := \{z \in \Omega \ : \ \mathrm{dist}(z, \Gamma_0) = \mathrm{dist}(z, \partial\Omega) = r\}$$

is compact, there exists $\tilde{z}_\omega \in \Gamma_r$ such that along some subsequence of $\{\tilde{z}_n\}_{n\geq 1}$, say $\{\tilde{z}_{n_m}\}_{m\geq 1}$, the following holds

$$\lim_{m\to\infty} \tilde{z}_{n_m} = \tilde{z}_\omega.$$

By construction, it becomes apparent that

$$B_r(\tilde{z}_\omega) \subset \Omega,$$

because $z \in B_r(\tilde{z}_\omega)$ implies that $z \in B_r(\tilde{z}_{n_m}) \subset \Omega$ for sufficiently large m. Moreover,

$$z_{n_m} \in B_r(\tilde{z}_{n_m}), \qquad |x - z_{n_m}| \leq \frac{1}{n_m}, \qquad m \geq 1,$$

and hence, letting $m \to \infty$ shows that

$$\lim_{m\to\infty} z_{n_m} = x, \qquad x \in \bar{B}_r(\tilde{z}_\omega).$$

Consequently, setting $z_x := \tilde{z}_\omega$, we conclude that

$$|x - z_x| = r, \qquad B_r(z_x) \subset \Omega,$$

because $x \in \Gamma_0$. Therefore, Ω satisfies the uniform interior sphere property on Γ_0 with parameter r.

Now, suppose that Γ_0 is of class $\mathcal{C}^1$, as discussed in Definition 1.5, and that Ω satisfies the uniform interior sphere property on Γ_0 with parameter r. Then, by Theorem 1.8, it is apparent that Ω has a tangent hyperplane at every $x_0 \in \Gamma_0$.

Let $z \in \Omega$ such that

$$\operatorname{dist}(z, \Gamma_0) \leq r$$

and pick a point $x \in \Gamma_0$ with

$$\operatorname{dist}(z, \Gamma_0) = |z - x| \leq r.$$

Necessarily, the tangent hyperplane to Γ_0 at x must be orthogonal to $z - x$. By Definition 1.2(a), there exists $z_x \in \Omega$ such that

$$|x - z_x| = r, \qquad B_r(z_x) \subset \Omega.$$

As $B_r(z_x) \subset \Omega$ touches Γ_0 at x, the vector $z_x - x$ must be orthogonal to $\partial\Omega$ at x and

$$\operatorname{dist}(z_x, \partial\Omega) = \operatorname{dist}(z_x, \Gamma_0) = |z_x - x| = r.$$

Consequently, z must lie in the line segment $[x, z_x]$ and hence,

$$z_x = x + r \frac{z - x}{|z - x|}.$$

Therefore,

$$\operatorname{dist}(z, \partial\Omega) = |z - x|, \qquad B_r\left(x + r \frac{z - x}{|z - x|}\right) \subset \Omega,$$

and, consequently, Ω satisfies the uniform interior sphere property in the strong sense on Γ_0 with parameter r. This ends the proof. $\qquad\square$

According to Lemma 1.1, in the special case when Ω is of class $\mathcal{C}^1$, Ω satisfies the uniform interior sphere property with parameter $r > 0$ if and only if it satisfies this property in the strong sense with the same parameter r. But, in the general case when Ω is not of class $\mathcal{C}^1$, the fact that Ω satisfies the uniform interior sphere property with parameter $r > 0$ and a uniform interior sphere property in the strong sense do not necessarily entail the strong condition to be satisfied with the same $r > 0$. As a counterexample, consider two points $z_0, z_1 \in \mathbb{R}^N$, $N \geq 2$, and $R > 0$ with

$$\mathrm{dist}(z_0, z_1) < 2R$$

for which $2R - \mathrm{dist}(z_0, z_1)$ is sufficiently small. Then, the union of the two balls

$$\Omega := B_R(z_0) \cup B_R(z_1),$$

is not of class $\mathcal{C}^1$, it satisfies the uniform interior sphere property with parameter $r = R$ (optimal), as well as the corresponding property in the strong sense for sufficiently small $r \in (0, R)$, but it does not satisfy the property in the strong sense with parameter $r = R$. Indeed, for every $x \in \partial\Omega$ there exists $i \in \{0, 1\}$ such that $x \in \partial B_R(z_i)$ and, hence, choosing $z_x := z_i$, we have that

$$|x - z_x| = |x - z_i| = R, \qquad B_R(z_x) = B_R(z_i) \subset \Omega.$$

Consequently, Ω satisfies the uniform interior sphere property with parameter $r = R$. Obviously, this property cannot be satisfied if $r > R$ and, consequently, $r = R$ is optimal. Now, pick two points $x_0, x_1 \in \partial B_R(z_0) \cap \partial B_R(z_1)$

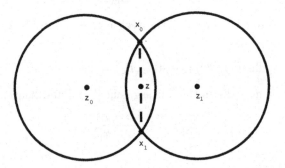

Fig. 1.4 The special case $\Omega := B_R(z_0) \cup B_R(z_1)$

such that

$$|x_0 - x_1| = \mathrm{diam}\,[\bar{B}_R(z_0) \cap \bar{B}_R(z_1)],$$

as illustrated by Figure 1.4. Note that $|x_0 - x_1| < 2R$. Then, setting

$$z := \frac{x_0 + x_1}{2},$$

we have that

$$|z - x_0| = |z - x_1| = \frac{|x_0 - x_1|}{2}$$

and it can be easily seen that Ω satisfies the uniform interior sphere property in the strong sense with

$$r := \frac{|x_0 - x_1|}{2} < R,$$

though, obviously, the property fails for greater values of r.

In general, a domain Ω satisfying the uniform interior sphere property does not necessarily satisfy it in the strong sense, as the following example shows. Let $\Omega \subset \mathbb{R}^2$ be the open set

$$\Omega := B_2(0) \setminus \{(0,0), (1/n,0) : n \geq 1\}, \tag{1.24}$$

which consists of the ball $B_2(0)$ perforated by $(0,0)$ and the sequence $(1/n,0)$, $n \geq 1$, which converges to $(0,0)$ as $n \to \infty$. As

$$\partial\Omega = \{x \in \mathbb{R}^2 : |x| = 2\} \cup \{(0,0), (1/n,0) : n \geq 1\},$$

Ω cannot satisfy a uniform interior sphere property in the strong sense, because of the special structure of its boundary around $(0,0)$. To prove this we will argue by contradiction. So, suppose Ω satisfies the uniform interior sphere property in the strong sense with parameter $r > 0$ and pick an integer $n \geq 1$ such that

$$\frac{1}{n} - \frac{1}{n+1} = \frac{1}{n(n+1)} < r,$$

and a real number $\epsilon \in \mathbb{R}$ satisfying

$$\frac{1}{n+1} < \epsilon < \frac{1}{n}, \qquad \epsilon - \frac{1}{n+1} < \frac{1}{n} - \epsilon.$$

Then, $z := (\epsilon, 0) \in \Omega$ and

$$\operatorname{dist}(z, \partial\Omega) = \epsilon - \frac{1}{n+1}.$$

Moreover, $x_z := (1/(n+1), 0)$ provides us with the unique point of $\partial\Omega$ such that

$$\operatorname{dist}(z, \partial\Omega) = |z - x_z|.$$

Thus, according to Definition 1.2(b), we should have that

$$B_r\left(x_z + r\frac{z - x_z}{|z - x_z|}\right) \subset \Omega,$$

which is impossible because

$$x_z + r\frac{z - x_z}{|z - x_z|} = \left(\frac{1}{n+1} + r, 0\right), \qquad \frac{1}{n+1} + r > \frac{1}{n},$$

and hence,

$$(1/n, 0) \in B_r \left(x_z + r \frac{z - x_z}{|z - x_z|} \right) \cap \partial\Omega = \emptyset.$$

Nevertheless, Ω satisfies the uniform interior sphere property with parameter $r := 1/2$. Indeed, for every $x \in \partial B_2(0)$ the point $z_x := \frac{3}{4}x$ satisfies

$$z_x \in \Omega, \quad |x - z_x| = \frac{1}{2}, \quad B_{\frac{1}{2}}(z_x) \subset \Omega.$$

Moreover, setting

$$z_{(0,0)} := (0, 1/2), \qquad z_{(1/n,0)} := (1/n, 1/2), \qquad n \geq 1,$$

it is apparent that

$$(0,0) \in \partial B_{\frac{1}{2}}(z_{(0,0)}), \qquad (1/n, 0) \in \partial B_{\frac{1}{2}}(z_{(1/n,0)}), \quad n \geq 1,$$

and

$$B_{\frac{1}{2}}(z_{(0,0)}) \cup \bigcup_{n \geq 1} B_{\frac{1}{2}}(z_{(1/n,0)}) \subset \Omega.$$

Therefore, Ω satisfies the uniform interior sphere condition with $r = 1/2$.

Theorem 1.9 in the Appendix of this chapter shows that Ω satisfies the uniform interior sphere property in the strong sense on Γ_0 whenever Γ_0 is of class $\mathcal{C}^2$. Undoubtedly, this is the most paradigmatic class of domains satisfying Definition 1.2(b).

Suppose Ω is of class $\mathcal{C}^2$ around $x_0 \in \partial\Omega$. Then, owing to the proof of Theorem 1.8, Ω satisfies a uniform interior sphere property around x_0 and any interior sphere $B_r(z)$ at x_0 must be tangent to the tangent hyperplane of $\partial\Omega$ at x_0. Thus, the vector $x_0 - z$ is orthogonal to $\partial\Omega$ at x_0 and, hence,

$$\mathbf{n} := \frac{x_0 - z}{|x_0 - z|}$$

is the **outward unit normal** at x_0. Note that the tangent hyperplane to any interior sphere $B_r(z)$ at x_0 does not vary with (r, z) and, hence, $\mathbf{n}$ is uniquely determined. In this context, a given $\nu \in \mathbb{R}^N \setminus \{0\}$ is said to be an **outward pointing vector** at x_0 whenever

$$\langle \nu, \mathbf{n} \rangle > 0 \tag{1.25}$$

(see Figure 1.5). As $\partial\Omega$ and $\mathbf{n}$ are orthogonal at x_0, (1.25) implies that

$$x_0 + t\nu \in \mathbb{R}^N \setminus \bar{\Omega} \quad \text{for sufficiently small} \quad t > 0.$$

Consequently, the nomenclature of *outward pointing vector* at $x_0 \in \partial\Omega$ is unambiguous. Figure 1.6 illustrates the existence of non-smooth boundaries

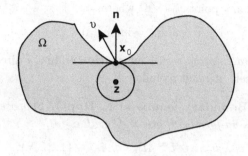

Fig. 1.5 Smooth boundary around $x_0 \in \partial\Omega$

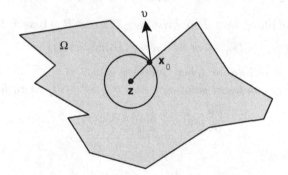

Fig. 1.6 Non-smooth boundary at $x_0 \in \partial\Omega$

satisfying an interior sphere property. In these cases, the previous concept of outward normal does not make sense, but yet a concept of outward pointing vector can be given through the next definition.

Definition 1.3. *Suppose Ω satisfies an interior sphere property at $x_0 \in \partial\Omega$. Then, $\nu \in \mathbb{R}^N \setminus \{0\}$ is said to be an **outward pointing vector** at x_0 if*

$$x_0 + t\nu \in \mathbb{R}^N \setminus \bar{\Omega} \quad \text{for sufficiently small } t > 0,$$

and there exist $z \in \Omega$ and $r > 0$ such that

$$|z - x_0| = r, \qquad B_r(z) \subset \Omega, \qquad \langle \nu, x_0 - z \rangle > 0.$$

1.4 Boundary lemma of E. Hopf

The main result of this section improves Theorem 1.2 by establishing that any non-constant superharmonic function $u \in \mathcal{C}^2(\Omega) \cap \mathcal{C}(\bar{\Omega})$ of $\mathfrak{L}$ in Ω must decay linearly at any point $x_0 \in \partial\Omega$ where

$$u(x_0) = \inf_{\bar{\Omega}} u \leq 0$$

along any *outward pointing vector* for which u admits a directional derivative. Most precisely, it reads as follows.

Theorem 1.3 (Boundary lemma of E. Hopf). *Suppose the differential operator $\mathfrak{L}$ is uniformly elliptic in Ω with $c \geq 0$,*

$$a_{ij}, \; b_j, \; c \in L^\infty(\Omega), \qquad i,j \in \{1, ..., N\}, \tag{1.26}$$

and $u \in \mathcal{C}^2(\Omega)$ is a non-constant function satisfying

$$\mathfrak{L}u \geq 0 \quad in \quad \Omega, \qquad and \quad m := \inf_{\Omega} u \in (-\infty, 0].$$

Assume, in addition, that there exist $x_0 \in \partial\Omega$ and $R > 0$ such that

$$u(x_0) = m, \qquad u \in \mathcal{C}(B_R(x_0) \cap \bar{\Omega}),$$

and Ω satisfies an interior sphere property at x_0.

Then, for any outward pointing vector $\nu \in \mathbb{R}^N \setminus \{0\}$ at x_0 for which

$$\frac{\partial u}{\partial \nu}(x_0) := \lim_{\substack{x \in \Omega \\ x \to x_0}} \langle \nu, \nabla u(x) \rangle$$

exists, necessarily

$$\frac{\partial u}{\partial \nu}(x_0) < 0. \tag{1.27}$$

Thanks to Theorem 1.2, under the assumptions of Theorem 1.3, we have that $u(x) > m$ for all $x \in \Omega$, as we are assuming that u is not a constant. Therefore, Theorem 1.3 establishes that any non-constant superharmonic function $u(x)$ decays linearly towards its minimum, $m = u(x_0)$, as $x \in \Omega$ approximates $x_0 \in \partial\Omega$, if $m \leq 0$. In Figure 1.7 we have represented the profile of u along the straight line passing through x_0 in the direction of the vector ν; we are denoting

$$\tan \alpha := \frac{\partial u}{\partial \nu}(x_0) < 0.$$

Proof. Let $z \in \Omega$ and a sufficiently small $r > 0$ such that

$$B_r(z) \subset \Omega, \qquad \bar{B}_r(z) \cap \partial\Omega = \{x_0\}, \qquad \langle \nu, x_0 - z \rangle > 0,$$

and

$$u \in \mathcal{C}^2(B_r(z)) \cap \mathcal{C}(\bar{B}_r(z)).$$

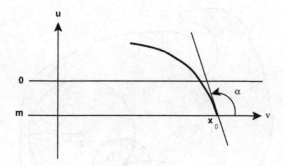

Fig. 1.7   Linear decay guaranteed by (1.27)

Now, consider the domain

$$D := B_{\frac{r}{2}}(x_0) \cap B_r(z)$$

and set

$$S_1 := \partial B_{\frac{r}{2}}(x_0) \cap \bar{B}_r(z), \qquad S_2 := B_{\frac{r}{2}}(x_0) \cap \partial B_r(z).$$

Then, $\partial D = S_1 \cup S_2$ and S_1 is a compact subset of Ω (see Figure 1.8).

By (1.26), the proof of (1.20) can be adapted to show that for sufficiently large $\alpha > 0$ the function

$$v(x) := e^{-\alpha |x-z|^2} - e^{-\alpha r^2}, \qquad x \in \mathbb{R}^N,$$

satisfies

$$\mathfrak{L}v(x) < 0 \quad \text{for all} \quad x \in B_r(z). \tag{1.28}$$

Throughout the remaining of the proof we suppose that α has been chosen to satisfy (1.28). Note that (1.26) is needed in order to obtain (1.28), because $x_0 \in \partial B_r(z) \cap \partial \Omega$; in the Proof of Theorem 1.2 the sign of $\mathfrak{L}v$ was fixed within a compact subset of Ω, where the coefficients of $\mathfrak{L}$ are always bounded under condition (1.2). Also, note that $v(x) > 0$ if $x \in B_r(z)$, while $v(x) = 0$ if $x \in \partial B_r(z)$ and $v(x) < 0$ for all $x \in \mathbb{R}^N \setminus \bar{B}_r(z)$.

Thanks to Theorem 1.2,

$$u(x) > m \quad \text{for each} \quad x \in \Omega. \tag{1.29}$$

Thus, since S_1 is a compact subset of Ω, there exists $\xi > 0$ such that

$$u(x) \geq \xi + m \quad \text{for all} \quad x \in S_1. \tag{1.30}$$

Now, fix $\epsilon > 0$ such that

$$0 < \epsilon < \frac{\xi}{1 - e^{-\alpha r^2}}$$

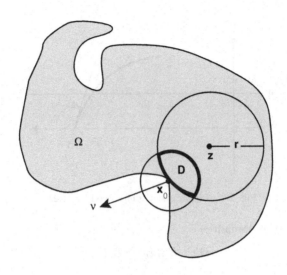

Fig. 1.8   The construction of D

and consider the auxiliary function

$$w(x) := u(x) - \epsilon v(x), \qquad x \in \Omega.$$

Note that $\frac{\partial w}{\partial \nu}(x_0)$ is well defined if $\frac{\partial u}{\partial \nu}(x_0)$ exists, because $v \in \mathcal{C}^\infty(\mathbb{R}^N)$. Moreover, in such case,

$$\frac{\partial w}{\partial \nu}(x_0) = \frac{\partial u}{\partial \nu}(x_0) - \epsilon \frac{\partial v}{\partial \nu}(x_0). \tag{1.31}$$

It follows from (1.28) that

$$\mathfrak{L}w(x) = \mathfrak{L}u(x) - \epsilon \mathfrak{L}v(x) \geq -\epsilon \mathfrak{L}v(x) > 0 \quad \text{for all} \quad x \in D,$$

since $D \subset B_r(z) \subset \Omega$. Moreover, for each $x \in S_1$ we have that

$$0 \leq v(x) < 1 - e^{-\alpha r^2}$$

and hence,

$$0 \leq \epsilon v(x) < \epsilon \left(1 - e^{-\alpha r^2}\right) < \xi.$$

Thus, we find from (1.30) that

$$w(x) = u(x) - \epsilon v(x) \geq m + \xi - \epsilon v(x) > m \quad \text{for all} \quad x \in S_1.$$

Further, since $v(x) = 0$ for each $x \in S_2 \subset \partial B_r(z)$, we have that

$$w(x) = u(x) - \epsilon v(x) = u(x).$$

Thus, by (1.29),

$$w(x) = u(x) > m \quad \text{for each} \quad x \in S_2 \setminus \{x_0\} \subset \Omega$$

while

$$w(x_0) = u(x_0) = m.$$

Consequently, $w \in \mathcal{C}^2(D) \cap \mathcal{C}(\bar{D})$ and it satisfies the following properties:

(1) $\mathfrak{L}w(x) > 0$ for all $x \in D$,
(2) $w(x) > m$ for all $x \in \partial D \setminus \{x_0\}$,
(3) $w(x_0) = m$.

Therefore, Theorem 1.2 implies that

$$w(x) > m \quad \text{for every} \quad x \in \bar{D} \setminus \{x_0\}$$

and, necessarily,

$$\frac{\partial w}{\partial \nu}(x_0) \leq 0. \tag{1.32}$$

It should be noted that $w(x) > m$, $x \in D$, cannot be obtained straight-away from the definition of w, as $v(x) > 0$ for all $x \in D$. By (1.31) and (1.32) it becomes apparent that

$$\frac{\partial u}{\partial \nu}(x_0) \leq \epsilon \frac{\partial v}{\partial \nu}(x_0) = \epsilon \langle \nu, \nabla v(x_0) \rangle.$$

On the other hand, by a direct calculation,

$$\nabla v(x_0) = -2\alpha(x_0 - z)e^{-\alpha r^2}$$

and, consequently,

$$\frac{\partial u}{\partial \nu}(x_0) \leq \epsilon \langle \nu, \nabla v(x_0) \rangle = -2\alpha\epsilon e^{-\alpha r^2} \langle \nu, x_0 - z \rangle < 0,$$

because $\langle \nu, x_0 - z \rangle > 0$. This shows (1.27) and ends the proof. $\qquad \square$

1.5 Positivity properties of super-harmonic functions

The following corollaries from Theorems 1.2 and 1.3, provide us with some important positivity properties of the superharmonic functions of $\mathfrak{L}$ in Ω.

Theorem 1.4. *Suppose Ω is bounded, $\mathfrak{L}$ is uniformly elliptic in Ω with $c \geq 0$, and $u \in \mathcal{C}^2(\Omega) \cap \mathcal{C}(\bar{\Omega}) \setminus \{0\}$ satisfies*

$$\begin{cases} \mathfrak{L}u \geq 0 & \text{in } \Omega, \\ u \geq 0 & \text{on } \partial\Omega. \end{cases}$$

Then,

$$u(x) > 0 \quad \text{for all} \quad x \in \Omega.$$

If, in addition, $u \in C^1(\bar{\Omega})$ and (1.26) is satisfied, then, for every

$$x \in \partial\Omega \cap u^{-1}(0)$$

where Ω satisfies the interior sphere property, and any outward pointing vector ν at x, we have that

$$\frac{\partial u}{\partial \nu}(x) < 0.$$

Proof.   Set

$$m := \inf_{\bar{\Omega}} u.$$

If $m > 0$, then $u(x) > 0$ for all $x \in \bar{\Omega}$ and we are done. So, suppose $m \leq 0$. Then, thanks to Theorem 1.2,

$$\inf_{\bar{\Omega}} u = \inf_{\partial\Omega} u = m \leq 0$$

and, hence, $m = 0$, since $u|_{\partial\Omega} \geq 0$. Moreover, $u(x) > 0$ for all $x \in \Omega$, because $u \neq 0$. The remaining assertions follow from Theorem 1.3.   □

Corollary 1.1. *Suppose Ω is bounded, $\mathfrak{L}$ is uniformly elliptic in Ω with $c \geq 0$, $f \in C(\Omega)$ and $g \in C(\partial\Omega)$. Then, the boundary value problem*

$$\begin{cases} \mathfrak{L}u = f & \text{in } \Omega, \\ u = g & \text{on } \partial\Omega, \end{cases} \tag{1.33}$$

admits, at most, a unique solution $u \in C^2(\Omega) \cap C(\bar{\Omega})$. In particular, $u = 0$ is the unique function $u \in C^2(\Omega) \cap C(\bar{\Omega})$ solving

$$\begin{cases} \mathfrak{L}u = 0 & \text{in } \Omega, \\ u = 0 & \text{on } \partial\Omega. \end{cases} \tag{1.34}$$

Proof.   On the contrary, assume that there exist $u, v \in C^2(\Omega) \cap C(\bar{\Omega})$, $u \neq v$, solving (1.33). Then, the auxiliary function

$$w := u - v \in C^2(\Omega) \cap C(\bar{\Omega}) \setminus \{0\}$$

provides us with a solution of (1.34) and, according to Theorem 1.4, $w(x) > 0$ for every $x \in \Omega$. Similarly, $-w$ provides us with a solution of (1.34) such that $-w(x) > 0$ for each $x \in \Omega$. This contradiction ends the proof.   □

1.6 Uniform decay property of E. Hopf

Essentially, the next result provides us with a lower estimate for the decay rate of all positive superharmonic functions of $\mathfrak{L}$ in Ω at the points of $\partial\Omega$ where they vanish.

Theorem 1.5 (Uniform decay property of E. Hopf). *Let* $\mathfrak{L}$ *be a uniformly elliptic operator in* Ω *whose coefficients satisfy* (1.26)*. Then, for each* $R > 0$ *there exists a constant*

$$M := M(\mathfrak{L}, R) > 0$$

such that for every $x_0 \in \Omega$ *with* $B_R(x_0) \subset \Omega$ *and any function* $u \in \mathcal{C}^2(B_R(x_0)) \cap \mathcal{C}(\bar{B}_R(x_0))$ *satisfying*

$$u(x) > 0 \qquad \text{for all} \quad x \in B_R(x_0) \tag{1.35}$$

and

$$\mathfrak{L}u(x) \geq 0 \qquad \text{for all} \quad x \in B_R(x_0) \tag{1.36}$$

the following estimate holds:

$$u(x) \geq \left(M \min_{\bar{B}_{\frac{R}{2}}(x_0)} u \right) \operatorname{dist}(x, \partial B_R(x_0)) \quad \forall \; x \in \bar{B}_R(x_0). \tag{1.37}$$

Proof. Let $R > 0$ and $x_0 \in \Omega$ such that $B_R(x_0) \subset \Omega$, and suppose $u \in \mathcal{C}^2(B_R(x_0)) \cap \mathcal{C}(\bar{B}_R(x_0))$ satisfies (1.35) and (1.36). Consider the auxiliary functions

$$E(x) := e^{\alpha(R^2 - |x - x_0|^2)}, \qquad v(x) := E(x) - 1, \qquad x \in \mathbb{R}^N,$$

where $\alpha > 0$ is a constant to be chosen later. The function v satisfies

$$\begin{cases} v(x) > 0 & \text{if and only if} \quad |x - x_0| < R, \\ v(x) = 0 & \text{if and only if} \quad |x - x_0| = R, \\ v(x) < 0 & \text{if and only if} \quad |x - x_0| > R. \end{cases}$$

Now, set

$$D := \left\{ x \in \mathbb{R}^N \;:\; \frac{R}{2} < |x - x_0| < R \right\} \subset \Omega$$

and

$$c^+ := \max\{c, 0\} \geq c, \qquad \mathfrak{L}^+ := \mathfrak{L} - c + c^+.$$

Then,

$$\mathfrak{L}^+ v = - \sum_{i,j=1}^N a_{ij} \frac{\partial^2 E}{\partial x_i \partial x_j} + \sum_{j=1}^N b_j \frac{\partial E}{\partial x_j} + c^+ v$$

and, differentiating and rearranging terms, shows that, for every $x \in \Omega$,

$$\mathfrak{L}^+ v(x) = \left\{ -4\alpha^2 \sum_{i,j=1}^N a_{ij}(x)(x_i - x_{0i})(x_j - x_{0j}) \right.$$
$$\left. + 2\alpha \sum_{j=1}^N [a_{jj}(x) - b_j(x)(x_j - x_{0j})] + c^+(x) \right\} E(x) - c^+(x)$$
$$= \left\{ -4\alpha^2 (x - x_0)^T A_x (x - x_0) + 2\alpha[\operatorname{tr} A_x - \langle b(x), x - x_0 \rangle] \right.$$
$$\left. + c^+(x) \right\} E(x) - c^+(x),$$

where A_x is the matrix of the principal coefficients of $\mathfrak{L}$ (see (1.3)), $\operatorname{tr} A_x$ stands for the trace of A_x, and $b := (b_1, ..., b_N)$. Let $\mu > 0$ be the ellipticity constant of $\mathfrak{L}$ in Ω (see Definition 1.1). Then, for every $x \in D$,

$$(x - x_0)^T A_x (x - x_0) \geq \mu |x - x_0|^2 > \mu \frac{R^2}{4}.$$

Moreover,

$$|\operatorname{tr} A_x - \langle b(x), x - x_0 \rangle| \leq |\operatorname{tr} A_x| + |b(x)| R$$

and hence, thanks to (1.26), there exists a constant

$$C := C \left(\|a_{jj}\|_{L^\infty(\Omega)}, \|b_j\|_{L^\infty(\Omega)}, R \right) = C(\mathfrak{L}, R) > 0$$

such that

$$|\operatorname{tr} A_x - \langle b(x), x - x_0 \rangle| \leq C \qquad \text{for all } x \in D,$$

independently of x_0. Thus, for every $x \in D$, we have that

$$\mathfrak{L}^+ v(x) \leq \left(-\alpha^2 \mu R^2 + 2\alpha C + \|c^+\|_{L^\infty(\Omega)} \right) E(x) - c^+(x)$$

and, therefore, there exists

$$\alpha := \alpha(\mathfrak{L}, R) > 0$$

such that

$$\mathfrak{L}^+ v(x) < 0 \qquad \text{for all } x \in D. \tag{1.38}$$

Throughout the rest of this proof, we suppose $\alpha > 0$ has been chosen to satisfy (1.38).

According to (1.35), $u(x) > 0$ for each $x \in \bar{B}_{\frac{R}{2}}(x_0)$ and hence, by the continuity of u,

$$u_R := \inf_{\bar{B}_{\frac{R}{2}}(x_0)} u > 0.$$

Subsequently, we consider the auxiliary function

$$w := u - \epsilon v, \qquad \epsilon := \frac{u_R}{e^{\alpha R^2} - 1}, \qquad (1.39)$$

which is well defined in $\bar{B}_R(x_0)$. Thanks to (1.39), it is apparent that

$$w \geq u_R - \epsilon(e^{\alpha R^2} - 1) = 0 \qquad \text{in} \quad \bar{B}_{\frac{R}{2}}(x_0),$$

because

$$u \geq u_R \qquad \text{and} \qquad v \leq v(x_0) = e^{\alpha R^2} - 1.$$

In particular,

$$w \geq 0 \qquad \text{on} \qquad \partial B_{\frac{R}{2}}(x_0).$$

Moreover, since $v = 0$ on $\partial B_R(x_0)$ and $u \geq 0$ in $\bar{B}_R(x_0)$, we have that

$$w \geq 0 \qquad \text{on} \qquad \partial B_R(x_0).$$

Consequently,

$$w \geq 0 \qquad \text{on} \qquad \partial D = \partial B_R(x_0) \cup \partial B_{\frac{R}{2}}(x_0). \qquad (1.40)$$

Furthermore, thanks to (1.38), we find that, in D,

$$\mathcal{L}^+ w = \mathcal{L}^+ u - \epsilon \mathcal{L}^+ v > \mathcal{L}^+ u = (\mathcal{L} - c + c^+)u \geq (c^+ - c)u \geq 0,$$

because

$$\mathcal{L} u \geq 0, \qquad c^+ \geq c, \qquad \text{and} \ u \geq 0.$$

Thus, w is a superharmonic function of $\mathcal{L}^+$ in D satisfying (1.40) and, therefore, Theorem 1.4 yields to

$$w = u - \epsilon v \geq 0 \qquad \text{in} \quad D. \qquad (1.41)$$

On the other hand, for every $x \in D$, we have that

$$v(x) = e^{\alpha(R^2 - |x - x_0|^2)} - 1 = e^{\alpha(R + |x - x_0|)(R - |x - x_0|)} - 1$$

$$\geq e^{\frac{3}{2}\alpha R(R - |x - x_0|)} - 1 \geq \frac{3}{2}\alpha R(R - |x - x_0|)$$

$$= \frac{3}{2}\alpha R \operatorname{dist}(x, \partial B_R(x_0)).$$

Thus, it follows from (1.39) and (1.41) that, for every $x \in D$,

$$u(x) \geq \epsilon v(x) \geq \frac{3\alpha R}{2 \left(e^{\alpha R^2} - 1\right)} u_R \, \mathrm{dist} \left(x, \partial B_R(x_0)\right). \qquad (1.42)$$

Finally, taking into account that, for any $x \in \bar{B}_{\frac{R}{2}}(x_0)$,

$$u(x) \geq u_R \geq \frac{u_R}{R} \, \mathrm{dist} \left(x, \partial B_R(x_0)\right),$$

it becomes apparent that (1.42) implies

$$u \geq \min \left\{ \frac{3\alpha R}{2 \left(e^{\alpha R^2} - 1\right)}, \frac{1}{R} \right\} u_R \, \mathrm{dist} \left(\cdot, \partial B_R(x_0)\right) \qquad \text{in } \bar{B}_R(x_0).$$

As the constant

$$M(\mathfrak{L}, R) := \min \left\{ \frac{3\alpha R}{2 \left(e^{\alpha R^2} - 1\right)}, \frac{1}{R} \right\}$$

satisfies (1.37) for all admissible u, the proof is complete. $\qquad \square$

For bounded domains satisfying the uniform interior sphere property in the strong sense, the following sharp version of Theorem 1.5 holds.

Theorem 1.6. *Suppose Ω is a bounded domain satisfying the uniform interior sphere property in the strong sense on a closed and open subset Γ_0 of $\partial\Omega$, and $\mathfrak{L}$ is a uniformly elliptic operator in Ω whose coefficients are measurable and bounded in Ω.*

Let $u \in \mathcal{C}^2(\Omega) \cap \mathcal{C}(\bar{\Omega})$ be a superharmonic function of $\mathfrak{L}$ in Ω such that

$$u(x) > 0 \qquad \text{for all } x \in \bar{\Omega} \setminus \Gamma_0.$$

Then, there exists $\delta > 0$ such that

$$u(x) \geq \delta \, \mathrm{dist} \left(x, \Gamma_0\right) \qquad \text{for all } x \in \Omega.$$

Proof. Suppose Ω satisfies the uniform interior sphere property in the strong sense on Γ_0 with parameter $R > 0$, and consider the compact subset of $\bar{\Omega}$ defined by

$$K_R := \left\{ x \in \bar{\Omega} \; : \; \mathrm{dist} \left(x, \Gamma_0\right) \geq R/2 \right\}.$$

Let $u \in \mathcal{C}^2(\Omega) \cap \mathcal{C}(\bar{\Omega})$ such that $\mathfrak{L}u \geq 0$ in Ω and $u(x) > 0$ for every $x \in \bar{\Omega} \setminus \Gamma_0$. Then,

$$u_L := \min_{K_R} u > 0.$$

Let $x \in \Omega$ with $\mathrm{dist} \left(x, \Gamma_0\right) \leq R$ and consider $y_x \in \Gamma_0$ such that

$$\mathrm{dist} \left(x, \Gamma_0\right) = |x - y_x| \quad \text{and} \quad B_R(x_0) \subset \Omega,$$

where
$$x_0 := y_x + R \frac{x - y_x}{|x - y_x|}.$$

Since $\bar{B}_{\frac{R}{2}}(x_0) \subset K_R$, we have that
$$\min_{\bar{B}_{\frac{R}{2}}(x_0)} u \geq u_L,$$

and hence, according to Theorem 1.5, there exists a constant $M = M(\mathfrak{L}, R)$ (independent of x) such that
$$u(z) \geq M u_L \, \text{dist}\,(z, \partial B_R(x_0)) \qquad \text{for all } z \in \bar{B}_R(x_0).$$

Therefore, for every $x \in \Omega \cap (\Gamma_0 + \bar{B}_R)$, we find that
$$u(x) \geq M u_L \, \text{dist}\,(x, \partial B_R(x_0)) = M u_L |x - y_x| = M u_L \, \text{dist}\,(x, \Gamma_0).$$

Finally, as
$$\eta := \inf_{x \in K_R} \frac{u(x)}{\text{dist}\,(x, \Gamma_0)} > 0,$$

it becomes apparent that making the choice
$$\delta := \min\{\eta, M u_L\}$$

concludes the proof. $\qquad\qquad\qquad\qquad\qquad\qquad\qquad\qquad\qquad\qquad\qquad\qquad$ □

As the class of domains for which Theorem 1.6 holds true will play a significant role throughout this book, it is appropriate to introduce the following concept.

Definition 1.4. *Suppose Ω is a bounded domain, Γ_0 is a closed and open subset of $\partial\Omega$, and $\mathfrak{L}$ is a uniformly elliptic operator in Ω satisfying (1.26). It is said that $(\mathfrak{L}, \Omega)$ satisfies the* **decay property of E. Hopf on Γ_0** *if for any superharmonic function $u \in \mathcal{C}^2(\Omega) \cap \mathcal{C}(\bar{\Omega})$ of $\mathfrak{L}$ in Ω with $u(x) > 0$ for all $x \in \bar{\Omega} \setminus \Gamma_0$ there exists $\delta > 0$ such that*
$$u(x) \geq \delta \, \text{dist}\,(x, \Gamma_0) \qquad \text{for all } x \in \Omega. \tag{1.43}$$

Using this concept, Theorem 1.6 can be equivalently stated as follows.

Corollary 1.2. *Suppose Ω is a bounded domain, Γ_0 is a closed and open subset of $\partial\Omega$, and $\mathfrak{L}$ is a uniformly elliptic operator in Ω whose coefficients are measurable and bounded in Ω. Then, $(\mathfrak{L}, \Omega)$ satisfies the decay property of E. Hopf on Γ_0 if Ω satisfies the uniform interior sphere property in the strong sense on Γ_0.*

1.7   The generalized minimum principle of M. H. Protter and H. F. Weinberger

The next result provides us with a sharp generalization of Theorems 1.2 and 1.3 to cover the general case when the coefficient function $c(x)$ is not necessarily non-negative.

Theorem 1.7 (Generalized minimum principle). *Suppose $\mathfrak{L}$ is a uniformly elliptic operator in Ω, and $(\mathfrak{L}, \Omega)$ admits a strictly positive superharmonic function $h \in \mathcal{C}^2(\Omega) \cap \mathcal{C}(\bar{\Omega})$, in the sense that*

i) $h(x) > 0$ *for all* $x \in \bar{\Omega}$,
ii) $\mathfrak{L}h \geq 0$ *in* Ω.

Then, for any superharmonic function $u \in \mathcal{C}^2(\Omega)$ of $\mathfrak{L}$ in Ω such that

$$m := \inf_{\Omega} \frac{u}{h} \in (-\infty, 0], \tag{1.44}$$

either

$$u(x) > mh(x) \qquad \text{for all} \quad x \in \Omega, \tag{1.45}$$

or

$$u = mh \qquad in \quad \Omega. \tag{1.46}$$

Further, suppose (1.26), (1.45), and the following three conditions:

(a) $h \in \mathcal{C}^1(\bar{\Omega})$;
(b) $u(x_0) = mh(x_0)$ *for some $x_0 \in \partial\Omega$, and there exists $R > 0$ such that $u \in \mathcal{C}(B_R(x_0) \cap \bar{\Omega})$;*
(c) Ω *satisfies the interior tangent sphere property at x_0 and there is an outward pointing vector $\nu \in \mathbb{R}^N \setminus \{0\}$ for which $\frac{\partial(u/h)}{\partial\nu}(x_0)$ exists.*

Then,

$$\frac{\partial(u/h)}{\partial\nu}(x_0) < 0.$$

Remark 1.2. When $c \geq 0$, the function $h := 1$ satisfies (i) and (ii) and, hence, it provides us with a strict positive superharmonic function of $(\mathfrak{L}, \Omega)$. Consequently, in this special case, Theorem 1.7 provides us with Theorems 1.2 and 1.3 simultaneously, for as $u/h = u$. Theorem 1.7 is substantially sharper than these results, because it does not impose any restriction on the sign of $c \in L^\infty(\Omega)$.

Proof. Subsequently, we set

$$v(x) := \frac{u(x)}{h(x)} \quad \text{for every} \quad x \in \Omega.$$

A direct calculation shows that, in Ω,

$$\mathfrak{L}u = \mathfrak{L}(hv) = -\sum_{i,j=1}^{N} a_{ij} \frac{\partial^2(hv)}{\partial x_i \partial x_j} + \sum_{j=1}^{N} b_j \frac{\partial(hv)}{\partial x_j} + chv$$

$$= -h \sum_{i,j=1}^{N} a_{ij} \frac{\partial^2 v}{\partial x_i \partial x_j} + h \sum_{j=1}^{N} \left(b_j - \frac{2}{h} \sum_{i=1}^{N} a_{ij} \frac{\partial h}{\partial x_i} \right) \frac{\partial v}{\partial x_j} + v\mathfrak{L}h$$

$$= h \left[-\sum_{i,j=1}^{N} a_{ij} \frac{\partial^2 v}{\partial x_i \partial x_j} + \sum_{j=1}^{N} \left(b_j - \frac{2}{h} \sum_{i=1}^{N} a_{ij} \frac{\partial h}{\partial x_i} \right) \frac{\partial v}{\partial x_j} + \frac{\mathfrak{L}h}{h} v \right]$$

and hence,

$$\mathfrak{L}u = h\mathfrak{L}_h v \quad \text{in} \quad \Omega, \tag{1.47}$$

where

$$\mathfrak{L}_h := -\sum_{i,j=1}^{N} a_{ij} \frac{\partial^2}{\partial x_i \partial x_j} + \sum_{j=1}^{N} \left(b_j - \frac{2}{h} \sum_{i=1}^{N} a_{ij} \frac{\partial h}{\partial x_i} \right) \frac{\partial}{\partial x_j} + \frac{\mathfrak{L}h}{h}. \tag{1.48}$$

As $h \in \mathcal{C}^2(\Omega) \cap \mathcal{C}(\bar{\Omega})$, we have that

$$\frac{\mathfrak{L}h}{h} \in L^\infty_{loc}(\Omega), \quad b_j - \frac{2}{h} \sum_{i=1}^{N} a_{ij} \frac{\partial h}{\partial x_i} \in L^\infty_{loc}(\Omega), \quad 1 \le j \le N.$$

Moreover,

$$\frac{\mathfrak{L}h}{h} \ge 0 \quad \text{in} \quad \Omega,$$

and, consequently, Theorems 1.2 and 1.3 can be applied to the operator $\mathfrak{L}_h$ defined by (1.48).

Since $\mathfrak{L}u \ge 0$ in Ω and $h(x) > 0$ for all $x \in \bar{\Omega}$, (1.47) implies that

$$\mathfrak{L}_h v \ge 0 \quad \text{in} \quad \Omega.$$

On the other hand, (1.44) can be equivalently expressed in the form

$$m = \inf_\Omega v \in (-\infty, 0].$$

Therefore, according to Theorem 1.2, either

$$v(x) > m \quad \text{for every} \quad x \in \Omega, \tag{1.49}$$

or

$$v = m \quad \text{in} \quad \Omega. \tag{1.50}$$

As (1.49) and (1.50) imply (1.45) and (1.46), respectively, the proof of the alternative of the first part of the theorem is complete.

Now, suppose (1.26), (1.45), and conditions (a), (b) and (c). Then, all the coefficients of $\mathfrak{L}_h$ live in $L^\infty(\Omega)$, and

$$v(x_0) = m, \quad v \in \mathcal{C}(B_R(x_0) \cap \bar{\Omega}).$$

Consequently, owing to Theorem 1.3, we find that

$$\frac{\partial v}{\partial \nu}(x_0) < 0,$$

which concludes the proof. □

1.8 Appendix: Smooth domains

The following definition fixes the concept of regularity for open sets.

Definition 1.5. *Let $\Omega \subset \mathbb{R}^N$, $N \geq 1$, be an open set with boundary $\partial\Omega$, $x_0 \in \partial\Omega$, and consider an integer $k \geq 1$. Then, Ω is said to be of class $\mathcal{C}^k$ at x_0 if there exist $R := R(x_0) > 0$, an open neighborhood D of zero in $\mathbb{R}^N$, and a bijection $\Phi : B_R(x_0) \to D$ such that $\Phi(x_0) = 0$ and*

i) $D\Phi(x_0) \in \mathrm{Iso}(\mathbb{R}^N)$,

ii) $\Phi(B_R(x_0) \cap \Omega) = \{(x_1, ..., x_N) \in D \; : \; x_N > 0\}$,

iii) $\Phi(B_R(x_0) \cap \partial\Omega) = \{(x_1, ..., x_N) \in D \; : \; x_N = 0\}$,

iv) $\Phi \in \mathcal{C}^k(B_R(x_0); D)$ *and* $\Phi^{-1} \in \mathcal{C}^k(D; B_R(x_0))$.

In such case, Φ is said to be a diffeomorphism straightening $\partial\Omega$ at x_0 into the hyperplane $x_N = 0$.

Given an open subset $\Gamma \subset \partial\Omega$, Γ is said to be of class $\mathcal{C}^k$ if Ω is of class $\mathcal{C}^k$ at each $x_0 \in \Gamma$. The whole domain Ω is said to be of class $\mathcal{C}^k$ when $\partial\Omega$ is of class $\mathcal{C}^k$.

Given an open subset $\Gamma \subset \partial\Omega$ of class $\mathcal{C}^k$, a function $\psi : \Gamma \to \mathbb{R}$ is said to be of class $\mathcal{C}^k$ if, for every $x_0 \in \Gamma$ and any diffeomorphism Φ straightening $\partial\Omega$ at x_0, the following condition holds

$$\psi \circ \Phi^{-1} \in \mathcal{C}^k(D \cap \{(x_1, ..., x_N) \in \mathbb{R}^N : x_N = 0\}).$$

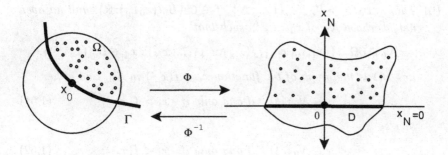

Fig. 1.9   The straightening diffeomorphism Φ ($\Gamma := \partial\Omega$)

Remark 1.3.

(a) According to the inverse function theorem, Property (i) implies Property (iv) if $\Phi \in \mathcal{C}^k(B_R(x_0); \mathbb{R}^N)$.

(b) Suppose $\partial\Omega$ is of class $\mathcal{C}^k$ for some integer $k \geq 1$. Then, $\psi \in \mathcal{C}^k(\bar{\Omega})$ implies $\psi|_{\partial\Omega} \in \mathcal{C}^k(\partial\Omega)$. Conversely, according to the H. F. F. Tietze extension theorems, every function $\psi \in \mathcal{C}^k(\partial\Omega)$ admits an extension to a function of class $\mathcal{C}^k(\bar{\Omega})$.

The next result shows that Ω is of class $\mathcal{C}^k$ at $x_0 \in \partial\Omega$ if and only if in a neighborhood of x_0 the boundary $\partial\Omega$ is the graph of a $\mathcal{C}^k$ function of $N-1$ variables. Subsequently, the coordinates of a vector $x \in \mathbb{R}^N$ are denoted by

$$x = (x_1, ..., x_N),$$

and, for a given $i \in \{1, ..., N-1\}$, we will denote by $x_{[i]}$ the vector

$$x_{[i]} := (x_1, ..., x_{i-1}, x_{i+1}, ..., x_N) \in \mathbb{R}^{N-1},$$

while

$$x_{[N]} := (x_1, ..., x_{N-1}) \in \mathbb{R}^{N-1}.$$

Also, for every $\eta > 0$ and $x_0 \in \mathbb{R}^N$, we denote by $B_\eta((x_0)_{[i]})$ the ball of $\mathbb{R}^{N-1}$ of radius η centered at $(x_0)_{[i]} \in \mathbb{R}^{N-1}$. Using these notations, the following result holds.

Theorem 1.8. *Let* $\Omega \subset \mathbb{R}^N$, $N \geq 1$, *be an open set with boundary* $\partial\Omega$, $x_0 \in \partial\Omega$, *and* $k \in \mathbb{N}$, $k \geq 1$. *Then, the following conditions are equivalent:*

(a) Ω *is of class* $\mathcal{C}^k$ *at* x_0.

(b) *There exist $\eta > 0$, $i \in \{1, ..., N\}$, $f \in C^k(\bar{B}_\eta((x_0)_{[i]}); \mathbb{R})$, and an open neighborhood $\mathcal{N}_0$ of x_0 in $\mathbb{R}^N$ such that*

$$\mathcal{N}_0 \cap \partial\Omega = \{(x_1, ..., x_{i-1}, f(x_{[i]}), x_{i+1}, ..., x_N) : x_{[i]} \in B_\eta((x_0)_{[i]})\},$$

i.e., $\partial\Omega$ is the graph of the function $x_i = f(x_{[i]})$ in $\mathcal{N}_0$, and either

$$x \in \mathcal{N}_0 \cap \Omega \quad \text{if and only if} \quad x_i > f(x_{[i]}), \tag{1.51}$$

or else

$$x \in \mathcal{N}_0 \cap \Omega \quad \text{if and only if} \quad x_i < f(x_{[i]}). \tag{1.52}$$

Proof. Suppose (b) with, e.g., (1.51). Then, by Remark 1.3(a), it is easy to see that the map

$$\Phi(x) := \left((x - x_0)_{[i]}, x_i - f(x_{[i]})\right), \qquad x \in B_R(x_0),$$

provides us with a diffeomorphism straightening $\partial\Omega$ at x_0 for sufficiently small $R > 0$, as discussed in Definition 1.5. Therefore, (b) implies (a).

Suppose (a) and let Φ be a diffeomorphism straightening $\partial\Omega$ at x_0, as discussed in Definition 1.5. Let Φ_N denote its N-th coordinate map. Necessarily, $\nabla\Phi_N(x_0) \neq 0$, because $D\Phi(x_0) \in \text{Iso}(\mathbb{R}^N)$. Thus,

$$\frac{\partial\Phi_N}{\partial x_i}(x_0) \neq 0 \quad \text{for some} \quad i \in \{1, ..., N\}. \tag{1.53}$$

Moreover, by definition,

$$B_R(x_0) \cap \partial\Omega = \Phi_N^{-1}(0).$$

Consequently, (b) should follow by applying the implicit function theorem to the equation $\Phi_N = 0$ at x_0, by expressing x_i as a function of the remaining variables, $x_i = f(x_{[i]})$. Indeed, consider the one-parametric family of equations

$$\Phi_N(x_1, ..., x_N) - c = 0, \qquad c \sim 0.$$

According to the implicit function theorem, we find from (1.53) that there exist an open neighborhood $\mathcal{M}$ of

$$(c, x_{[i]}) = (0, (x_0)_{[i]}) \in \mathbb{R}^N$$

and a (unique) function $F \in C^k(\mathcal{M}; \mathbb{R})$ such that

$$F(0, (x_0)_{[i]}) = (x_0)_i$$

and, for every $(c, x_{[i]}) \in \mathcal{M}$,

$$\Phi_N(x_1, ..., x_{i-1}, F(c, x_{[i]}), x_{i+1}, ..., x_N) - c = 0. \tag{1.54}$$

By the uniqueness obtained as an application of the implicit function theorem, it is apparent that $\partial\Omega$ is given through the graph of the function

$$f := F(0, \cdot).$$

Moreover, differentiating (1.54) with respect to c, yields to

$$1 = \frac{\partial \Phi_N}{\partial x_i}(x_0)\frac{\partial F}{\partial c}(0, (x_0)_{[i]}). \tag{1.55}$$

Thus, if, e.g., $\frac{\partial \Phi_N}{\partial x_i}(x_0) > 0$, then, by continuity, (1.55) implies that

$$\frac{\partial F}{\partial c}(c, x_{[i]}) > 0$$

for $c \sim 0$ and $x_{[i]} \sim (x_0)_{[i]}$. According to Definition 1.5 (ii), for every $x \in \Omega$ sufficiently close to x_0, we have that

$$c := \Phi_N(x) > 0$$

and hence,

$$x_i = F(c, x_{[i]}) > F(0, x_{[i]}) = f(x_{[i]}),$$

whereas if $x \notin \bar{\Omega}$, then $c := \Phi_N(x) < 0$ and, so,

$$x_i = F(c, x_{[i]}) < F(0, x_{[i]}) = f(x_{[i]}).$$

Consequently, in such case, condition (1.51) holds. Similarly, (1.52) holds when $\frac{\partial \Phi_N}{\partial x_i}(x_0) < 0$. This concludes the proof. $\square$

The following result shows that Ω satisfies the uniform interior sphere property in the strong sense if it is of class $\mathcal{C}^2$.

Theorem 1.9. *Let $\Omega \subset \mathbb{R}^N$, $N \geq 1$, be an open set with boundary $\partial\Omega$, and Γ_0 a closed and open subset of $\partial\Omega$ of class $\mathcal{C}^2$. Then, Ω satisfies the uniform interior sphere property in the strong sense on Γ_0.*

Proof. By Lemma 1.1(b), it suffices to prove that Ω satisfies the uniform interior sphere property on Γ_0.

Pick $x_0 \in \Gamma_0$, and let $\eta > 0$, $i \in \{1, ..., N\}$, and $f \in \mathcal{C}^2(\bar{B}_\eta((x_0)_{[i]}); \mathbb{R})$, satisfying Theorem 1.8(b) with $k = 2$. Subsequently, we set

$$K := \bar{B}_\eta((x_0)_{[i]}) \times \bar{B}_\eta((x_0)_{[i]})$$

and consider the function $R \in \mathcal{C}^2(K; \mathbb{R})$ defined through

$$R(\tilde{x}, \tilde{y}) := f(\tilde{x}) - f(\tilde{y}) - Df(\tilde{y})(\tilde{x} - \tilde{y}) - \frac{1}{2}D^2 f(\tilde{y})(\tilde{x} - \tilde{y}, \tilde{x} - \tilde{y}),$$

for all $(\tilde{x}, \tilde{y}) \in K$, where Df and $D^2 f$ stand for the first and the second order differentials of f. It is well known that $Df(\tilde{y})$ is given by the gradient $\nabla f(\tilde{y})$, while $D^2 f(\tilde{y})$ is given through the Hessian matrix of f at $\tilde{y}$. By our regularity assumptions on f, we have that

$$\lim_{\tilde{x} \to \tilde{y}} \frac{R(\tilde{x}, \tilde{y})}{|\tilde{x} - \tilde{y}|^2} = 0, \qquad |\tilde{y} - (x_0)_{[i]}| \le \eta.$$

Therefore, the quotient function Q defined by

$$Q(\tilde{x}, \tilde{y}) := \begin{cases} \frac{R(\tilde{x}, \tilde{y})}{|\tilde{x} - \tilde{y}|^2}, & \text{if } \tilde{x} \neq \tilde{y}, \\ 0, & \text{if } \tilde{x} = \tilde{y}, \end{cases}$$

is continuous in K. As K is compact, Q is uniformly continuous in K and, therefore, for every $h > 0$ there exists $\delta > 0$ such that

$$|\tilde{x} - \tilde{y}| \le \delta \quad \Longrightarrow \quad |R(\tilde{x}, \tilde{y})| \le h|\tilde{x} - \tilde{y}|^2 \qquad (1.56)$$

for every $\tilde{x}, \tilde{y} \in K$. The importance of this estimate will become apparent later.

By the choice of f and $i \in \{1, ..., N\}$, the tangent hyperplane to $\partial \Omega$ at x_0 is given through

$$x_i = x_{0i} + \nabla f((x_0)_{[i]}) \left(x_{[i]} - (x_0)_{[i]} \right)$$

and hence,

$$\nu := \left(\frac{\partial f}{\partial x_1}, ..., \frac{\partial f}{\partial x_{i-1}}, -1, \frac{\partial f}{\partial x_{i+1}}, ..., \frac{\partial f}{\partial x_N} \right) \Bigg|_{((x_0)_{[i]})}$$

is the normal vector to $\partial \Omega$ at x_0 if $i < N$, whereas

$$\nu := \left(\frac{\partial f}{\partial x_1}, ..., \frac{\partial f}{\partial x_{N-1}}, -1 \right) \Bigg|_{((x_0)_{[i]})}$$

if $i = N$. Subsequently, the sign $\pm$ is chosen so that the unit normal vector

$$\mathbf{n} := \pm \frac{\nu}{|\nu|}$$

satisfy $x_0 - \epsilon \mathbf{n} \in \Omega$ for sufficiently small $\epsilon > 0$. For this choice, $\mathbf{n}$ provides us with the *outward unit normal to* Ω at $x_0 \in \partial \Omega$. Next, we will show that

$$B_\epsilon(x_0 - \epsilon \mathbf{n}) \subset \Omega$$

for sufficiently small $\epsilon > 0$. Setting $\mathbf{n} := (n_1, ..., n_N)$, it is easy to see that this holds from the estimate

$$D_0 := \left[f(x_{[i]}) + \epsilon n_i - x_{0i} \right]^2 + \sum_{\substack{j=1 \\ j \neq i}}^{N} (x_j + \epsilon n_j - x_{0j})^2 \ge \epsilon^2. \qquad (1.57)$$

Indeed, by Theorem 1.8(b), we have that

$$(x_1, ..., x_{i-1}, f(x_{[i]}), x_{i+1}, ..., x_N) \in \partial\Omega$$

for $x_{[i]} \sim (x_0)_{[i]}$, and (1.57) implies that

$$(x_1, ..., x_{i-1}, f(x_{[i]}), x_{i+1}, ..., x_N) \in \mathbb{R}^N \setminus B_\epsilon(x_0 - \epsilon\mathbf{n}).$$

Therefore, $B_\epsilon(x_0 - \epsilon\mathbf{n}) \subset \Omega$ because $x_0 - \epsilon\mathbf{n} \in \Omega$ for sufficiently small $\epsilon > 0$.

The estimate (1.57) can be derived from the next chain of identities

$$D_0 = \left[\epsilon n_i + \nabla f\left((x_0)_{[i]}\right)\left(x_{[i]} - (x_0)_{[i]}\right) \right.$$
$$\left. + \frac{1}{2} D^2 f\left((x_0)_{[i]}\right)\left(x_{[i]} - (x_0)_{[i]}, x_{[i]} - (x_0)_{[i]}\right) + R\left(x_{[i]}, (x_0)_{[i]}\right) \right]^2$$

$$+ \sum_{\substack{j=1 \\ j \neq i}}^{N} (x_j - x_{0j})^2 + \sum_{\substack{j=1 \\ j \neq i}}^{N} \epsilon^2 n_j^2 + 2\epsilon \sum_{\substack{j=1 \\ j \neq i}}^{N} n_j(x_j - x_{0j})$$

$$= \epsilon^2 \sum_{j=1}^{N} n_j^2 + 2\epsilon n_i \nabla f\left((x_0)_{[i]}\right)\left(x_{[i]} - (x_0)_{[i]}\right) + 2\epsilon \sum_{\substack{j=1 \\ j \neq i}}^{N} n_j(x_j - x_{0j})$$

$$+ \left[\nabla f\left((x_0)_{[i]}\right)\left(x_{[i]} - (x_0)_{[i]}\right) \right]^2 + \sum_{\substack{j=1 \\ j \neq i}}^{N} (x_j - x_{0j})^2$$

$$+ \epsilon n_i D^2 f\left((x_0)_{[i]}\right)\left(x_{[i]} - (x_0)_{[i]}, x_{[i]} - (x_0)_{[i]}\right) + \mathfrak{R}\left(x_{[i]}, (x_0)_{[i]}\right),$$

where $\mathfrak{R}$ is a certain C^2 function satisfying

$$\lim_{x_{[i]} \to (x_0)_{[i]}} \frac{\mathfrak{R}\left(x_{[i]}, (x_0)_{[i]}\right)}{\left| x_{[i]} - (x_0)_{[i]} \right|^2} = 0, \tag{1.58}$$

whose explicit knowledge is not important for our purposes here. Since

$$\sum_{j=1}^{N} n_j^2 = |\mathbf{n}|^2 = 1, \qquad \left[\nabla f\left((x_0)_{[i]}\right)\left(x_{[i]} - (x_0)_{[i]}\right) \right]^2 \geq 0,$$

and

$$n_i \frac{\partial f}{\partial x_j}\left((x_0)_{[i]}\right) + n_j = \pm \left(\frac{-1}{|\nu|} \frac{\partial f}{\partial x_j}\left((x_0)_{[i]}\right) + n_j \right) = 0, \qquad j \neq i,$$

we find from the previous identity that

$$D_0 \geq \epsilon^2 + \sum_{\substack{j=1 \\ j \neq i}}^{N} (x_j - x_{0j})^2 + \mathfrak{R}\left(x_{[i]}, (x_0)_{[i]}\right)$$

$$+ \epsilon n_i D^2 f\left((x_0)_{[i]}\right)\left(x_{[i]} - (x_0)_{[i]}, x_{[i]} - (x_0)_{[i]}\right).$$

Therefore, owing to (1.58), in a neighborhood of x_0 we obtain that

$$D_0 \geq \epsilon^2 + \frac{1}{2}|x_{[i]} - (x_0)_{[i]}|^2 \geq \epsilon^2$$

for sufficiently small $\epsilon > 0$. Actually, by (1.56), this estimate is satisfied uniformly in a neighborhood of x_0, because f is of class C^2. Consequently, (1.57) holds and, therefore, Ω satisfies the uniform interior sphere property on Γ_0. This completes the proof. $\square$

1.9 Comments on Chapter 1

Most of the contents covered in this chapter were elaborated from the classical textbook of M. H. Protter and H. F. Weinberger [183], as well as many of the subsequent historical remarks; it is the most paradigmatic book about the *maximum principle* for elliptic operators, and, undoubtedly, it counts among the best monographs on Partial Differential Equations of the 20th Century. However, the reader should pay special attention to all necessary changes for re-stating all classical results in terms of minimum principles, instead of maximum principles, as it has been done in the present monograph.

In the special case when $\mathfrak{L} = -\Delta$, Theorem 1.1 goes back to C. F. Gauss [75] and S. Earnshaw [58], and to A. Paraf [174] when $N = 2$ (two dimensions) and $c > 0$. The version of A. Paraf was extended by E. Picard [178] and L. Lichtenstein [125], [126], to cover the more general case when $c \geq 0$. Later, T. Moutard [169] generalized the Paraf result to more than two dimensions. These results were used by M. Picone [179] to obtain a generalized minimum principle. Theorem 1.2 of Section 1.2 goes back to E. Hopf [103]; it was the first result where the continuity assumptions on the coefficients of $\mathfrak{L}$ were removed. The proof of Theorem 1.2 given in this chapter is based upon the proof of Theorems 5 and 6 of Section 3 in M. H. Protter and H. F. Weinberger [183].

Section 1.3 is a non-trivial re-elaboration of the contents of the two paragraphs after Theorem 1 of W. Walter [224]. The example (1.24) is due to R. Redheffer [187].

Under some additional continuity properties on the coefficients of the operator, Theorem 1.3 goes back to G. Giraud [80], [81]. The version included in this chapter is attributable to E. Hopf [104] and O. A. Oleinik [172]. It is a re-elaboration from Theorems 7 and 8 of Section 3 in M. H. Protter and H. F. Weinberger [183].

Theorems 1.5 and 1.6 are based on Lemma 1 of W. Walter [224]. The relevance of these uniform versions of Theorem 1.3 will become apparent in Chapter 2.

Theorem 1.7 goes back to Theorem 10 of Chapter 2 of M. H. Protter and H. F. Weinberger [183]; it generalizes substantially Theorems 1.2 and 1.3. In Chapter 7 we will show that the existence of h satisfying conditions (i) and (ii) is, actually, equivalent to the positivity of the principal eigenvalue of $\mathfrak{L}$ in Ω under Dirichlet boundary conditions. This characterization will provide us with a sharp substantial generalization of all classical results of this chapter.

Section 1.8 reviews some very basic concepts and results, which, however, might be difficult to document in the available literature. For example, though L. C. Evans in the Remark on p. 330 of [60] claimed that " the interior ball condition automatically holds if $\partial\Omega$ is C^2 ", the reader should recognize that, being certainly elementary, the proof of Theorem 1.9 is far from ' automatic '.

Chapter 2

Classifying supersolutions

This chapter will be developed under the following general hypotheses:

H1 Ω is a bounded domain of $\mathbb{R}^N$, $N \geq 1$, whose boundary consists of two disjoint open and closed subsets, Γ_0 and Γ_1,

$$\partial\Omega := \Gamma_0 \cup \Gamma_1,$$

and it satisfies the interior sphere property at every $x \in \partial\Omega$. Necessarily, Γ_0 and Γ_1 must possess finitely many components. Either Γ_0, or Γ_1, might be empty.

H2 The differential operator

$$\mathfrak{L} := -\sum_{i,j=1}^{N} a_{ij} \frac{\partial^2}{\partial x_i \partial x_j} + \sum_{j=1}^{N} b_j \frac{\partial}{\partial x_j} + c \qquad (2.1)$$

is uniformly elliptic in Ω and

$$a_{ij},\, b_j,\, c \in L^\infty(\Omega), \qquad i, j \in \{1, ..., N\}.$$

H3 $\beta \in \mathcal{C}(\Gamma_1; \mathbb{R})$ is a continuous function and $\nu = (\nu_1, ..., \nu_N) \in \mathcal{C}(\partial\Omega; \mathbb{R}^N)$ is an outward pointing nowhere tangent vector field, in the sense that $\nu(x)$ is an outward pointing vector at x for all $x \in \partial\Omega$.

Under these assumptions, we can introduce the boundary operator

$$\mathfrak{B} \,:\, \mathcal{C}(\Gamma_0) \otimes \mathcal{C}^1(\Gamma_1) \to C(\partial\Omega)$$

defined by

$$\mathfrak{B}\psi := \begin{cases} \psi & \text{on} \ \ \Gamma_0, \\ \frac{\partial\psi}{\partial\nu} + \beta\psi & \text{on} \ \ \Gamma_1, \end{cases} \qquad \text{for all} \ \ \psi \in \mathcal{C}(\Gamma_0) \otimes \mathcal{C}^1(\Gamma_1). \qquad (2.2)$$

Note that $\mathfrak{B}$ is the *Dirichlet boundary operator* on Γ_0, subsequently denoted by $\mathfrak{D}$, the *Neumann* boundary operator on Γ_1 if $\nu = \mathbf{n}$ and $\beta = 0$, denoted by $\mathfrak{N}$, and a *first order regular oblique derivative boundary operator* on Γ_1 in any other case.

The following concepts are pivotal.

Definition 2.1. *Suppose $u \in \mathcal{C}^2(\Omega) \cap \mathcal{C}^1(\Omega \cup \Gamma_1) \cap \mathcal{C}(\bar{\Omega})$. Then:*

(a) *u is said to be a **supersolution** of $(\mathfrak{L}, \mathfrak{B}, \Omega)$ if*

$$\begin{cases} \mathfrak{L}u \geq 0 & \text{in } \Omega, \\ \mathfrak{B}u \geq 0 & \text{on } \partial\Omega. \end{cases}$$

*When some of these inequalities is strict, u is said to be a **strict supersolution** of $(\mathfrak{L}, \mathfrak{B}, \Omega)$.*

(b) *u is said to be a **subsolution** of $(\mathfrak{L}, \mathfrak{B}, \Omega)$ if*

$$\begin{cases} \mathfrak{L}u \leq 0 & \text{in } \Omega, \\ \mathfrak{B}u \leq 0 & \text{on } \partial\Omega. \end{cases}$$

*When some of these inequalities is strict, u is said to be a **strict subsolution** of $(\mathfrak{L}, \mathfrak{B}, \Omega)$.*

According to Definition 2.1, the main assumption of Theorem 1.7 can be expressed by simply saying that $h \in \mathcal{C}^2(\Omega) \cap \mathcal{C}(\bar{\Omega})$ is a supersolution of $(\mathfrak{L}, \mathfrak{D}, \Omega)$ everywhere positive in $\bar{\Omega}$. Consequently, it must be a positive strict supersolution of $(\mathfrak{L}, \mathfrak{D}, \Omega)$, because $\mathfrak{D}h > 0$ on $\Gamma_0 = \partial\Omega$.

The main goal of this chapter is ascertaining all admissible *canonical behaviors* of the supersolutions of $(\mathfrak{L}, \mathfrak{B}, \Omega)$ from the existence of a single positive supersolution. This objective will be accomplished through Theorems 2.1 and 2.4. Later, we will derive from Theorem 2.1 the strong positivity of the resolvent operator of the linear boundary value problem (2.31) associated to $(\mathfrak{L} + \omega, \mathfrak{B}, \Omega)$ for sufficiently large ω, which will be established by Theorem 2.2. In Chapter 7, using the theory of positive operators developed in Chapter 6, we will infer from these results the existence and uniqueness of the *principal eigenvalue* of $(\mathfrak{L} + \omega, \mathfrak{B}, \Omega)$.

2.1　First classification theorem

The following result provides us with all admissible *canonical behaviors* of the supersolutions of $(\mathfrak{L}, \mathfrak{B}, \Omega)$ in the presence of a supersolution everywhere positive in $\bar{\Omega}$. It should be remembered that the existence of such a supersolution was the main assumption of Theorem 1.7 in the special case when $\Gamma_1 = \emptyset$ ($\mathfrak{B} = \mathfrak{D}$).

Theorem 2.1. *Suppose* $(\mathfrak{L}, \mathfrak{B}, \Omega)$ *admits a supersolution* $h \in \mathcal{C}^2(\Omega) \cap \mathcal{C}^1(\bar{\Omega})$ *such that*

$$h(x) > 0 \qquad \text{for all } x \in \bar{\Omega}. \tag{2.3}$$

Then, any supersolution $u \in \mathcal{C}^2(\Omega) \cap \mathcal{C}^1(\bar{\Omega})$ *of* $(\mathfrak{L}, \mathfrak{B}, \Omega)$ *must satisfy some of the following alternatives:*

A1 $u = 0$ *in* Ω.
A2 $u(x) > 0$ *for every* $x \in \Omega \cup \Gamma_1$, *and*

$$\frac{\partial u}{\partial \nu}(x) < 0 \qquad \text{for all } \quad x \in u^{-1}(0) \cap \Gamma_0.$$

A3 *There exists a constant* $m < 0$ *such that*

$$u = mh \quad \text{in} \quad \bar{\Omega}.$$

In such case, $u(x) < 0$ *for all* $x \in \bar{\Omega}$, $\Gamma_0 = \emptyset$, *and*

$$\begin{cases} \mathfrak{L}h = 0 & \text{in } \Omega, \\ \mathfrak{B}h = 0 & \text{on } \partial\Omega. \end{cases} \tag{2.4}$$

In other words, $\tau = 0$ *must be an eigenvalue to a positive eigenfunction* (h *itself) of the linear eigenvalue problem*

$$\begin{cases} \mathfrak{L}\psi = \tau\psi & \text{in } \Omega, \\ \mathfrak{B}\psi = 0 & \text{on } \partial\Omega. \end{cases} \tag{2.5}$$

Proof. Let $u \in \mathcal{C}^2(\Omega) \cap \mathcal{C}^1(\bar{\Omega})$ be a supersolution of $(\mathfrak{L}, \mathfrak{B}, \Omega)$. Then, thanks to (2.3),

$$v := \frac{u}{h} \in \mathcal{C}^2(\Omega) \cap \mathcal{C}^1(\bar{\Omega}).$$

Moreover, arguing as in the proof of Theorem 1.7, we find that

$$\mathfrak{L}u = h\mathfrak{L}_h v, \tag{2.6}$$

where

$$\mathfrak{L}_h := -\sum_{i,j=1}^{N} a_{ij} \frac{\partial^2}{\partial x_i \partial x_j} + \sum_{j=1}^{N} b_{hj} \frac{\partial}{\partial x_j} + c_h, \tag{2.7}$$

with

$$c_h := \frac{\mathfrak{L}h}{h}, \qquad b_{hj} := b_j - \frac{2}{h} \sum_{i=1}^{N} a_{ij} \frac{\partial h}{\partial x_i}, \qquad 1 \le j \le N.$$

As $h \in \mathcal{C}^1(\bar{\Omega})$, we find from the hypothesis H2 that

$$b_{hj}, \quad c_h \in L^{\infty}(\Omega), \qquad 1 \le j \le N.$$

Moreover,

$$c_h \geq 0$$

because h is a supersolution of $(\mathfrak{L}, \mathfrak{B}, \Omega)$ and, hence, $\mathfrak{L}h \geq 0$ in Ω, by Definition 2.1.

As u is a supersolution of $(\mathfrak{L}, \mathfrak{B}, \Omega)$, (2.3) and (2.6) imply

$$\mathfrak{L}_h v \geq 0 \quad \text{in} \quad \Omega. \tag{2.8}$$

Moreover, $\mathfrak{B}u \geq 0$ on $\partial\Omega$, and, in particular, $\mathfrak{B}u = u \geq 0$ on Γ_0. Thus,

$$v \geq 0 \quad \text{on} \quad \Gamma_0. \tag{2.9}$$

Similarly, on Γ_1 we have that

$$0 \leq \mathfrak{B}u = \mathfrak{B}(hv) = \frac{\partial(hv)}{\partial\nu} + \beta hv = h\frac{\partial v}{\partial\nu} + v\frac{\partial h}{\partial\nu} + \beta hv$$

$$= h\frac{\partial v}{\partial\nu} + \left(\frac{\partial h}{\partial\nu} + \beta h\right)v = h\left(\frac{\partial v}{\partial\nu} + \frac{\mathfrak{B}h}{h}v\right)$$

and, consequently,

$$0 \leq \mathfrak{B}u = h\mathfrak{B}_h v \quad \text{on} \quad \Gamma_1, \tag{2.10}$$

where $\mathfrak{B}_h$ stands for the boundary operator

$$\mathfrak{B}_h\psi := \begin{cases} \psi & \text{on} \quad \Gamma_0, \\ \frac{\partial\psi}{\partial\nu} + \beta_h\psi & \text{on} \quad \Gamma_1, \end{cases} \qquad \psi \in \mathcal{C}(\Gamma_0) \otimes \mathcal{C}^1(\Gamma_1), \tag{2.11}$$

with

$$\beta_h := \frac{\mathfrak{B}h}{h} \geq 0 \quad \text{on} \quad \Gamma_1, \tag{2.12}$$

because $\mathfrak{B}h \geq 0$ on $\partial\Omega$. Note that $\beta_h \in \mathcal{C}(\Gamma_1)$. Incidentally, (2.12) holds independently of the sign of $\beta \in \mathcal{C}(\Gamma_1)$.

Combining (2.8), (2.9) and (2.10), it becomes apparent that v provides us with a supersolution of $(\mathfrak{L}_h, \mathfrak{B}_h, \Omega)$.

Now, we will show that Alternative A3 occurs if $u(x_0) < 0$ for some $x_0 \in \Omega$. Indeed, in this case we have that

$$v(x_0) = \frac{u(x_0)}{h(x_0)} < 0$$

and hence,

$$m := \min_{\bar{\Omega}} v < 0.$$

Thus, as $c_h \geq 0$ and v is a superharmonic function of $\mathfrak{L}_h$ in Ω, it follows from Theorem 1.2 that either

$$v(x) > m \quad \text{for all} \quad x \in \Omega, \tag{2.13}$$

or

$$v = m \quad \text{in} \quad \Omega. \tag{2.14}$$

Suppose (2.13). Then, since $m < 0$, (2.9) implies that

$$v(x) > m \quad \text{for all} \quad x \in \Omega \cup \Gamma_0. \tag{2.15}$$

Let $x_1 \in \bar{\Omega}$ be such that

$$m = v(x_1).$$

By (2.15), $x \in \Gamma_1$. Consequently, owing to Theorem 1.3, we have that

$$\frac{\partial v}{\partial \nu}(x_1) < 0.$$

Thus, according to (2.10) and (2.11),

$$0 \le \mathfrak{B}_h v(x_1) = \frac{\partial v}{\partial \nu}(x_1) + \beta_h(x_1)v(x_1) < \beta_h(x_1)v(x_1) = \beta_h(x_1)m$$

and, consequently,

$$\beta_h(x_1) < 0,$$

which contradicts (2.12). Therefore, (2.14) holds, and hence,

$$u = mh \quad \text{in} \quad \Omega. \tag{2.16}$$

By continuity, (2.16) must be satisfied in $\bar{\Omega}$, and, so, Alternative A3 occurs. The remaining assertions of Alternative A3 follow very easily from (2.16). Indeed, suppose $\Gamma_0 \ne \emptyset$ and pick $x_0 \in \Gamma_0$. Then, $h(x_0) > 0$ and hence,

$$u(x_0) = mh(x_0) < 0,$$

which is impossible, for as $u \ge 0$ on Γ_0. Thus, $\Gamma_0 = \emptyset$. Moreover, the estimates

$$0 \le \mathfrak{L}u = m\mathfrak{L}h \le 0 \quad \text{in} \quad \Omega$$

imply

$$\mathfrak{L}u = 0,$$

while the estimates

$$0 \le \mathfrak{B}u = m\mathfrak{B}h \le 0 \quad \text{on} \quad \partial\Omega = \Gamma_1,$$

entail

$$\mathfrak{B}u = 0 \quad \text{on} \quad \partial\Omega,$$

and hence (2.4) holds.

To complete the proof of the theorem, it suffices to show that some of the first two alternatives occurs if $u \geq 0$ in $\bar{\Omega}$. So, suppose

$$u \geq 0 \quad \text{in} \quad \bar{\Omega}.$$

Then, either

$$u(x_0) = 0 \quad \text{for some} \quad x_0 \in \Omega, \tag{2.17}$$

or

$$u(x) > 0 \quad \text{for all} \quad x \in \Omega. \tag{2.18}$$

Suppose (2.17). Then, $v \geq 0$ in $\bar{\Omega}$ and $v(x_0) = 0$. Thus, as v is a supersolution of $(\mathfrak{L}_h, \mathfrak{B}_h, \Omega)$, we find from Theorem 1.2 that $v = 0$ in $\bar{\Omega}$. Therefore, $u = 0$ in $\bar{\Omega}$ and Alternative A1 holds.

In case (2.18), $v(x) > 0$ for all $x \in \Omega$, and hence, since v is a supersolution of $(\mathfrak{L}_h, \mathfrak{B}_h, \Omega)$, it follows from Theorem 1.3 that

$$\frac{\partial v}{\partial \nu}(x) < 0 \quad \text{for all} \quad x \in v^{-1}(0) \cap \partial\Omega. \tag{2.19}$$

Suppose $\Gamma_1 \neq \emptyset$ and $v(x_1) = 0$ for some $x_1 \in \Gamma_1$. Then,

$$0 \leq \mathfrak{B}_h v(x_1) = \frac{\partial v}{\partial \nu}(x_1) + \beta_h(x_1)v(x_1) = \frac{\partial v}{\partial \nu}(x_1),$$

which contradicts (2.19). Thus, $v(x_1) > 0$, and so, $u(x_1) > 0$ for all $x_1 \in \Gamma_1$. Moreover, for every

$$x_0 \in \Gamma_0 \cap u^{-1}(0),$$

we have that $v(x_0) = 0$ and hence, it follows from (2.19) that

$$\frac{\partial u}{\partial \nu}(x_0) = \frac{\partial(hv)}{\partial \nu}(x_0)$$

$$= h(x_0)\frac{\partial v}{\partial \nu}(x_0) + v(x_0)\frac{\partial h}{\partial \nu}(x_0)$$

$$= h(x_0)\frac{\partial v}{\partial \nu}(x_0) < 0.$$

Therefore, Alternative A2 holds. The proof is complete. $\qquad\square$

As an immediate consequence from Theorem 2.1, the following result holds. Essentially, it shows the *strong positivity* of any smooth supersolution $u \neq 0$ of $(\mathfrak{L}, \mathfrak{B}, \Omega)$ from the existence of a strict supersolution h satisfying (2.3).

Corollary 2.1. *Suppose $(\mathfrak{L}, \mathfrak{B}, \Omega)$ possesses a strict supersolution $h \in \mathcal{C}^2(\Omega) \cap \mathcal{C}^1(\bar{\Omega})$ such that $h(x) > 0$ for all $x \in \bar{\Omega}$. Then, any supersolution $u \in \mathcal{C}^2(\Omega) \cap \mathcal{C}^1(\bar{\Omega})$, $u \neq 0$, of $(\mathfrak{L}, \mathfrak{B}, \Omega)$ (in particular, any strict supersolution) satisfies $u(x) > 0$ for every $x \in \Omega \cup \Gamma_1$, and*

$$\frac{\partial u}{\partial \nu}(x) < 0 \quad \text{for all} \quad x \in u^{-1}(0) \cap \Gamma_0.$$

Proof. Since h is a strict supersolution of $(\mathfrak{L}, \mathfrak{B}, \Omega)$, (2.4) cannot be satisfied and therefore, Alternative A3 of Theorem 2.1 cannot occur. As Alternative A1 is excluded, because $u \neq 0$, Alternative A2 holds. □

As a consequence from Corollary 2.1, the next uniqueness result holds.

Corollary 2.2. *Suppose* $(\mathfrak{L}, \mathfrak{B}, \Omega)$ *possesses a strict supersolution* $h \in \mathcal{C}^2(\Omega) \cap \mathcal{C}^1(\bar{\Omega})$ *such that* $h(x) > 0$ *for all* $x \in \bar{\Omega}$. *Then,* $u = 0$ *is the unique function* $u \in \mathcal{C}^2(\Omega) \cap \mathcal{C}^1(\bar{\Omega})$ *satisfying*

$$\begin{cases} \mathfrak{L}u = 0 & \text{in } \Omega, \\ \mathfrak{B}u = 0 & \text{on } \partial\Omega. \end{cases} \tag{2.20}$$

Therefore, for every $(f, g) \in \mathcal{C}(\bar{\Omega}) \times \mathcal{C}(\partial\Omega)$, *the boundary value problem*

$$\begin{cases} \mathfrak{L}u = f & \text{in } \Omega, \\ \mathfrak{B}u = g & \text{on } \partial\Omega, \end{cases} \tag{2.21}$$

admits, at most, a unique solution $u \in \mathcal{C}^2(\Omega) \cap \mathcal{C}^1(\bar{\Omega})$.

Proof. Suppose $u \in \mathcal{C}^2(\Omega) \cap \mathcal{C}^1(\bar{\Omega})$, $u \neq 0$, solves (2.20). Then, according to Corollary 2.1, $u(x) > 0$ for all $x \in \Omega \cup \Gamma_1$. Similarly, since $-u \neq 0$ provides us with another solution of (2.20), we have that $-u(x) > 0$ for all $x \in \Omega \cup \Gamma_1$, which is impossible. Consequently, $u = 0$ is the unique function of $\mathcal{C}^2(\Omega) \cap \mathcal{C}^1(\bar{\Omega})$ solving (2.20). The uniqueness of the solution of (2.21) follows easily from the previous uniqueness result. □

2.2 Existence of positive strict supersolutions

The following result provides us with a sufficient condition for the existence of a strict supersolution satisfying (2.3).

Proposition 2.1. *Suppose there are* $\psi \in \mathcal{C}^2(\bar{\Omega})$ *and* $\gamma > 0$ *such that*

$$\frac{\partial\psi}{\partial\nu}(x) \geq \gamma \quad \text{for all} \quad x \in \Gamma_1. \tag{2.22}$$

Then, there exists $\omega_0 \in \mathbb{R}$ *such that, for every* $\omega > \omega_0$, $(\mathfrak{L} + \omega, \mathfrak{B}, \Omega)$ *possesses a strict supersolution* $h \in \mathcal{C}^2(\bar{\Omega})$ *with* $h(x) > 0$ *for all* $x \in \bar{\Omega}$.

Proof. Suppose $h \in \mathcal{C}^2(\bar{\Omega})$, with $h(x) > 0$ for all $x \in \bar{\Omega}$, is a strict supersolution of $(\mathfrak{L} + \omega_0, \mathfrak{B}, \Omega)$ for some $\omega_0 \in \mathbb{R}$. Then, for every $\omega > \omega_0$,

$$(\mathfrak{L} + \omega)h = (\mathfrak{L} + \omega_0)h + (\omega - \omega_0)h \geq (\omega - \omega_0)h > 0$$

and hence, h is a strict supersolution of $(\mathfrak{L}+\omega, \mathfrak{B}, \Omega)$. Therefore, it suffices to show that there are $\omega_0 \in \mathbb{R}$ and $M > 0$ for which the function h defined through

$$h(x) := e^{M\psi(x)}, \qquad x \in \bar{\Omega}, \tag{2.23}$$

is a strict supersolution of $(\mathfrak{L}+\omega_0, \mathfrak{B}, \Omega)$. Indeed, by definition, $h(x) > 0$ for all $x \in \bar{\Omega}$. In particular,

$$\mathfrak{B}h(x) = h(x) > 0 \qquad \text{for all } x \in \Gamma_0.$$

Moreover, on Γ_1 we have that

$$\mathfrak{B}h = Mh\frac{\partial\psi}{\partial\nu} + \beta h = \left(M\frac{\partial\psi}{\partial\nu} + \beta\right)h$$

and hence, according to (2.22), we find that

$$\mathfrak{B}h \geq (M\gamma + \beta)h \geq 0 \qquad \text{on} \quad \Gamma_1$$

for sufficiently large $M > 0$. Suppose $M > 0$ has been chosen in this way. Then, since $\mathfrak{L}h \in L^\infty(\Omega)$ and $h(x) > 0$ for all $x \in \bar{\Omega}$, it becomes apparent that

$$(\mathfrak{L}+\omega_0)h > 0 \qquad \text{in } \Omega$$

for sufficiently large $\omega_0 \in \mathbb{R}$. This concludes the proof. $\qquad\square$

The next result provides us with a simple sufficient condition for the existence of a function ψ satisfying the requirements of Proposition 2.1.

Lemma 2.1. *Suppose Ω is of class $\mathcal{C}^2$. Then, there exist $\psi \in \mathcal{C}^2(\bar{\Omega})$ and $\gamma > 0$ satisfying (2.22).*

Proof. Fix $x_0 \in \partial\Omega$. Then, by Definition 1.5, there exist $R > 0$, an open subset $D \subset \mathbb{R}^N$, and a bijection

$$\Phi \; : \; B_R(x_0) \to D = \Phi(B_R(x_0))$$

such that:

i) $\Phi(x_0) = 0$ and $D\Phi(x_0) \in \text{Iso}(\mathbb{R}^N)$,

ii) $\Phi(B_R(x_0) \cap \Omega) = \{(x_1, ..., x_N) \in D \; : \; x_N > 0\}$,

iii) $\Phi(B_R(x_0) \cap \partial\Omega) = \{(x_1, ..., x_N) \in D \; : \; x_N = 0\}$,

iv) $\Phi \in \mathcal{C}^2(B_R(x_0); D)$ and $\Phi^{-1} \in \mathcal{C}^2(D; B_R(x_0))$.

As $\nu(x_0)$ is an outward pointing vector at $x_0 \in \partial\Omega$,

$$x_0 + t\nu(x_0) \in B_R(x_0) \setminus \bar{\Omega}$$

for sufficiently small $t > 0$, and hence

$$\Phi(x_0 + t\nu(x_0)) \in \{(x_1, \ldots, x_N) \in D \ : \ x_N < 0\}. \tag{2.24}$$

Moreover,

$$\langle \nu(x_0), \mathbf{n} \rangle > 0,$$

where $\mathbf{n}$ stands for the outward unit normal to Ω at $x_0 \in \partial\Omega$.

Let $\mathbf{T}_{x_0}$ denote the tangent hyperplane to $\partial\Omega$ at x_0. Then,

$$\mathbf{T}_{x_0} = x_0 + \{v \in \mathbb{R}^N \ : \ \langle v, \mathbf{n} \rangle = 0\}$$

and

$$\mathbb{R}^N = \mathbf{T}_{x_0} \oplus \text{span}\,[\nu(x_0)].$$

Subsequently, we denote by P_N the projection of $\mathbb{R}^N$ onto the N-th coordinate. According to (i)–(iii), it becomes apparent that

$$D\Phi(x_0)\,(\mathbf{T}_{x_0}) = \{v \in \mathbb{R}^N \ : \ P_N v = 0\} \tag{2.25}$$

(see Figure 2.1) and

$$\mathbb{R}^N = D\Phi(x_0)\,(\mathbf{T}_{x_0}) \oplus \text{span}\,[\tilde{\nu}], \quad \tilde{\nu} := D\Phi(x_0)\nu(x_0). \tag{2.26}$$

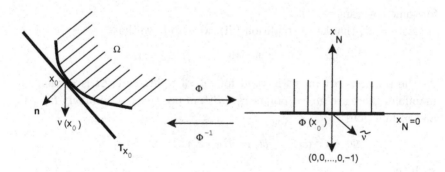

Fig. 2.1 Sketch of the construction

Also, since $\Phi \in \mathcal{C}^2(B_R(x_0); D)$ and $\Phi(x_0) = 0$, we have that

$$\Phi(x_0 + t\nu(x_0)) = t\tilde{\nu} + o(t) \qquad \text{as } t \downarrow 0. \tag{2.27}$$

Due to (2.24),

$$P_N \Phi(x_0 + t\nu(x_0)) < 0$$

for sufficiently small $t > 0$. Thus, (2.27) implies that $P_N \tilde{\nu} \leq 0$. On the other hand, thanks to (2.25) and (2.26), it is apparent that $P_N \tilde{\nu} \neq 0$ and, therefore,

$$P_N \tilde{\nu} < 0. \tag{2.28}$$

Consequently, $\tilde{\nu}$ points outward the open set $\Phi(B_R(x_0) \cap \Omega)$ at $\Phi(x_0) = 0$, as illustrated by Figure 2.1.

Now, consider the function

$$\Psi(x) := -P_N \Phi(x), \qquad x \in B_R(x_0).$$

Clearly, $\Psi \in \mathcal{C}^2(B_R(x_0))$ and, for every $x \in \partial\Omega$ sufficiently close to x_0, we have that

$$\frac{\partial \Psi}{\partial \nu(x)}(x) = D\Psi(x)\nu(x) = -P_N D\Phi(x)\nu(x).$$

Consequently, by (2.26), (2.28), and the continuity of the map

$$x \mapsto D\Phi(x)\nu(x),$$

it becomes apparent that R can be shortened, if necessary, so that

$$\frac{\partial \Psi}{\partial \nu(x)}(x) \geq \gamma \qquad \text{for all} \quad x \in B_R(x_0) \cap \partial\Omega,$$

for some constant $\gamma > 0$.

Note that, thanks to condition (iii), we also have that

$$\Psi(x) = 0 \qquad \text{for all} \quad x \in B_R(x_0) \cap \partial\Omega.$$

As the previous argument is valid for all $x_0 \in \partial\Omega$ and $\partial\Omega$ is a compact manifold, there are $m \geq 1$ points $x_0^j \in \partial\Omega$, m positive real numbers $R_j > 0$, and m functions

$$\Psi_j \in \mathcal{C}^2(\mathcal{U}_j), \quad \mathcal{U}_j := B_{R_j}(x_0^j), \qquad 1 \leq j \leq m,$$

such that

$$\bigcup_{j=1}^m (\mathcal{U}_j \cap \partial\Omega) = \partial\Omega,$$

$$\Psi_j = 0 \quad \text{in} \quad \mathcal{U}_j \cap \partial\Omega, \qquad 1 \leq j \leq m, \tag{2.29}$$

and, for some positive constant $\gamma > 0$,

$$\frac{\partial \Psi_j}{\partial \nu(x)}(x) \geq \gamma \quad \text{for all } x \in \mathcal{U}_j \cap \partial \Omega, \quad 1 \leq j \leq m. \tag{2.30}$$

Clearly, there exists $\epsilon > 0$ such that

$$\{x \in \Omega \; : \; \text{dist}\,(x, \partial \Omega) \leq \epsilon\} \subset \bigcup_{j=1}^{m} \mathcal{U}_j$$

and hence, setting

$$\mathcal{U}_{m+1} := \{x \in \Omega \; : \; \text{dist}\,(x, \partial \Omega) > \epsilon\},$$

the open neighborhoods $\mathcal{U}_j$, $1 \leq j \leq m + 1$, provide us with a covering of $\bar{\Omega}$. By Proposition 2.2 in the Appendix, there exist

$$\psi_j \in \mathcal{C}_0^\infty(\mathbb{R}^N), \qquad 1 \leq j \leq m+1,$$

with

$$\text{supp}\, \psi_j \subset \bar{\mathcal{U}}_j, \qquad 1 \leq j \leq m+1,$$

such that

$$\sum_{j=1}^{m+1} \psi_j = 1 \quad \text{in} \quad \bar{\Omega}.$$

It remains to check that the function ψ defined through

$$\psi(x) := \sum_{j=1}^{m} \Psi_j(x)\psi_j(x) + \psi_{m+1}(x), \qquad x \in \bar{\Omega},$$

satisfies the requirements of the lemma. Indeed, $\psi \in \mathcal{C}^2(\bar{\Omega})$ and

$$\frac{\partial \psi}{\partial \nu(x)}(x) = \sum_{j=1}^{m} \frac{\partial \Psi_j}{\partial \nu(x)}(x)\psi_j(x) + \sum_{j=1}^{m} \Psi_j(x)\frac{\partial \psi_j}{\partial \nu(x)}(x)$$

for all $x \in \partial \Omega$, because $\psi_{m+1} = 0$ on a neighborhood of $\partial \Omega$. Thus, by (2.29) and (2.30), we have that

$$\frac{\partial \psi}{\partial \nu(x)}(x) = \sum_{j=1}^{m} \frac{\partial \Psi_j}{\partial \nu(x)}(x)\psi_j(x) \geq \gamma \sum_{j=1}^{m} \psi_j(x) = \gamma$$

for all $x \in \partial \Omega$, since $\psi_{m+1} = 0$ on $\partial \Omega$. This ends the proof. $\qquad \square$

Essentially, the function $\psi(x)$ constructed in the proof of Lemma 2.1 behaves like $-\text{dist}\,(x, \partial \Omega)$ for all $x \in \Omega$ sufficiently close to $\partial \Omega$.

2.3 Positivity of the resolvent operator

As an immediate consequence from Corollary 2.2 and Proposition 2.1, the following result holds.

Theorem 2.2. *Assume that there exist* $\psi \in \mathcal{C}^2(\bar{\Omega})$ *and* $\gamma > 0$ *satisfying* (2.22). *Then, there exists* $\omega_0 \in \mathbb{R}$ *such that, for every* $\omega > \omega_0$, $u = 0$ *is the unique function* $u \in \mathcal{C}^2(\Omega) \cap \mathcal{C}^1(\bar{\Omega})$ *solving the problem*

$$\begin{cases} (\mathfrak{L} + \omega)u = 0 & \text{in } \Omega, \\ \mathfrak{B}u = 0 & \text{on } \partial\Omega. \end{cases}$$

Therefore, for every $\omega > \omega_0$ *and* $(f, g) \in \mathcal{C}(\bar{\Omega}) \times \mathcal{C}(\partial\Omega)$, *the problem*

$$\begin{cases} (\mathfrak{L} + \omega)u = f & \text{in } \Omega, \\ \mathfrak{B}u = g & \text{on } \partial\Omega, \end{cases} \tag{2.31}$$

has, at most, a unique solution $u \in \mathcal{C}^2(\Omega) \cap \mathcal{C}^1(\bar{\Omega})$. *Moreover, if such solution exists and* $f \geq 0$, $g \geq 0$, *with* $(f, g) \neq (0, 0)$, *then*

$$u(x) > 0 \qquad \text{for all } x \in \Omega \cup \Gamma_1 \tag{2.32}$$

and

$$\frac{\partial u}{\partial \nu}(x) < 0 \qquad \text{for all } x \in u^{-1}(0) \cap \Gamma_0. \tag{2.33}$$

Proof. According to Proposition 2.1, there exists $\omega_0 \in \mathbb{R}$ such that, for every $\omega > \omega_0$, $(\mathfrak{L} + \omega, \mathfrak{B}, \Omega)$ possesses a strict supersolution $h \in \mathcal{C}^2(\bar{\Omega})$ such that $h(x) > 0$ for all $x \in \bar{\Omega}$. Consequently, the uniqueness follows from Corollary 2.2.

Suppose $\omega > \omega_0$, $f \geq 0$, $g \geq 0$, $(f, g) \neq (0, 0)$, and (2.31) admits a solution $u \in \mathcal{C}^2(\Omega) \cap \mathcal{C}^1(\bar{\Omega})$. Then, u provides us with a strict supersolution of $(\mathfrak{L} + \omega, \mathfrak{B}, \Omega)$ and, therefore, by Corollary 2.1, (2.32) and (2.33) are satisfied. The proof is complete. □

2.4 Behavior of the positive supersolutions on Γ_0

The following result provides us with the point-wise behavior of the positive supersolutions of $(\mathfrak{L}, \mathfrak{B}, \Omega)$.

Theorem 2.3. *Suppose there are* $\psi \in \mathcal{C}^2(\bar{\Omega})$ *and* $\gamma > 0$ *satisfying* (2.22). *Let* $u \in \mathcal{C}^2(\Omega) \cap \mathcal{C}^1(\bar{\Omega})$, $u > 0$, *be a supersolution of* $(\mathfrak{L}, \mathfrak{B}, \Omega)$. *Then,* u *satisfies* (2.32) *and* (2.33).

If, in addition, $(\mathfrak{L}, \Omega)$ satisfies the decay property of E. Hopf on Γ_0, as discussed in Definition 1.4, then, there exists a constant $\delta := \delta(u) > 0$ such that

$$u(x) \geq \delta \operatorname{dist}(x, \partial\Omega) \qquad \text{for all } x \in \Omega. \tag{2.34}$$

Proof. By Proposition 2.1, there exist $\omega > 0$ and $h \in \mathcal{C}^2(\bar{\Omega})$ such that $h(x) > 0$ for all $x \in \bar{\Omega}$ and h is a supersolution of $(\mathfrak{L} + \omega, \mathfrak{B}, \Omega)$. As u is a supersolution of $(\mathfrak{L}, \mathfrak{B}, \Omega)$, we find that

$$(\mathfrak{L} + \omega)u = \mathfrak{L}u + \omega u \geq \omega u > 0 \qquad \text{in} \quad \Omega,$$

since $u > 0$ in Ω. Also, $\mathfrak{B}u \geq 0$ on $\partial\Omega$. Thus, $u \in \mathcal{C}^2(\Omega) \cap \mathcal{C}^1(\bar{\Omega})$, $u > 0$, is a supersolution of $(\mathfrak{L} + \omega, \mathfrak{B}, \Omega)$, and, therefore, according to Corollary 2.1, u must satisfy (2.32) and (2.33). In particular, u is a superharmonic function of $(\mathfrak{L}, \Omega)$ such that $u(x) > 0$ for all $x \in \bar{\Omega} \setminus \Gamma_0$. Consequently, by Definition 1.4, there exists $\delta > 0$ satisfying (2.34). $\square$

By the hypotheses H1, H2 and H3, it follows from Corollary 1.2 that $(\mathfrak{L}, \Omega)$ satisfies the decay property of E. Hopf on Γ_0 if Ω satisfies the uniform interior sphere property in the strong sense on Γ_0. Moreover, according to Theorem 1.9, this occurs if Ω is of class $\mathcal{C}^2$. Consequently, owing to Lemma 2.1, the next result holds.

Corollary 2.3. *Suppose Ω is of class $\mathcal{C}^2$ and $u \in \mathcal{C}^2(\Omega) \cap \mathcal{C}^1(\bar{\Omega})$, $u > 0$, is a supersolution of $(\mathfrak{L}, \mathfrak{B}, \Omega)$. Then, u satisfies (2.32), (2.33), and (2.34) for some $\delta = \delta(u) > 0$.*

2.5 Second classification theorem

The main result of this section is the following sharp improvement of Theorem 2.1, where the positive supersolution h can vanish on Γ_0.

Theorem 2.4. *Suppose:*

i) *$(\mathfrak{L}, \Omega)$ satisfies the decay property of E. Hopf on Γ_0;*

ii) *There exist $\psi \in \mathcal{C}^2(\bar{\Omega})$ and $\gamma > 0$ satisfying (2.22);*

iii) *$(\mathfrak{L}, \mathfrak{B}, \Omega)$ has a positive supersolution $h \in \mathcal{C}^2(\Omega) \cap \mathcal{C}^1(\bar{\Omega})$, $h > 0$.*

Then, any supersolution $u \in \mathcal{C}^2(\Omega) \cap \mathcal{C}^1(\bar{\Omega})$ of $(\mathfrak{L}, \mathfrak{B}, \Omega)$ must satisfy some of the Alternatives A1, A2, or A3, of Theorem 2.1.

Remark 2.1. By Corollary 1.2, Theorem 1.9, and Lemma 2.1, conditions (i) and (ii) are satisfied when Ω is of class $\mathcal{C}^2$.

Proof.   Thanks to Theorem 2.3, the supersolution h satisfies

$$h(x) > 0 \qquad \text{for all } x \in \Omega \cup \Gamma_1, \tag{2.35}$$

$$\frac{\partial h}{\partial \nu}(x) < 0 \qquad \text{for all } x \in h^{-1}(0) \cap \Gamma_0, \tag{2.36}$$

and there exists a constant $\delta := \delta(h) > 0$ such that

$$h(x) \geq \delta \operatorname{dist}(x, \partial\Omega) \qquad \text{for all } x \in \Omega. \tag{2.37}$$

Let $u \in \mathcal{C}^2(\Omega) \cap \mathcal{C}^1(\bar{\Omega})$ be a supersolution of $(\mathfrak{L}, \mathfrak{B}, \Omega)$. Then, $\mathfrak{L}u \geq 0$ in Ω and $\mathfrak{B}u \geq 0$ on $\partial\Omega$. Suppose $u > 0$. Then, thanks again to Theorem 2.3, u satisfies (2.32) and (2.33), and, therefore, Alternative A2 holds. Consequently, in case $u \geq 0$ some of the first two alternatives must occur. To complete the proof of the theorem it remains to show that Alternative A3 is satisfied if u is somewhere negative. So, suppose

$$u(x_-) < 0 \qquad \text{for some } x_- \in \Omega. \tag{2.38}$$

Subsequently, for every $\lambda \geq 0$, we consider the function v_λ defined by

$$v_\lambda(x) := u(x) + \lambda h(x), \qquad x \in \bar{\Omega}.$$

We claim that $v_\lambda \geq 0$ in Ω for sufficiently large $\lambda > 0$. On the contrary, assume that, for each integer $k \geq 1$, there exists $x_k \in \Omega$ such that

$$v_k(x_k) = u(x_k) + k h(x_k) < 0. \tag{2.39}$$

As $\bar{\Omega}$ is compact, there exist $x_0 \in \bar{\Omega}$ and a subsequence of $\{k\}_{k \geq 1}$, say $\{k_m\}_{m \geq 1}$, such that

$$\lim_{m \to \infty} x_{k_m} = x_0.$$

Thanks to (2.39), we have that

$$\frac{1}{k_m} u(x_{k_m}) + h(x_{k_m}) < 0, \qquad m \geq 1. \tag{2.40}$$

Moreover, by the continuity of u in $\bar{\Omega}$,

$$\lim_{m \to \infty} \frac{1}{k_m} u(x_{k_m}) = 0.$$

Thus, letting $m \to \infty$ in (2.40) gives $h(x_0) \leq 0$, and hence

$$h(x_0) = 0,$$

because we are assuming that $h \geq 0$ in Ω. Therefore, owing to (2.35),

$$x_0 \in h^{-1}(0) \cap \Gamma_0.$$

This entails $\Gamma_0 \neq \emptyset$. Consequently, if $\Gamma_0 = \emptyset$, then $v_\lambda \geq 0$ for sufficiently large $\lambda > 0$. So, suppose $\Gamma_0 \neq \emptyset$, and, for each $m \geq 1$, let

$$y_{k_m} \in \Gamma_0$$

such that

$$\text{dist}\,(x_{k_m}, \Gamma_0) = |x_{k_m} - y_{k_m}|.$$

As u is a supersolution of $(\mathfrak{L}, \mathfrak{B}, \Omega)$, we have that $u \geq 0$ on Γ_0 and hence,

$$-u(x_{k_m}) \leq u(y_{k_m}) - u(x_{k_m}), \qquad m \geq 1.$$

Thus, since $u \in \mathcal{C}^1(\bar\Omega)$, this implies that

$$-u(x_{k_m}) \leq L|y_{k_m} - x_{k_m}| = L\,\text{dist}\,(x_{k_m}, \Gamma_0), \qquad m \geq 1, \qquad (2.41)$$

where $L \geq 0$ is the Lipschitz constant of u in $\bar\Omega$.

Now, combining (2.39) and (2.41) yields

$$k_m h(x_{k_m}) < -u(x_{k_m}) \leq L\,\text{dist}\,(x_{k_m}, \Gamma_0), \qquad m \geq 1,$$

and, therefore, (2.37) implies that

$$\delta k_m \text{dist}\,(x_{k_m}, \partial\Omega) < L\,\text{dist}\,(x_{k_m}, \Gamma_0), \qquad m \geq 1. \qquad (2.42)$$

Finally, as for sufficiently large m

$$\text{dist}\,(x_{k_m}, \partial\Omega) = \text{dist}\,(x_{k_m}, \Gamma_0),$$

because $x_{k_m} \to x_0$ as $m \to \infty$, we find from (2.42) that

$$\delta k_m < L$$

for sufficiently large $m \geq 1$, which is impossible, because

$$\lim_{m \to \infty} k_m = \infty.$$

This contradiction shows that $v_\lambda \geq 0$ for sufficiently large $\lambda > 0$.

Let Λ denote the set of λ's, $\lambda > 0$, for which $v_\lambda \geq 0$. We have just proven that $\Lambda \neq \emptyset$. According to (2.38), $\lambda \notin \Lambda$ for sufficiently small $\lambda > 0$. Moreover, since $h \geq 0$, we have that $[\lambda, \infty) \subset \Lambda$ for all $\lambda \in \Lambda$. In particular,

$$\mu := \inf \Lambda > 0.$$

Clearly,

$$v_\mu \geq 0.$$

Indeed, if $v_\mu(x_\mu) < 0$ for some $x_\mu \in \Omega$, then, for sufficiently small $\epsilon > 0$, there exists $x_{\mu+\epsilon} \in \Omega$, $x_{\mu+\epsilon} \sim x_\mu$, such that

$$v_{\mu+\epsilon}(x_{\mu+\epsilon}) < 0,$$

which is impossible, by the definition of μ. Also,

$$\mathfrak{L}v_\mu = \mathfrak{L}u + \mu\mathfrak{L}h \geq 0 \qquad \text{in} \quad \Omega,$$

and

$$\mathfrak{B}v_\mu = \mathfrak{B}u + \mu\mathfrak{B}h \geq 0 \qquad \text{on} \quad \partial\Omega.$$

Thus, $v_\mu \in \mathcal{C}^2(\Omega) \cap \mathcal{C}^1(\bar\Omega)$ is a non-negative supersolution of $(\mathfrak{L}, \mathfrak{B}, \Omega)$. Consequently, according to Theorem 2.3, either

$$v_\mu = 0, \tag{2.43}$$

or

$$\begin{cases} v_\mu(x) > 0 & \forall\, x \in \Omega \cup \Gamma_1, \\[2mm] \dfrac{\partial v_\mu}{\partial \nu}(x) < 0 & \forall\, x \in v_\mu^{-1}(0) \cap \Gamma_0. \end{cases} \tag{2.44}$$

Suppose (2.43). Then, setting $m := -\mu < 0$, we obtain that

$$u = mh$$

and Alternative A3 holds; (2.4) follows easily from this identity taking into account that u and h are supersolutions of $(\mathfrak{L}, \mathfrak{B}, \Omega)$ with $m < 0$. Consequently, to complete the proof it suffices to show that (2.44) contradicts the minimality of μ. Indeed, by the definition of μ, for every $k \geq 1$ there exists a point $x_k \in \Omega$ such that

$$v_{\mu-\frac{1}{k}}(x_k) = u(x_k) + \left(\mu - \frac{1}{k}\right)h(x_k) = v_\mu(x_k) - \frac{h(x_k)}{k} < 0. \tag{2.45}$$

Arguing as above, there exist $x_0 \in \bar\Omega$ and a subsequence of $\{k\}_{k\geq 1}$, say $\{k_m\}_{m\geq 1}$, such that

$$\lim_{m\to\infty} x_{k_m} = x_0.$$

Thanks to (2.45), we have that

$$v_\mu(x_{k_m}) < \frac{h(x_{k_m})}{k_m}, \qquad m \geq 1. \tag{2.46}$$

Moreover, by the continuity of h in $\bar\Omega$,

$$\lim_{m\to\infty} \frac{h(x_{k_m})}{k_m} = 0.$$

Thus, letting $m \to \infty$ in (2.46) shows that

$$v_\mu(x_0) \leq 0$$

and, therefore, according to (2.44), we find that

$$x_0 \in v_\mu^{-1}(0) \cap \Gamma_0.$$

As above, this entails $\Gamma_0 \neq \emptyset$ and ends the proof of the theorem when $\Gamma_0 = \emptyset$. So, suppose $\Gamma_0 \neq \emptyset$, and, for every $m \geq 1$, let $y_{k_m} \in \Gamma_0$ such that

$$\text{dist}\,(x_{k_m}, \Gamma_0) = |x_{k_m} - y_{k_m}|.$$

The same argument used above shows that (2.41) holds.

On the other hand, by (2.45), we have that

$$-u(x_{k_m}) > \left(\mu - \frac{1}{k_m}\right) h(x_{k_m}), \qquad m \geq 1,$$

and hence (2.41) yields

$$\left(\mu - \frac{1}{k_m}\right) h(x_{k_m}) < L\,\text{dist}\,(x_{k_m}, \Gamma_0), \qquad m \geq 1.$$

Thus, since

$$\lim_{m \to \infty} k_m = \infty,$$

we have that

$$h(x_{k_m}) < \frac{L k_m}{\mu k_m - 1}\,\text{dist}\,(x_{k_m}, \Gamma_0)$$

for sufficiently large m. Hence, going back to (2.45) we find that, for sufficiently large m,

$$0 > v_\mu(x_{k_m}) - \frac{h(x_{k_m})}{k_m}$$

$$\geq \delta_\mu\,\text{dist}\,(x_{k_m}, \Gamma_0) - \frac{L}{\mu k_m - 1}\,\text{dist}\,(x_{k_m}, \Gamma_0)$$

$$= \left(\delta_\mu - \frac{L}{\mu k_m - 1}\right)\,\text{dist}\,(x_{k_m}, \Gamma_0),$$

because $(\mathfrak{L}, \Omega)$ possesses the decay property of E. Hopf on Γ_0 and v_μ is a positive supersolution of $(\mathfrak{L}, \mathfrak{B}, \Omega)$. We have denoted by $\delta_\mu > 0$ the constant of the decay property corresponding to v_μ. The previous inequality cannot be satisfied, because it entails

$$\text{dist}\,(x_{k_m}, \Gamma_0) < 0$$

for sufficiently large m. Therefore, $v_\mu = 0$. This ends the proof. $\qquad\square$

2.6   Appendix: Partitions of the unity

Subsequently, we denote by $C_0^\infty(\mathbb{R}^N)$ the set of functions $\psi : \mathbb{R}^N \to \mathbb{R}$ of class C^∞ with compact support

$$\operatorname{supp}\psi := \overline{\psi^{-1}(\mathbb{R} \setminus \{0\})} \ .$$

Also, for every subset $A \subset \mathbb{R}^N$, we will denote by χ_A the *characteristic function* of A, i.e.,

$$\chi_A(x) = \begin{cases} 1, & \text{if } x \in A, \\ 0, & \text{if } x \in \mathbb{R}^N \setminus A. \end{cases}$$

Lemma 2.2. *Let D, Ω, be two bounded domains of $\mathbb{R}^N$, $N \geq 1$, with $\bar{D} \subset \Omega$. Then, there exists $\psi \in C_0^\infty(\mathbb{R}^N)$ such that $0 \leq \psi \leq 1$ in $\mathbb{R}^N$, and*

$$\psi = 1 \quad \text{in } \bar{D}, \qquad \operatorname{supp}\psi \subset \bar{\Omega}.$$

Proof.   For every $\delta > 0$, let E_δ denote the function defined through

$$E_\delta(x) := \begin{cases} e^{-\frac{\delta^2}{\delta^2 - |x|^2}}, & \text{if } |x| \leq \delta, \\ 0, & \text{if } |x| > \delta, \end{cases}$$

and set

$$C_\delta := \left(\int_{\mathbb{R}^N} E_\delta(x)\,dx \right)^{-1}.$$

Then, the function

$$\omega_\delta := C_\delta E_\delta$$

satisfies

$$\omega_\delta \in C_0^\infty(\mathbb{R}^N), \qquad \omega_\delta \geq 0 \text{ in } \mathbb{R}^N, \qquad \operatorname{supp}\omega_\delta \subset \bar{B}_\delta(0),$$

and

$$\int_{\mathbb{R}^N} \omega_\delta(x)\,dx = 1.$$

Subsequently, we set

$$\epsilon := \operatorname{dist}(\partial D, \partial \Omega), \qquad \delta := \epsilon/3,$$

and, for any $\eta > 0$,

$$D_\eta := D + B_\eta(0) = \{ x \in \mathbb{R}^N \ : \ \operatorname{dist}(x, D) < \eta \}.$$

It is easy to check that the function

$$\psi(x) := \int_{\mathbb{R}^N} \chi_{D_{2\delta}}(y)\,\omega_\delta(x - y)\,dy, \qquad x \in \mathbb{R}^N,$$

satisfies all the requirements of the statement. $\square$

The following result establishes the existence of a *partition of the unity* subordinated to each covering of $\bar{\Omega}$ by open connected neighborhoods.

Proposition 2.2. *Let Ω be a bounded domain of $\mathbb{R}^N$, $N \geq 1$, and $m \geq 1$ open connected subsets $\mathcal{U}_j \subset \mathbb{R}^N$, $1 \leq j \leq m$, such that*

$$\bar{\Omega} \subset \bigcup_{j=1}^{m} \mathcal{U}_j. \tag{2.47}$$

Then, there exist $\psi_j \in \mathcal{C}_0^\infty(\mathbb{R}^N)$, $1 \leq j \leq m$, such that

$$\operatorname{supp} \psi_j \subset \bar{\mathcal{U}}_j, \qquad 1 \leq j \leq m,$$

and

$$\sum_{j=1}^{m} \psi_j = 1 \quad in \quad \bar{\Omega}.$$

Any set of m functions ψ_j, $1 \leq j \leq m$, satisfying all the requirements of Proposition 2.2 will be refereed to as a **partition of the unity** in Ω subordinated to the covering $\mathcal{U}_j$, $1 \leq j \leq m$.

Proof. According to (2.47), for sufficiently small $\epsilon > 0$, the open sets

$$D_j := \{x \in \mathcal{U}_j \ : \ \operatorname{dist}(x, \partial \mathcal{U}_j) > \epsilon\}, \qquad 1 \leq j \leq m,$$

satisfy

$$\bar{\Omega} \subset \bigcup_{j=1}^{m} D_j,$$

because $\bar{\Omega}$ is compact. By Lemma 2.2, for every $1 \leq j \leq m$, there exists $\Psi_j \in \mathcal{C}_0^\infty(\mathbb{R}^N)$ such that $0 \leq \Psi_j \leq 1$ in $\mathbb{R}^N$, and

$$\Psi_j = 1 \quad in \ D_j, \qquad \operatorname{supp} \Psi_j \subset \bar{\mathcal{U}}_j.$$

Let $\Psi \in \mathcal{C}_0^\infty(\mathbb{R}^N)$ be the function defined through

$$\Psi(x) := \sum_{j=1}^{m} \Psi_j(x), \qquad x \in \mathbb{R}^N.$$

By construction, the functions

$$\psi_j(x) := \frac{\Psi_j(x)}{\Psi(x)}, \qquad x \in \mathbb{R}^N, \qquad 1 \leq j \leq m,$$

satisfy

$$0 \leq \psi_j \leq 1 \quad in \ \mathbb{R}^N, \qquad \psi_j \in \mathcal{C}^\infty(\mathbb{R}^N), \qquad \operatorname{supp} \psi_j \subset \bar{\mathcal{U}}_j,$$

for all $1 \leq j \leq m$, and

$$\sum_{j=1}^{m} \psi_j = 1 \quad in \quad \bar{\Omega}.$$

This concludes the proof. $\qquad\qquad\qquad\qquad\qquad\qquad\qquad\qquad\square$

2.7 Comments on Chapter 2

The main results of this chapter are Theorems 2.1 and 2.4. Both establish that all non-zero supersolutions of $(\mathfrak{L}, \mathfrak{B}, \Omega)$ must have constant sign in Ω if $(\mathfrak{L}, \mathfrak{B}, \Omega)$ admits a positive supersolution h. Theorem 2.4 is an extremely sharp version of Theorem 2.1, because in Theorem 2.4 the supersolution h can vanish on some piece of $\partial\Omega$, while h must be separated away from zero in $\bar{\Omega}$ for the validity of Theorem 2.1, which is the classical condition of M. H. Protter and H. F. Weinberger [183] in their Theorem 10 of Chapter 2, which is Theorem 1.7 of Chapter 1.

Theorem 2.2, which is a straightforward consequence from Theorem 2.1, establishes the *positivity of the inverse* for the boundary value problem (2.31) associated to $(\mathfrak{L}+\omega, \mathfrak{B}, \Omega)$ for sufficiently large ω. Theorem 2.2 goes back to Theorem 6.1 of H. Amann [9] in the special case when Ω is of class $\mathcal{C}^2$, as no sign restriction for β was imposed therein. Note that, previously, M. H. Protter and H. F. Weinberger in Chapter 2 of [183] assumed that $\beta \geq 0$. But, according to the last paragraph of the proof of Theorem 6.1 on page 239 of H. Amann [9], where it was claimed that

"the assertion follows by an obvious combination of the generalized maximum principle of Protter and Weinberger [183] with Bony's maximum principle [28]".

the reader might believe that Theorem 2.1 is a direct consequence of Theorem 1.7. Although this is certainly true, we estimate that Theorem 2.1 is far from being an obvious consequence of Theorem 1.7. Nevertheless, it should be remarked that the last assertion of H. Amann in the statement of Theorem 6.1 of [9], where it is asserted that

$$\frac{\partial u}{\partial \nu}(x) < 0 \qquad \text{for} \quad x \in \Gamma_0,$$

does not seem to be true unless $u(x) = 0$.

Theorem 2.4 goes back to Theorem 2 of W. Walter [224] in the special case when $\Gamma_1 = \emptyset$. It was substantially refined by J. López-Gómez in Theorem 5.2 of [144] to cover the general case treated in this chapter. Theorem 2.1 should be weighted against Theorem 1 of W. Walter [224], whose proof is, according to Section 4 of W. Walter [224],

"closely related to the idea of families of upper solutions which goes back to A. McNabb [160] and is also known under the name Serrin's sweeping principle."

Some of the regularity assumptions imposed in the results of this chapter can be substantially relaxed. For example, in Lemma 2.1 the regularity of $\partial\Omega$ can be relaxed up to assume that Γ_1 is of class $\mathcal{C}^2$. The regularity of Γ_0 is far from necessary, because ψ can be taken to be positive on Γ_0.

The exposition of this chapter has been elaborated from the materials of J. López-Gómez [144].

In Chapter 7, using the maximum principle of J. M. Bony [28], it will become apparent that most of the results of Chapters 1 and 2 are still valid in the weak sense within the context of the Sobolev spaces.

Chapter 3

Representation theorems

Let H be a real vector space. A *scalar product* $\langle \cdot, \cdot \rangle$ in H is a *bilinear form*

$$\langle \cdot, \cdot \rangle \; : \; H \times H \to \mathbb{R}$$

symmetric and *positive definite*. Every scalar product $\langle \cdot, \cdot \rangle$ induces a canonical norm in H through

$$\|u\| := \sqrt{\langle u, u \rangle}, \qquad u \in H,$$

which satisfies the Cauchy–Schwarz inequality

$$|\langle u, v \rangle| \le \|u\| \, \|v\|, \qquad u, v \in H. \tag{3.1}$$

Indeed, for every $u, v \in H$ and $t \in \mathbb{R}$, we have that

$$0 \le \|tu + v\|^2 = \langle tu + v, tu + v \rangle = t^2 \|u\|^2 + 2t\langle u, v \rangle + \|v\|^2.$$

Thus, the polynomial

$$P(t) := \|u\|^2 t^2 + 2\langle u, v \rangle t + \|v\|^2, \qquad t \in \mathbb{R},$$

admits at most a unique real root. Consequently,

$$\langle u, v \rangle^2 \le \|u\|^2 \|v\|^2, \qquad u, v \in H,$$

and, extracting square roots in this inequality, (3.1) holds. Moreover, the *parallelogram identity*

$$\left\| \frac{u+v}{2} \right\|^2 + \left\| \frac{u-v}{2} \right\|^2 = \frac{\|u\|^2 + \|v\|^2}{2}, \qquad u, v \in H, \tag{3.2}$$

is satisfied. Indeed,

$$
\begin{aligned}
\|u + v\|^2 + \|u - v\|^2 &= \langle u + v, u + v \rangle + \langle u - v, u - v \rangle \\
&= \|u\|^2 + \|v\|^2 + 2\langle u, v \rangle + \|u\|^2 + \|v\|^2 - 2\langle u, v \rangle \\
&= 2(\|u\|^2 + \|v\|^2).
\end{aligned}
$$

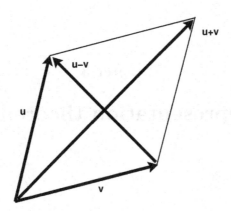

Fig. 3.1   The parallelogram of sides u and v

Geometrically, (3.2) establishes that the sum of the squares of the lengths of the diagonals of a parallelogram equals the sum of the squares of the lengths of all sides, as illustrated by Figure 3.1.

The real vector space H with the scalar product $\langle \cdot, \cdot \rangle$ is said to be a *Hilbert space* if the normed vector space $(H, \| \cdot \|)$ is complete, i.e., if it is a Banach space. Throughout this chapter, we will suppose that H is a Hilbert space with scalar product $\langle \cdot, \cdot \rangle$ and associated norm $\| \cdot \|$.

In this chapter we are going to study the theorem of G. Stampacchia [212], which has been a milestone for the development of the calculus of variations and its applications. As a byproduct, we will derive from it the representation theorem of P. D. Lax and A. N. Milgram [122], which is a substantial extension of the representation theorem of F. Riesz [191]. Essentially, the theorem of G. Stampacchia [212] is an abstract nonlinear counterpart of the representation theorem of P. D. Lax and A. N. Milgram [122], which is utterly linear.

This chapter is distributed as follows. Section 3.1 establishes the existence of the projection operator associated to each closed and convex subset K of H, Section 3.2 shows how the projection operator equals the orthogonal projection when K is a linear subspace of H, Section 3.3 studies the representation theorem of F. Riesz, Section 3.4 introduces some basic concepts of the theory of bilinear forms, Section 3.5 studies the theorem of G. Stampacchia, Section 3.6 derives from it the theorem of P. D. Lax and A. N. Milgram, and, finally, Section 3.7 establishes the existence of continuous projections on any convex and closed subset of a uniformly convex Banach space.

3.1 The projection on a closed convex set

The following result constructs the projection operator on any closed and convex subset of the Hilbert space H.

Theorem 3.1. *Let $K \subset H$ be a closed and convex set. Then, for every $u \in H$, there exists a unique $\mathcal{P}_K u \in K$ such that*

$$\|u - \mathcal{P}_K u\| = \operatorname{dist}(u, K) = \min_{k \in K} \|u - k\|.$$

Necessarily,

$$\mathcal{P}_K k = k \qquad \text{if } k \in K,$$

and, for every $u \in H$, $\mathcal{P}_K u$ is the unique element of K for which

$$\langle u - \mathcal{P}_K u, k - \mathcal{P}_K u \rangle \leq 0 \qquad \text{for all } k \in K. \tag{3.3}$$

The underlying map

$$H \xrightarrow{\mathcal{P}_K} K$$
$$u \mapsto \mathcal{P}_K u$$

*is said to be the **projection of H on K**.*

As illustrated by Figure 3.2, Theorem 3.1 has an obvious geometrical meaning. For every $u \in H \setminus K$, the orthogonal hyperplane to the vector $u - \mathcal{P}_K u$ through the point $\mathcal{P}_K u$, given by

$$T := \{ h \in H \ : \ \langle u - \mathcal{P}_K u, h - \mathcal{P}_K u \rangle = 0 \},$$

divides the whole space H into the two closed and convex subspaces

$$H_+ := \{ h \in H \ : \ \langle u - \mathcal{P}_K u, h - \mathcal{P}_K u \rangle \geq 0 \},$$
$$H_- := \{ h \in H \ : \ \langle u - \mathcal{P}_K u, h - \mathcal{P}_K u \rangle \leq 0 \},$$

in such a way that, according to (3.3), $K \subset H_-$ and $u \in H_+$. Consequently, (3.3) does actually provide us with a separation property. As the closed convex K is the set of fixed points of $\mathcal{P}_K$, K and $\mathcal{P}_K$ determine each other.

Proof. Fix $u \in H$. Then, the quantity

$$d := \operatorname{dist}(u, K) = \inf_{k \in K} \|u - k\| \geq 0$$

is well defined. Actually, $d = 0$ if and only if $u \in K$, since K is closed. In such case, we define

$$\mathcal{P}_K u := u.$$

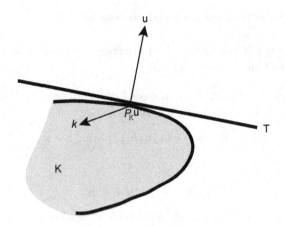

Fig. 3.2 The projection operator

Now, suppose $u \in H \setminus K$. Then, $d > 0$ and there exists a sequence $\{k_n\}_{n \geq 1}$ in K such that

$$d := \lim_{n \to \infty} \|u - k_n\|. \tag{3.4}$$

Setting

$$d_n := \|u - k_n\|, \qquad n \geq 1,$$

we find from (3.2) that, for every n, $m \geq 1$,

$$\left\| \frac{u - k_n + u - k_m}{2} \right\|^2 + \left\| \frac{u - k_n - (u - k_m)}{2} \right\|^2 = \frac{\|u - k_n\|^2 + \|u - k_m\|^2}{2}$$

and hence,

$$\left\| u - \frac{k_n + k_m}{2} \right\|^2 + \left\| \frac{k_m - k_n}{2} \right\|^2 = \frac{d_n^2 + d_m^2}{2}.$$

As K is convex, we have that

$$k_n, k_m \in K \quad \Longrightarrow \quad \frac{k_n + k_m}{2} \in K$$

for all n, $m \geq 1$, and hence,

$$d^2 = \text{dist}^2(u, K) \leq \left\| u - \frac{k_n + k_m}{2} \right\|^2.$$

Thus,

$$d^2 + \left\| \frac{k_m - k_n}{2} \right\|^2 \leq \frac{d_n^2 + d_m^2}{2}, \qquad n, m \geq 1,$$

and, therefore,

$$\left\|\frac{k_m - k_n}{2}\right\|^2 \leq \frac{d_n^2 + d_m^2}{2} - d^2 \to 0 \qquad \text{as} \quad n, m \to \infty.$$

This shows that $\{k_n\}_{n \geq 1}$ is a Cauchy sequence in K and, since K is closed, there exists $\mathcal{P}_K u \in K$ such that

$$\lim_{n \to \infty} k_n = \mathcal{P}_K u.$$

By (3.4), it is apparent that

$$d = \|u - \mathcal{P}_K u\|$$

and, consequently, d is attained at $\mathcal{P}_K u$.

Now, we will show that, for any $m \in K$, the next identity holds

$$\|u - m\| = \text{dist}\,(u, K) = \min_{k \in K} \|u - k\| \tag{3.5}$$

if and only if

$$\langle u - m, k - m \rangle \leq 0 \qquad \text{for all } k \in K. \tag{3.6}$$

Suppose $m \in K$ satisfies (3.5) and let $k \in K$. As K is convex,

$$(1 - t)m + tk \in K \qquad \text{for all } t \in [0, 1].$$

Thus, owing to (3.5),

$$\|u - m\| \leq \|u - (1 - t)m - tk\| = \|u - m + t(m - k)\|$$

for every $t \in [0, 1]$, and hence,

$$\|u - m\|^2 \leq \|u - m + t(m - k)\|^2$$
$$= \|u - m\|^2 + t^2\|m - k\|^2 + 2t\langle u - m, m - k \rangle.$$

Consequently,

$$t^2\|m - k\|^2 + 2t\langle u - m, m - k \rangle \geq 0 \qquad \text{for all } t \in [0, 1],$$

and, therefore, for every $t \in (0, 1]$,

$$2\langle u - m, k - m \rangle \leq t\|m - k\|^2.$$

Letting $t \downarrow 0$ in this estimate, (3.6) holds.

Conversely, let $m \in K$ satisfying (3.6) and pick a $k \in K$. Then,

$$\|u - m\|^2 - \|u - k\|^2 = \langle u - m, u - m \rangle - \langle u - k, u - k \rangle$$
$$= \|u\|^2 + \|m\|^2 - 2\langle u, m \rangle - \left(\|u\|^2 + \|k\|^2 - 2\langle u, k \rangle\right)$$
$$= \|m\|^2 - \|k\|^2 + 2\langle u, k - m \rangle$$
$$= 2\langle u - m, k - m \rangle + 2\langle m, k - m \rangle + \|m\|^2 - \|k\|^2$$
$$= 2\langle u - m, k - m \rangle + 2\langle m, k \rangle - \|m\|^2 - \|k\|^2$$
$$= 2\langle u - m, k - m \rangle - \|m - k\|^2.$$

Thus, according to (3.6), we obtain that

$$\|u - m\|^2 - \|u - k\|^2 \le -\|m - k\|^2 \le 0$$

and, consequently,

$$\|u - m\| \le \|u - k\| \qquad \text{for all } k \in K.$$

Equivalently,

$$\|u - m\| = \min_{k \in K} \|u - k\|.$$

Therefore, (3.5) and (3.6) are indeed equivalent.

It remains to prove that $\mathcal{P}_K u$ is the unique $k \in K$ for which

$$d = \|u - k\|.$$

Let $u \in H$ and $m_1, m_2 \in K$ such that

$$\|u - m_1\| = \|u - m_2\| = \min_{k \in K} \|u - k\|.$$

Then, by the equivalence of (3.5) and (3.6), we have that

$$\langle u - m_j, k - m_j \rangle \le 0 \qquad \text{for all } (k, j) \in K \times \{1, 2\},$$

and, in particular,

$$\langle u - m_1, m_2 - m_1 \rangle \le 0 \quad \text{and} \quad \langle u - m_2, m_1 - m_2 \rangle \le 0.$$

Thus, adding these inequalities, yields

$$
\begin{aligned}
0 &\ge \langle u - m_1, m_2 - m_1 \rangle + \langle u - m_2, m_1 - m_2 \rangle \\
&= \langle u, m_2 - m_1 \rangle - \langle m_1, m_2 - m_1 \rangle + \langle u, m_1 - m_2 \rangle - \langle m_2, m_1 - m_2 \rangle \\
&= \|m_2 - m_1\|^2
\end{aligned}
$$

and, therefore, $m_1 = m_2$, which concludes the proof. $\qquad\square$

The next result reveals that the projection operator

$$\mathcal{P}_K : H \to K$$

is globally Lipschitz with constant one.

Proposition 3.1. *Let $K \subset H$ be a closed and convex subset of H, and $\mathcal{P}_K : H \to K$ the projection operator of H on K. Then,*

$$\|\mathcal{P}_K u - \mathcal{P}_K v\| \le \|u - v\| \qquad \text{for all } u, v \in H. \tag{3.7}$$

As $\mathcal{P}_K k = k$ for all $k \in K$, the Lipschitz constant of $\mathcal{P}_K$ equals 1.

Proof. According to Theorem 3.1, and, most precisely, due to (3.3), we have that

$$\langle u - \mathcal{P}_K u, k - \mathcal{P}_K u \rangle \leq 0 \quad \text{and} \quad \langle v - \mathcal{P}_K v, k - \mathcal{P}_K v \rangle \leq 0$$

for all $u, v \in H$ and $k \in K$. In particular, for every $u, v \in H$,

$$\langle u - \mathcal{P}_K u, \mathcal{P}_K v - \mathcal{P}_K u \rangle \leq 0, \qquad \langle v - \mathcal{P}_K v, \mathcal{P}_K u - \mathcal{P}_K v \rangle \leq 0,$$

and, hence, adding these inequalities shows that

$$\begin{aligned}
0 &\geq \langle u - \mathcal{P}_K u, \mathcal{P}_K v - \mathcal{P}_K u \rangle + \langle v - \mathcal{P}_K v, \mathcal{P}_K u - \mathcal{P}_K v \rangle \\
&= \langle u - v, \mathcal{P}_K v - \mathcal{P}_K u \rangle + \langle \mathcal{P}_K v - \mathcal{P}_K u, \mathcal{P}_K v - \mathcal{P}_K u \rangle \\
&= \langle u - v, \mathcal{P}_K v - \mathcal{P}_K u \rangle + \| \mathcal{P}_K u - \mathcal{P}_K v \|^2.
\end{aligned}$$

Consequently, for every $u, v \in H$, we find that

$$\| \mathcal{P}_K u - \mathcal{P}_K v \|^2 \leq \langle u - v, \mathcal{P}_K u - \mathcal{P}_K v \rangle$$

and, therefore, according to (3.1), we conclude that

$$\| \mathcal{P}_K u - \mathcal{P}_K v \|^2 \leq \| \mathcal{P}_K u - \mathcal{P}_K v \| \, \| u - v \|.$$

This shows (3.7) and ends the proof. □

3.2 The orthogonal projection on a closed subspace

Subsequently, given a closed subspace N of H, we denote by $N^\perp$ the orthogonal of N in H

$$N^\perp := \{ u \in H \ : \ \langle u, n \rangle = 0 \text{ for all } n \in N \}.$$

As any subspace of H is convex, Theorem 3.1 guarantees that the projection $\mathcal{P}_N : H \to N$ is well defined. The next result identifies $\mathcal{P}_N$ as the *orthogonal projection* of H on N.

Theorem 3.2. *Let N be a closed subspace of H, and denote by $\mathcal{P}_N$ the projection of H on N. Then,*

$$\mathcal{P}_N \in \mathcal{L}(H, N), \qquad \| \mathcal{P}_N \|_{\mathcal{L}(H,N)} = 1,$$

and

$$\mathcal{P}_N |_N = I_N, \qquad R[I_H - \mathcal{P}_N] \subset N^\perp.$$

Consequently, $\mathcal{P}_N$ is the **orthogonal projection** *of H on N.*

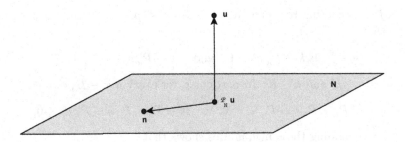

Fig. 3.3   The orthogonal projection

Proof.   Let $u \in H$. By Theorem 3.1, we have that

$$\langle u - \mathcal{P}_N u, n - \mathcal{P}_N u \rangle \leq 0 \qquad \text{for all } n \in N,$$

and hence, for every $n \in N$ and $\lambda > 0$,

$$0 \geq \langle u - \mathcal{P}_N u, \lambda n - \mathcal{P}_N u \rangle = \lambda \langle u - \mathcal{P}_N u, n \rangle - \langle u - \mathcal{P}_N u, \mathcal{P}_N u \rangle,$$

since $\lambda n \in N$ for all $\lambda \in \mathbb{R}$. Thus, for every $n \in N$ and $\lambda > 0$,

$$\langle u - \mathcal{P}_N u, n \rangle \leq \frac{1}{\lambda} \langle u - \mathcal{P}_N u, \mathcal{P}_N u \rangle, \qquad (3.8)$$

and hence, letting $\lambda \uparrow \infty$, (3.8) implies that

$$\langle u - \mathcal{P}_N u, n \rangle \leq 0 \qquad \text{for all } n \in N.$$

As N is a linear subspace, we also have that

$$\langle u - \mathcal{P}_N u, -n \rangle \geq 0 \qquad \text{for all } n \in N.$$

Consequently,

$$\langle u - \mathcal{P}_N u, n \rangle = 0 \qquad \text{for all } n \in N \qquad (3.9)$$

and, therefore,

$$R[I_H - \mathcal{P}_N] \subset N^\perp.$$

The fact that

$$\mathcal{P}_N n = n, \qquad n \in N,$$

is a direct consequence from Theorem 3.1.

By Proposition 3.1, to conclude the proof it suffices to prove that $\mathcal{P}_N$ is a linear operator. Indeed, due to (3.9), for every $\lambda, \mu \in \mathbb{R}$, $u, v \in H$, and $n \in N$, we have that

$$\langle \lambda u + \mu v - \mathcal{P}_N (\lambda u + \mu v), n \rangle = 0.$$

Similarly,

$$\langle \lambda u + \mu v - \lambda \mathcal{P}_N u - \mu \mathcal{P}_N v, n \rangle = \lambda \langle u - \mathcal{P}_N u, n \rangle + \mu \langle v - \mathcal{P}_N v, n \rangle = 0.$$

Thus, subtracting both identities shows that

$$\langle \mathcal{P}_N(\lambda u + \mu v) - \lambda \mathcal{P}_N u - \mu \mathcal{P}_N v, n \rangle = 0$$

for all $n \in N$. In particular, making the choice

$$n := \mathcal{P}_N(\lambda u + \mu v) - \lambda \mathcal{P}_N u - \mu \mathcal{P}_N v$$

it becomes apparent that, for every $u, v \in H$ and $\lambda, \mu \in \mathbb{R}$,

$$\| \mathcal{P}_N(\lambda u + \mu v) - \lambda \mathcal{P}_N u - \mu \mathcal{P}_N v \|^2 = 0,$$

which implies

$$\mathcal{P}_N(\lambda u + \mu v) = \lambda \mathcal{P}_N u + \mu \mathcal{P}_N v$$

and concludes the proof. $\square$

3.3 The representation theorem of F. Riesz

In this section, we will obtain the representation theorem of F. Riesz from the theory already developed in Sections 3.2 and 3.3. It identifies any Hilbert space H with its dual space

$$H' = \mathcal{L}(H, \mathbb{R}).$$

Consequently, any real Hilbert space is *reflexive*, in the sense that

$$H'' = H.$$

The abstract theory of Hilbert spaces relies on this feature.

Theorem 3.3 (of representation of F. Riesz). *Let H be a real Hilbert space. Then, for every $\varphi \in H'$ there exists a unique $u_\varphi \in H$ such that*

$$\varphi(u) = \langle u_\varphi, u \rangle \qquad \text{for all } u \in H. \tag{3.10}$$

Moreover,

$$\| \varphi \|_{H'} = \| u_\varphi \| \qquad \text{for all } \varphi \in H',$$

and, actually, the map

$$H' \xrightarrow{\mathfrak{D}} H$$
$$\varphi \mapsto \mathfrak{D}(\varphi) := u_\varphi \tag{3.11}$$

establishes an isometric isomorphism between H' and H.

Proof.   If $\varphi = 0$, then (3.10) is satisfied if and only if $u_\varphi = 0$. So, suppose $\varphi \in H' \setminus \{0\}$ and consider

$$N := N[\varphi] = \{u \in H \ : \ \varphi(u) = 0\}.$$

Then, N is a closed proper subspace of H.

Next, we will show that there exists $p \in N^\perp$ with $\|p\| = 1$ such that

$$H = N \oplus \operatorname{span} [p]. \tag{3.12}$$

Indeed, pick $u \in H \setminus N$, and let $\mathcal{P}_N$ be the orthogonal projection of H on N. Then, by the definition of N, $\varphi(u) \neq 0$ and, hence,

$$\varphi(u - \mathcal{P}_N u) = \varphi(u) - \varphi(\mathcal{P}_N u) = \varphi(u) \neq 0.$$

Thus,

$$q := \frac{u - \mathcal{P}_N u}{\varphi(u - \mathcal{P}_N u)} \in H \setminus N.$$

On the other hand, owing to Theorem 3.2,

$$u - \mathcal{P}_N u \in R[I_H - \mathcal{P}_N] \subset N^\perp$$

and, consequently,

$$\langle q, n \rangle = 0 \qquad \text{for all } n \in N.$$

Therefore, the vector

$$p := \frac{q}{\|q\|} \in H \setminus N$$

satisfies

$$\|p\| = 1, \qquad p \in N^\perp. \tag{3.13}$$

Moreover, it satisfies (3.12). Indeed, for every $v \in H$, we have that

$$v = v - \frac{\varphi(v)}{\varphi(p)} p + \frac{\varphi(v)}{\varphi(p)} p$$

and, obviously,

$$\varphi(v - \frac{\varphi(v)}{\varphi(p)} p) = 0.$$

Thus,

$$H = N + \operatorname{span} [\varphi].$$

Moreover, if

$$n + \lambda p = m + \mu p$$

for some $\lambda, \mu \in \mathbb{R}$ and $n, m \in N$, then

$$n - m = (\mu - \lambda)p$$

and hence, it follows from (3.13) that

$$\|n - m\|^2 = (\mu - \lambda)\langle n - m, p \rangle = 0.$$

Consequently,

$$n = m \quad \text{and} \quad (\mu - \lambda)p = 0,$$

which implies $\lambda = \mu$. Therefore, (3.12) holds. The algebraic direct sum is actually topological because the orthogonal projections of H on N and span $[p]$ are continuous.

Next, we will show that

$$u_\varphi := \varphi(p)p$$

satisfies (3.10). Indeed, according to (3.12), for each $u \in H$ there exist $n \in N$ and $\lambda \in \mathbb{R}$ (unique) such that

$$u = n + \lambda p.$$

Thus, since $\varphi(n) = 0$, we find from (3.13) that

$$\begin{aligned}
\langle u_\varphi, u \rangle &= \varphi(p)\langle p, n + \lambda p \rangle \\
&= \varphi(p)\langle p, n \rangle + \lambda\varphi(p)\|p\|^2 \\
&= \lambda\varphi(p) = \varphi(n + \lambda p) \\
&= \varphi(u),
\end{aligned}$$

which concludes the proof of (3.10).

To prove the uniqueness of u_φ, assume that there are $u_\varphi^1, u_\varphi^2 \in H$ such that

$$\langle u_\varphi^1, u \rangle = \langle u_\varphi^2, u \rangle \qquad \text{for all } u \in H.$$

Then, for every $u \in H$,

$$\langle u_\varphi^1 - u_\varphi^2, u \rangle = 0$$

and, in particular,

$$0 = \langle u_\varphi^1 - u_\varphi^2, u_\varphi^1 - u_\varphi^2 \rangle = \|u_\varphi^1 - u_\varphi^2\|^2,$$

which implies $u_\varphi^1 = u_\varphi^2$ and shows the uniqueness of u_φ. Consequently, the map (3.11) is well defined.

To complete the proof of the theorem it remains to show that the map $\mathfrak{D}$ defined by (3.11) is an isometric isomorphism.

Let $\lambda, \mu \in \mathbb{R}$ and $\varphi, \psi \in H'$. Then, for every $u \in H$, we have that

$$(\lambda\varphi + \mu\psi)(u) = \lambda\varphi(u) + \mu\psi(u)$$
$$= \lambda\langle u_\varphi, u\rangle + \mu\langle u_\psi, u\rangle$$
$$= \langle \lambda u_\varphi + \mu u_\psi, u\rangle$$

and hence,

$$u_{\lambda\varphi+\mu\psi} = \lambda u_\varphi + \mu u_\psi,$$

by uniqueness. In other words,

$$\mathfrak{D}(\lambda\varphi + \mu\psi) = \lambda\mathfrak{D}(\varphi) + \mu\mathfrak{D}(\psi),$$

and, consequently, $\mathfrak{D}$ is linear.

Now, suppose

$$\mathfrak{D}(\varphi) = u_\varphi = 0.$$

Then, $\varphi(u) = 0$ for all $u \in H$ and, so, $\varphi = 0$. Therefore, $\mathfrak{D}$ is injective. To show that it is surjective, for a given $u \in H$, let $\varphi_u : H \to \mathbb{R}$ denote the map defined by

$$\varphi_u(v) = \langle u, v\rangle, \qquad v \in H.$$

The map φ_u is linear and continuous, because

$$|\varphi_u(v)| \le \|u\|\,\|v\| \qquad \text{for all } v \in H.$$

Consequently, $\varphi_u \in H'$,

$$\|\varphi_u\|_{H'} \le \|u\|,$$

and, necessarily, by uniqueness,

$$\mathfrak{D}(\varphi_u) = u.$$

Therefore, $\mathfrak{D}$ establishes an isomorphism between H and H'.

Finally, owing to (3.10), we have that, for every $\varphi \in H'$ and $u \in H$,

$$|\varphi(u)| \le \|u_\varphi\|\,\|u\|, \qquad |\varphi(u_\varphi)| = \|u_\varphi\|^2.$$

Thus,

$$\|\varphi\|_{H'} = \|u_\varphi\| = \|\mathfrak{D}(\varphi)\|,$$

and, therefore, $\mathfrak{D}$ is isometric. The proof is complete. $\qquad\square$

3.4 Continuity and coercivity of bilinear forms

The following concept has shown to be pivotal for the development of the theory of partial differential equations.

Definition 3.1. *Let H be a real Hilbert space and $\mathfrak{a} : H \times H \to \mathbb{R}$ a bilinear form. Then:*

i) *$\mathfrak{a}$ is said to be **continuous** if there is a constant $\beta > 0$ such that*

$$|\mathfrak{a}(u, v)| \leq \beta \, \|u\| \, \|v\| \qquad \text{for all } u, v \in H. \tag{3.14}$$

ii) *$\mathfrak{a}$ is said to be **coercive** if there is a constant $\alpha > 0$ such that*

$$\mathfrak{a}(u, u) \geq \alpha \, \|u\|^2 \qquad \text{for all } u \in H. \tag{3.15}$$

If the bilinear form $\mathfrak{a} : H \times H \to \mathbb{R}$ is continuous and coercive, then

$$\alpha \, \|u\|^2 \leq \mathfrak{a}(u, u) \leq \beta \, \|u\|^2$$

for all $u \in H$. Actually, the following result holds.

Proposition 3.2. *Let $\mathfrak{a} : H \times H \to \mathbb{R}$ be a continuous and coercive bilinear form satisfying (3.14) and (3.15). Then, there exists $A \in \mathrm{Iso}(H)$ such that*

$$\alpha \, \|u\| \leq \|Au\| \leq \beta \|u\| \qquad \text{for all } u \in H, \tag{3.16}$$

and

$$\mathfrak{a}(u, v) = \langle Au, v \rangle \qquad \text{for all } u, v \in H. \tag{3.17}$$

Proof. Fix $u \in H$ and consider the map $\varphi_u : H \to \mathbb{R}$ defined through

$$\varphi_u(v) := \mathfrak{a}(u, v), \qquad v \in H.$$

As $\mathfrak{a}$ is bilinear, φ_u is a linear map. Moreover, by (3.14), we have that

$$|\varphi_u(v)| = |\mathfrak{a}(u, v)| \leq \beta \, \|u\| \, \|v\| \qquad \text{for every } v \in H$$

and hence $\varphi_u \in H'$. Therefore, according to Theorem 3.3, there exists a unique $\tilde{u} \in H$ such that

$$\varphi_u(v) = \langle \tilde{u}, v \rangle \qquad \text{for all } v \in H.$$

Let $A : H \to H$ be the map defined by

$$Au := \tilde{u}, \qquad u \in H.$$

By construction, A satisfies (3.17). To complete the proof it remains to show that $A \in \mathrm{Iso}(H)$. Indeed, for every $u, v, w \in H$ and $\lambda, \mu \in \mathbb{R}$,

$$
\begin{aligned}
\langle A(\lambda u + \mu v), w \rangle &= \mathfrak{a}(\lambda u + \mu v, w) \\
&= \lambda \mathfrak{a}(u, w) + \mu \mathfrak{a}(v, w) \\
&= \lambda \langle Au, w \rangle + \mu \langle Av, w \rangle \\
&= \langle \lambda Au + \mu Av, w \rangle
\end{aligned}
$$

and hence,

$$
\langle A(\lambda u + \mu v) - (\lambda Au + \mu Av), w \rangle = 0.
$$

Thus,

$$
A(\lambda u + \mu v) = \lambda Au + \mu Av
$$

for all $u, v \in H$ and $\lambda, \mu \in \mathbb{R}$. Therefore, A is linear.

Thanks to (3.15) and (3.17), it becomes apparent that

$$
\alpha \|u\|^2 \leq |\mathfrak{a}(u, u)| = |\langle Au, u \rangle| \leq \|Au\| \, \|u\|
$$

for all $u \in H$, and, consequently,

$$
\alpha \|u\| \leq \|Au\|,
$$

which provides us with the lower estimate of (3.16). Moreover, it follows from (3.14) that, for every $u \in H$,

$$
\|Au\|^2 = |\langle Au, Au \rangle| = |\mathfrak{a}(u, Au)| \leq \beta \|u\| \, \|Au\|
$$

and, therefore,

$$
\|Au\| \leq \beta \|u\| \qquad \text{for all } u \in H,
$$

which concludes the proof of (3.16). This ends the proof, because (3.16) entails that $A \in \mathrm{Iso}(H)$. $\qquad\qquad \square$

3.5 The theorem of G. Stampacchia

The main result of this section reads as follows.

Theorem 3.4 (of G. Stampacchia). *Suppose* $\mathfrak{a} : H \times H \to \mathbb{R}$ *is a continuous and coercive bilinear form, and K is a closed and convex subset of H. Then, for each $\varphi \in H'$ there exists a unique $f := f(\varphi) \in K$ such that*

$$
\mathfrak{a}(f, k - f) \geq \varphi(k - f) \qquad \text{for all } k \in K.
$$

Proof. Fix $\varphi \in H'$. Then, according to Theorem 3.3, there exists a unique $u_\varphi \in H$ such that

$$\varphi(u) = \langle u_\varphi, u \rangle \qquad \text{for all } u \in H.$$

Consequently, it suffices to prove that there is a unique $f \in K$ for which

$$\mathfrak{a}(f, k - f) \geq \langle u_\varphi, k - f \rangle \qquad \text{for all } k \in K. \tag{3.18}$$

According to Proposition 3.2, there exists $A \in \text{Iso}(H)$ for which (3.17) holds. Obviously, owing to (3.17), the inequality (3.18) can be expressed as

$$\langle Af, k - f \rangle \geq \langle u_\varphi, k - f \rangle \qquad \text{for all } k \in K.$$

Equivalently, for some $\rho > 0$,

$$\langle \rho u_\varphi - \rho Af + f - f, k - f \rangle \leq 0, \qquad \text{for all } k \in K. \tag{3.19}$$

Let $\mathcal{P}_K$ denote the projection of H on K. According to Theorem 3.1, for every $u \in H$, $\mathcal{P}_K u$ is characterized through

$$\langle u - \mathcal{P}_K u, k - \mathcal{P}_K u \rangle \leq 0 \qquad \text{for all } k \in K.$$

Thus, condition (3.19) holds if

$$f = \mathcal{P}_K (\rho u_\varphi - \rho Af + f).$$

Therefore, it suffices to prove that there exists $\rho > 0$ for which the operator $\mathfrak{S} : K \to K$ defined by

$$\mathfrak{S}(k) := \mathcal{P}_K (\rho u_\varphi - \rho Ak + k), \qquad k \in K,$$

has a fixed point. By Proposition 3.1, we have that, for every $k_1, k_2 \in K$ and $\rho > 0$,

$$\begin{aligned}
\|\mathfrak{S}(k_1) - \mathfrak{S}(k_2)\| &= \|\mathcal{P}_K (\rho u_\varphi - \rho Ak_1 + k_1) - \mathcal{P}_K (\rho u_\varphi - \rho Ak_2 + k_2)\| \\
&\leq \|\rho u_\varphi - \rho Ak_1 + k_1 - \rho u_\varphi + \rho Ak_2 - k_2\| \\
&= \|k_1 - k_2 - \rho A(k_1 - k_2)\| \\
&\leq \|I_H - \rho A\|_{\mathcal{L}(H)} \|k_1 - k_2\|.
\end{aligned}$$

Thus, thanks to the contracting mapping theorem, it suffices to show that there exists $\rho > 0$ such that

$$\|I_H - \rho A\|_{\mathcal{L}(H)} < 1. \tag{3.20}$$

Indeed, for every $\rho > 0$ and $u \in H$, we have that

$$\|(I_H - \rho A)u\|^2 = \langle u - \rho Au, u - \rho Au \rangle = \|u\|^2 + \rho^2 \|Au\|^2 - 2\rho \langle Au, u \rangle,$$

and hence, it becomes apparent from (3.15)–(3.17) that

$$\begin{aligned}
\|(I_H - \rho A)u\|^2 &= \|u\|^2 + \rho^2 \|Au\|^2 - 2\rho \mathfrak{a}(u, u) \\
&\leq (1 + \rho^2 \beta^2 - 2\rho \alpha) \|u\|^2.
\end{aligned}$$

Consequently,

$$\|I_H - \rho A\|_{\mathcal{L}(H)} \leq \sqrt{1 + \rho^2 \beta^2 - 2\rho \alpha}$$

for every $\rho > 0$, and, therefore, (3.20) holds provided

$$0 < \rho < 2\alpha/\beta^2.$$

The proof is complete. $\qquad\qquad\qquad\qquad\qquad\qquad\qquad\qquad\qquad\square$

3.6 The theorem of P. D. Lax and A. N. Milgram

When $K = H$, Theorem 3.4 provides us with a very celebrated result of P. D. Lax and A. N. Milgram, which is a substantial improvement of Riesz' representation Theorem 3.3, as the underlying bilinear form is not required to be symmetric and, so, it might not define a scalar product.

Theorem 3.5 (of P. D. Lax and A. N. Milgram). *Let $\mathfrak{a} : H^2 \to \mathbb{R}$ be a continuous and coercive bilinear form. Then, for every $\varphi \in H'$ there exists a unique $u_\varphi \in H$ such that*

$$\varphi(u) = \mathfrak{a}(u_\varphi, u) \qquad \text{for all } u \in H. \tag{3.21}$$

Proof. Choose $K := H$ and let $\varphi \in H'$. Then, according to Theorem 3.4, there exists a unique $u_\varphi \in H$ such that

$$\mathfrak{a}(u_\varphi, u - u_\varphi) \geq \varphi(u - u_\varphi) \qquad \text{for all } \quad u \in H.$$

Consequently,

$$\mathfrak{a}(u_\varphi, v) \geq \varphi(v), \qquad \text{for all } \quad v \in H,$$

and hence,

$$-\mathfrak{a}(u_\varphi, v) = \mathfrak{a}(u_\varphi, -v) \geq \varphi(-v) = -\varphi(v).$$

Therefore, (3.21) holds. The proof is complete. $\qquad\square$

If, in addition, the bilinear form $\mathfrak{a}(\cdot, \cdot)$ is symmetric, i.e.,

$$\mathfrak{a}(u, v) = \mathfrak{a}(v, u) \qquad \text{for every } u, v \in H,$$

then, Theorem 3.5 follows directly from Theorem 3.3, as the map

$$H \times H \xrightarrow{\langle \cdot, \cdot \rangle_\mathfrak{a}} \mathbb{R}$$
$$(u, v) \;\mapsto\; \langle u, v \rangle_\mathfrak{a} := \mathfrak{a}(u, v)$$

provides us with a scalar product in H and hence we can apply Theorem 3.3 to get Theorem 3.5. Therefore, the importance of Theorem 3.5 relies on the fact that it does not require the symmetry of $\mathfrak{a}$.

3.7 Projecting on a closed convex set of a u.c. B-space

Essentially, this section generalizes Theorem 3.1 to the more general context of uniformly convex Banach spaces (u.c. B-spaces). But the projection on a closed subspace of a u.c. B-space, though continuous, is not necessarily linear, unless the subspace is a hyperplane. This is in strong contrast with the situation already described in the context of Hilbert spaces.

This section is divided into four parts. The first one introduces some basic concepts and preliminaries. The second one shows the existence and continuity of the projection operator on any closed and convex subset of a u.c. B-space. The third section shows some *linear properties* of the projection on a closed subspace. Among them, it proves that it is homogeneous of degree one. Finally, the fourth section shows that the projection on a closed hyperplane is linear. It remains an open problem to ascertain whether this property holds in the context of u.c. B-spaces. Throughout this section, X will stand for a real Banach space with norm $\| \cdot \|$.

3.7.1 *Basic concepts and preliminaries*

This section collects some basic concepts and results whose proofs will not be given here, as they are outside the general scope of this book.

The next concept involves a geometrical property of the unit ball that might not be satisfied for any equivalent norm. Basically, it establishes that the unit ball is sufficiently curved.

Definition 3.2 (Uniformly convex B-space). *The Banach space X is said to be uniformly convex if for every $\epsilon > 0$ there exists $\delta = \delta(\epsilon) > 0$ such that*

$$\left. \begin{array}{r} x, y \in X \\ \|x\| \leq 1, \ \|y\| \leq 1 \\ \|x - y\| > \epsilon \end{array} \right\} \implies \left\| \frac{x+y}{2} \right\| < 1 - \delta. \tag{3.22}$$

To simplify notations, X is said to be a u.c. B-space if it is a uniformly convex real Banach space. Suppose $X = \mathbb{R}^N$. Then, though all norms in X are equivalent, and, in particular, so are the norms

$$\|x\|_p = \left(\sum_{j=1}^{N} |x_j|^p \right)^{\frac{1}{p}}, \qquad x = (x_1, \cdots, x_n) \in \mathbb{R}^N,$$

for all $p \in [1, \infty)$, it is easily seen that $(X, \| \cdot \|_p)$ is uniformly convex if $p \in (1, \infty)$, while this property fails for $p = 1$. More generally, any Hilbert space is a u.c. B-space.

The concept of uniform convexity goes back to J. A. Clarkson [42], where it was shown that $L^p(\Omega)$ is a u.c. B-space for all $p \in (1, \infty)$. As a by-product, the spaces $W^{2,p}(\Omega)$, $1 < p < \infty$, which are going to be introduced in Chapter 4, are u.c. B-spaces, because there is a natural isometry between $W^{2,p}(\Omega)$ and a certain product of $L^p(\Omega)$'s. A short time later, the following result was established by D. P. Milman, [162] and, independently, by B. J. Pettis [176].

Theorem 3.6 (of D. P. Milman and B. J. Pettis). *Every u.c. B-space X is reflexive, i.e., $X'' = X$.*

This result establishes an astonishing connection between a geometrical property, the uniform convexity of the unit ball, and a rather topological property, the reflexivity. Other proofs of this result, some of them very short, were given by S. Kakutani [110], J. R. Ringrose [192], J. Lindenstrauss and L. Tzafriri [128] (see the bottom of p. 127 in [128]), and H. Brézis [29] (see the proof of Theorem III.29). According to M. M. Day [48], there are reflexive B-spaces which are not equivalent to any u.c. B-space.

We now collect some classical results on weak topologies that will be used in the proof of the main theorem of this section. Subsequently, for any $f \in X'$, we denote by $\varphi_f : X \to \mathbb{R}$ the map defined by

$$\varphi_f(x) = f(x), \qquad x \in X.$$

Then, the **weak topology** $\sigma(X, X')$ is defined as the weakest topology in X for which every functional φ_f, $f \in X'$, is continuous.

The next result establishes that, for convex sets, the concepts of strong and weak closeness coincide; it is Theorem III.7 of H. Brézis [29]. A set $C \subset X$ is said to be *weakly closed* if it is closed in the weak topology $\sigma(X, X')$.

Theorem 3.7. *Let $C \subset X$ be a convex subset of the Banach space X. Then, C is weakly closed if and only if C is closed in the strong topology.*

The next result is a consequence from Theorem 3.7; it is Corollary III.8 of H. Brézis [29]. Subsequently, given a subset $Y \subset X$ and a functional $\Phi : Y \to (-\infty, \infty]$, Φ is said to be **lower semi continuous (l.s.c.)** if, for every $y_n, y \in Y$, $n \geq 1$,

$$\lim_{n \to \infty} y_n = y \qquad \text{implies} \qquad \Phi(y) \leq \liminf_{n \to \infty} \Phi(y_n).$$

Obviously, Φ is l.s.c. if it is continuous.

Corollary 3.1. *Let $\Phi : X \to (-\infty, \infty]$ be a l.s.c. convex functional. Then,*

$$\lim_{n \to \infty} x_n = x \quad \text{in } \sigma(X, X') \qquad \text{implies} \qquad \Phi(x) \leq \liminf_{n \to \infty} \Phi(x_n).$$

The following result provides us with a pivotal characterization of the reflexivity attributable to W. F. Eberlein [59] and V. L. Shmulyan [202]; it is Theorem III.16 of H. Brézis [29]. The interested reader might wish to have a look at p. 141 of K. Yosida [227] too.

Theorem 3.8 (of W. F. Eberlein and V. L. Shmulyan). *A B-space* X *is reflexive if and only if the unit ball*

$$B := \{x \in X \ : \ \|x\| \leq 1\}$$

is compact in the weak topology $\sigma(X, X')$.

The next theorem is a classical pivotal result in the calculus of variations; it is Corollary III.20 of H. Brézis [29].

Theorem 3.9. *Let* X *be a reflexive B-space,* $K \subset X$ *a non-empty closed and convex set, and* $\Phi : K \longrightarrow (-\infty, \infty]$ *a convex l.s.c. functional,* $\Phi \neq \infty$, *such that*

$$\lim_{\substack{k \in K \\ \|k\| \to \infty}} \Phi(k) = \infty \tag{3.23}$$

if K *is unbounded. Then, there exists* $k_{\min} \in K$ *such that*

$$\Phi(k_{\min}) = \min_{k \in K} \Phi(k). \tag{3.24}$$

Any point $k_{\min} \in K$ satisfying (3.24) is called a *minimizer* of Φ in K. In general, the minimizer is not unique. In the context of the calculus of variations, any functional Φ satisfying (3.23) is said to be *coercive* on K.

Finally, we need the following useful property of the u.c. B-spaces; it is Proposition III.30 of H. Brézis [29]

Proposition 3.3. *Let* X *be a u.c. B-space, and* $\{x_n\}_{n \geq 1}$ *a sequence of* X *such that*

$$\lim_{n \to \infty} x_n = x \quad in \ \sigma(X, X') \qquad and \quad \limsup_{n \to \infty} \|x_n\| \leq \|x\|.$$

Then,

$$\lim_{n \to \infty} \|x_n - x\| = 0.$$

Naturally, the reader is referred to Chapter III of H. Brézis [29] for the proofs of all the previous results.

3.7.2   The projection theorem

The main result of this section reads as follows.

Theorem 3.10. *Let X be a u.c. B-space and $K \subset X$ a closed and convex proper subset. Then, for every $x \in X$, there exists a unique $\mathcal{P}_K x \in K$ such that*

$$\|x - \mathcal{P}_K x\| = \operatorname{dist}(x, K) = \min_{k \in K} \|x - k\|.$$

Necessarily, $\mathcal{P}_K k = k$ for all $k \in K$. Moreover, the associated map

$$X \xrightarrow{\mathcal{P}_K} K$$
$$x \mapsto \mathcal{P}_K x$$

*is continuous. The operator $\mathcal{P}_K$ is called the **projection of X on K**.*

Proof.   Fixed $x \in X$, let $\Phi : K \to [0, \infty)$ denote the distance map

$$\Phi(k) := \|x - k\| \qquad \text{for all } k \in K.$$

As, for every $k_1, k_2 \in K$ and $t \in [0, 1]$, we have that

$$\begin{aligned}
\Phi(tk_1 + (1 - t)k_2) &= \|x - tk_1 - (1 - t)k_2\| \\
&= \|t(x - k_1) + (1 - t)(x - k_2)\| \\
&\leq t\|x - k_1\| + (1 - t)\|x - k_2\| \\
&= t\Phi(k_1) + (1 - t)\Phi(k_2),
\end{aligned}$$

the map Φ is convex. Moreover, since

$$|\Phi(k_1) - \Phi(k_2)| = |\|x - k_1\| - \|x - k_2\|| \leq \|k_1 - k_2\|,$$

Φ is continuous and hence lower semi-continuous. The coercivity of Φ is a direct consequence from the estimate

$$|\|k\| - \|x\|| \leq \|k - x\| = \Phi(k),$$

which implies (3.23).

According to Theorem 3.6, X is reflexive. Therefore, by Theorem 3.9, Φ admits a minimizer in K.

It should be noted that the existence of the minimizer is guaranteed as soon as X is reflexive. The uniform convexity of X is necessary for the uniqueness of the minimizer, as it will become apparent soon. The uniqueness is obvious if $x \in K$. So, suppose that $x \in X \setminus K$ and there are $k_1, k_2 \in K$, $k_1 \neq k_2$, such that

$$0 < \|x - k_1\| = \|x - k_2\| = \operatorname{dist}(x, K) \leq \|x - k\| \qquad (3.25)$$

for all $k \in K$. Then, setting

$$\tilde{x} := \frac{x}{\|x - k_j\|}, \qquad \tilde{k}_j := \frac{k_j}{\|x - k_j\|}, \qquad j \in \{1, 2\},$$

and dividing (3.25) by $\|x - k_1\|$, we are led to the identity

$$1 = \|\tilde{x} - \tilde{k}_1\| = \|\tilde{x} - \tilde{k}_2\| \leq \|\tilde{x} - \frac{k}{\|x - k_1\|}\|$$

for all $k \in K$. In particular, as K is convex, we obtain that

$$1 = \|\tilde{x} - \tilde{k}_1\| = \|\tilde{x} - \tilde{k}_2\| \leq \|\tilde{x} - \frac{tk_1 + (1-t)k_2}{\|x - k_1\|}\|$$

for all $t \in [0, 1]$. Consequently,

$$1 \leq \|\tilde{x} - t\tilde{k}_1 - (1-t)\tilde{k}_2\| = \|t(\tilde{x} - \tilde{k}_1) + (1-t)(\tilde{x} - \tilde{k}_2)\|$$
$$\leq t\|\tilde{x} - \tilde{k}_1\| + (1-t)\|\tilde{x} - \tilde{k}_2\| = t + 1 - t = 1$$

and, therefore,

$$\|\tilde{x} - t\tilde{k}_1 - (1-t)\tilde{k}_2\| = 1 \qquad \forall\, t \in [0, 1].$$

In particular,

$$\|\frac{\tilde{x} - \tilde{k}_1}{2} + \frac{\tilde{x} - \tilde{k}_2}{2}\| = 1. \tag{3.26}$$

By construction, we already know that

$$\|\frac{\tilde{x} - \tilde{k}_1}{2}\| = \|\frac{\tilde{x} - \tilde{k}_2}{2}\| = \frac{1}{2}$$

and

$$\|\tilde{x} - \tilde{k}_1 - \tilde{x} + \tilde{k}_2\| = \|\tilde{k}_1 - \tilde{k}_2\| > 0,$$

because $k_1 \neq k_2$. Thus, as X is a u.c. B-space, we should have

$$\|\frac{\tilde{x} - \tilde{k}_1}{2} + \frac{\tilde{x} - \tilde{k}_2}{2}\| \leq 1 - \delta$$

for some $\delta > 0$. Obviously, this contradicts (3.26) and, consequently, it shows the uniqueness of the minimizer. Subsequently, for every $x \in X$, we denote by $\mathcal{P}_K x \in K$ the unique minimizer of Φ in K.

To conclude the proof of the theorem, it remains to prove that $\mathcal{P}_K$ is continuous. Let $\{x_n\}_{n \geq 1}$ be a sequence in X such that

$$\lim_{n \to \infty} \|x_n - x\| = 0$$

for some $x \in X$. Then, as the map

$$\text{dist}\,(\cdot, K) \ : \ X \longrightarrow [0, \infty)$$

is continuous, we have that

$$\lim_{n \to \infty} \|x_n - \mathcal{P}_K x_n\| = \lim_{n \to \infty} \text{dist}\,(x_n, K)$$
$$= \text{dist}\,(x, K) = \|x - \mathcal{P}_K x\|. \tag{3.27}$$

As a by-product, the sequence $\{\mathcal{P}_K x_n\}_{n \geq 1}$ is bounded, because, for sufficiently large $n \geq 1$,

$$\|\mathcal{P}_K x_n\| \leq \|x_n\| + \|x - \mathcal{P}_K x\| + 1$$

and. moreover, $\{x_n\}_{n \geq 1}$ is bounded.

By Theorem 3.6, X is reflexive. Thus, owing to Theorem 3.8, there exists $w \in X$ and a subsequence of $\{x_n\}_{n \geq 1}$, say $\{x_{n_m}\}_{m \geq 1}$, such that

$$\lim_{m \to \infty} \mathcal{P}_K x_{n_m} = w \quad \text{in} \quad \sigma(X, X').$$

As, due to Theorem 3.7, K is closed for the weak topology $\sigma(X, X')$, necessarily $w \in K$ and

$$\lim_{m \to \infty} (x_{n_m} - \mathcal{P}_K x_{n_m}) = x - w \quad \text{in} \quad \sigma(X, X'). \tag{3.28}$$

Consequently, we obtain from Corollary 3.1 that

$$\|x - w\| \leq \liminf_{m \to \infty} \|x_{n_m} - \mathcal{P}_K x_{n_m}\|, \tag{3.29}$$

because $\|\cdot\| : X \to [0, \infty)$ is l.s.c. and convex. Thus, we find from (3.27) and (3.29) that

$$\|x - w\| \leq \|x - \mathcal{P}_K x\|$$

and, hence, $w = \mathcal{P}_K x$, by the definition of $\mathcal{P}_K$. Therefore, (3.28) does actually provide us with

$$\lim_{m \to \infty} (x_{n_m} - \mathcal{P}_K x_{n_m}) = x - \mathcal{P}_K x \quad \text{in} \quad \sigma(X, X'). \tag{3.30}$$

Finally, by (3.27) and (3.30), Proposition 3.3 implies that

$$\lim_{m \to \infty} (x_{n_m} - \mathcal{P}_K x_{n_m}) = x - \mathcal{P}_K x \quad \text{in} \quad X,$$

and, consequently,

$$\lim_{m \to \infty} \mathcal{P}_K x_{n_m} = \mathcal{P}_K x \quad \text{in} \quad X.$$

As this scheme can be repeated along any subsequence, we find that

$$\lim_{n \to \infty} \mathcal{P}_K x_n = \mathcal{P}_K x \quad \text{in} \quad X.$$

The proof is complete. $\square$

3.7.3 The projection on a closed linear subspace

Let N be a closed subspace of a u.c. B-space X. According to Theorem 3.2, when X is a Hilbert space, $\mathcal{P}_N$ is a linear operator and hence N admits a topological complement in X. Consequently, a natural question arises. Should $\mathcal{P}_N$ be a linear operator for any closed subspace N of a u.c. B-space X? The next theorem by J. Lindenstrauss and L. Tzafriri [127] shows that, in general, $\mathcal{P}_N$ is non-linear. It actually shows that all the projections $\mathcal{P}_N$ are linear if and only if X is Hilbertizable.

Theorem 3.11 (of J. Lindenstrauss and L. Tzafriri). *A B-space X is Hilbertizable, i.e., it admits an equivalent norm associated to an inner product, if and only if every closed subspace possesses a topological complement.*

Nevertheless, when X is an arbitrary u.c. B-space and N is a closed subspace of X, the projection $\mathcal{P}_N$ constructed by Theorem 3.10 satisfies the properties collected in the next result.

Proposition 3.4. *Suppose X is a u.c. B-space and N is a closed subspace of X. Then,*

$$\mathcal{P}_N(\lambda x) = \lambda \mathcal{P}_N x \tag{3.31}$$

for all $\lambda \in \mathbb{R}$ and $x \in X$. Moreover,

$$\mathcal{P}_N(x - \mathcal{P}_N x) = 0 \qquad \text{for all } x \in X, \tag{3.32}$$

and hence, $\mathcal{P}_N$ is linear if and only if, for every $x, y \in X$,

$$\mathcal{P}_N x = \mathcal{P}_N y = 0 \qquad \text{implies} \qquad \mathcal{P}_N(x + y) = 0. \tag{3.33}$$

Proof. According to Theorem 3.10,

$$\|x - \mathcal{P}_N x\| < \|x - n\| \tag{3.34}$$

for all $x \in X$ and $n \in N \setminus \{\mathcal{P}_N x\}$. Pick $\lambda > 0$. Then, (3.34) implies that

$$\|\lambda x - \lambda \mathcal{P}_N x\| < \|\lambda x - n\| \qquad \forall\, x \in X,\, n \in N \setminus \{\mathcal{P}_N x\}.$$

Thus,

$$\mathcal{P}_N(\lambda x) = \lambda \mathcal{P}_N x \tag{3.35}$$

for all $x \in X$ and $\lambda > 0$. Moreover, (3.34) also shows that

$$\|-x + \mathcal{P}_N x\| \leq \|-x - n\| \qquad \forall\, x \in X,\, n \in N.$$

Hence,

$$\mathcal{P}_N(-x) = -\mathcal{P}_N x \qquad (3.36)$$

for all $x \in X$. Thanks to (3.35) and (3.36), we find that, for every $\lambda < 0$ and $x \in X$,

$$\mathcal{P}_N(\lambda x) = \mathcal{P}_N(-\lambda(-x)) = -\lambda \mathcal{P}_N(-x) = \lambda \mathcal{P}_N x.$$

Moreover, $\mathcal{P}_N 0 = 0$, by definition. Consequently, (3.31) holds.

Now, we will prove (3.32). By definition, we have that

$$\|x - \mathcal{P}_N x - \mathcal{P}_N(x - \mathcal{P}_N x)\| \leq \|x - \mathcal{P}_N x - n\| \qquad (3.37)$$

for all $x \in X$ and $n \in N$. Similarly,

$$\|x - \mathcal{P}_N x\| \leq \|x - \tilde{n}\| \qquad (3.38)$$

for all $x \in X$ and $\tilde{n} \in N$. Thus, particularizing (3.37) at $n = 0$ and (3.38) at

$$\tilde{n} = \mathcal{P}_N x + \mathcal{P}_N(x - \mathcal{P}_N x) \in N,$$

it becomes apparent that

$$\|x - \mathcal{P}_N x - \mathcal{P}_N(x - \mathcal{P}_N x)\| \leq \|x - \mathcal{P}_N x\|$$
$$\leq \|x - \mathcal{P}_N x - \mathcal{P}_N(x - \mathcal{P}_N x)\|$$

and, therefore,

$$\|x - \mathcal{P}_N x - \mathcal{P}_N(x - \mathcal{P}_N x)\| = \|x - \mathcal{P}_N x\|,$$

which implies (3.32), by the uniqueness of $\mathcal{P}_N(x - \mathcal{P}_N x)$.

Obviously, (3.33) holds if $\mathcal{P}_N$ is linear. To show the converse, suppose (3.33) holds and let $x, y \in X$. Then, owing to (3.32),

$$0 = \mathcal{P}_N(x - \mathcal{P}_N x) = \mathcal{P}_N(y - \mathcal{P}_N y) = \mathcal{P}_N(x + y - \mathcal{P}_N(x + y)).$$

Thus, it follows from (3.33) that

$$0 = \mathcal{P}_N(x + y - \mathcal{P}_N x - \mathcal{P}_N y) = \mathcal{P}_N(x + y - \mathcal{P}_N(x + y))$$

and hence, by (3.31) and (3.33), we find that

$$\mathcal{P}_N(x + y - \mathcal{P}_N x - \mathcal{P}_N y - x - y + \mathcal{P}_N(x + y)) = 0.$$

Equivalently,

$$\mathcal{P}_N(\mathcal{P}_N(x + y) - \mathcal{P}_N x - \mathcal{P}_N y) = 0,$$

and, therefore, since

$$\mathcal{P}_N(x + y) - \mathcal{P}_N x - \mathcal{P}_N y \in N,$$

we obtain that

$$\mathcal{P}_N(x + y) = \mathcal{P}_N x + \mathcal{P}_N y.$$

Combining this identity with (3.31) shows that $\mathcal{P}_N$ is linear and ends the proof. $\square$

3.7.4 *The projection on a closed hyperplane*

The next result shows that $\mathcal{P}_N$ is indeed linear when N is a closed hyperplane.

Theorem 3.12. *Let X be a u.c. B-space and $N \subset X$ a closed hyperplane. Then, $\mathcal{P}_N \in \mathcal{L}(X, N)$.*

The proof of this result is based on the following

Lemma 3.1. *Let X be a u.c. B-space and $N \subset X$ a closed proper subspace. Then, there exists $z \in X$ such that*

$$\|z\| = 1 \qquad and \qquad \mathcal{P}_N z = 0. \tag{3.39}$$

Proof. Let $x \in X \setminus N$. Then,

$$x \neq \mathcal{P}_N x \in N$$

and hence,

$$z := \frac{x - \mathcal{P}_N x}{\|x - \mathcal{P}_N x\|}$$

is well defined and it satisfies $\|z\| = 1$. Moreover, according to (3.31) and (3.32), we have that

$$\mathcal{P}_N z = \frac{1}{\|x - \mathcal{P}_N x\|} \mathcal{P}_N (x - \mathcal{P}_N x) = 0.$$

This concludes the proof. □

Proof. [Proof of Theorem 3.12] By Theorem 3.10 and Proposition 3.4, it remains to prove that $\mathcal{P}_N$ satisfies (3.33). Let $z \in X$ satisfying (3.39). As N is a closed hyperplane,

$$X = N \oplus \operatorname{span}[z]. \tag{3.40}$$

Let $x, y \in X$ be such that

$$\mathcal{P}_N x = \mathcal{P}_N y = 0.$$

According to (3.40), there exist $\lambda, \mu \in \mathbb{R}$ and $n_x, n_y \in N$ (unique) such that

$$x = \lambda z + n_x, \qquad y = \mu z + n_y.$$

As $\mathcal{P}_N x = 0$, we have that

$$\|x\| = \|\lambda z + n_x\| \leq \|\lambda z + n\|$$

for all $n \in N$. In particular,

$$\|\lambda z + n_x\| \leq \|\lambda z\|.$$

On the other hand, by (3.31) and (3.39),

$$\mathcal{P}_N(\lambda z) = \lambda \mathcal{P}_N z = 0.$$

Thus,

$$\|\lambda z\| = \|\lambda z + n_x\|$$

and, consequently, $n_x = 0$, by the definition of $\mathcal{P}_N$. As this is valid for all $x \in X$, we also have that $n_y = 0$. So,

$$x + y = (\lambda + \mu)z$$

and, therefore, (3.31) implies that

$$\mathcal{P}_N(x + y) = \mathcal{P}_N((\lambda + \mu)z) = (\lambda + \mu)\mathcal{P}_N z = 0.$$

This shows (3.33) and concludes the proof. □

3.8 Comments on Chapter 3

The abstract concept of Hilbert space goes back to J. Von Neumann [171], where an axiomatic definition of Hilbert space was given for the first time in the context of separable spaces. The original *Hilbert space* of D. Hilbert [99] was nothing more than ℓ^2! In 1935, P. Jordan and J. Von Neumann [109] proved that the parallelogram identity (3.2) characterizes all Hilbertian norms.

Theorem 3.3 goes back to F. Riesz [191], where, rather remarkably, the separability requirement of J. Von Neumann [171] and predecessors was overcome. In his original paper, published in 1934, F. Riesz stressed that the theory of Hilbert spaces should be founded upon his representation theorem.

Twenty years later, in 1954, P. D. Lax and A. N. Milgram [122] formulated the variant of the Riesz' representation theorem established by Theorem 3.5. As it will become apparent in Chapter 4, Theorem 3.5 is an extremely useful device to get the existence of weak solutions in wide classes of linear boundary value problems of elliptic type.

Naturally, the importance of the concepts introduced in Definition 3.1 was revealed by P. D. Lax and A. N. Milgram [122] themselves, though, apparently, the word *coercive* was coined later (see p. 92 of K. Yosida [227], whose first edition goes back to 1964, and p. 298 of L. C. Evans [60]).

Ten years later, in 1964, G. Stampacchia [212] generalized Theorem 3.5 up to obtain Theorem 3.4, which has shown to be a milestone for the development of the calculus of variations (see D. Kinderlehrer and G. Stampacchia [113]). The proof of Theorem 3.4 given here goes back to J. L. Lions and G. Stampacchia [131]. It substantially differs from the original one given by G. Stampacchia [212].

The existence of the orthogonal projection on a closed subspace is a very classical result attributable to E. Schmidt [198], which allowed him to tidy up and generalize some abstract previous results of D. Hilbert on linear systems.

The existence of the projection operator on a closed and convex subset of a Hilbert space goes back, at least, to Th. Motzkin [168], who, in 1935, characterized the convexity of a closed set K in $\mathbb{R}^N$ by establishing that *K is convex if and only if to each point in $\mathbb{R}^N$ there corresponds a unique nearest point in K*. His demonstrations adapt *mutatis mutandis* to cover the general infinite dimensional case. Among weakly closed sets, convex sets are the unique ones admitting projections (see Theorem 4.8 of F. A. Valentine [221]). These existence results can be applied to analyze many problems of the calculus of variations, and, in particular, to solve a number of obstacle problems (see p. 4 of D. Kinderlehrer and G. Stampacchia [113]). Theorem 3.1 was the key tool of J. L. Lions and G. Stampacchia [131] to prove Theorem 3.4. The variational characterization of the projection operator established by (3.3) is attributable to them.

Theorems 3.10, 3.12 and Proposition 3.4 might be new. Anyway, they should be weighted against other related results on geometry of Banach spaces, as, e.g., B. Beauzamy [24], J. Diestel [53], J. Lindenstrauss and L. Tzafriri [128], and L. Schwartz [200]. The fact that the projection on a general closed subspace of a Banach space might not be linear had been observed independently by T. Kato [112] (see Lemma 2.3 therein).

Lemma 3.1 is a sharp version, for uniformly convex Banach spaces, of a classical result of F. Riesz [190] about the existence of "nearly orthogonal" projections in general Banach spaces (e.g., see Section III.2 of K. Yosida [227], Lemma 5.4 of D. Gilbarg and N. Trudinger [79], and Lemma VI.1 of H. Brézis [29]) with a number of important applications in functional analysis. Indeed, according to (3.39),

$$\text{dixt}\,(z, N) = \|z - \mathcal{P}_N z\| = \|z\| = 1$$

and, therefore, for any $\epsilon \in (0, 1)$, there exists an $x_\epsilon \in X$ such that

$$\|x_\epsilon\| = 1 \quad \text{and} \quad \text{dist}\,(x_\epsilon, N) \geq 1 - \epsilon.$$

This chapter has been strongly inspired by Chapter VI of H. Brézis [29], though most of its contents, are, certainly, folklore and, consequently, very well documented in many textbooks.

The author expresses his deepest gratitude to Professor F. Bombal, at Complutense University, for a very fruitful telephone discussion about the materials of Section 3.7.3.

Chapter 4

Existence of weak solutions

This chapter considers a second order uniformly elliptic differential operator of the form

$$\mathfrak{L} := - \operatorname{div}(A\nabla \cdot) + \langle b, \nabla \cdot \rangle + c \qquad (4.1)$$

in a bounded domain Ω of $\mathbb{R}^N$, $N \geq 1$, where 'div' stands for the divergence operator

$$\operatorname{div}(u_1, \ldots, u_N) = \sum_{j=1}^{N} \frac{\partial u_j}{\partial x_j},$$

$\langle \cdot, \cdot \rangle$ is the Euclidean inner product of $\mathbb{R}^N$, and

$$\begin{cases} A = (a_{ij})_{1 \leq i,j \leq N} \in \mathcal{M}_N^{\mathrm{sym}}(W^{1,\infty}(\Omega)), \\[2mm] b = (b_1, ..., b_N) \in (L^\infty(\Omega))^N, \quad c \in L^\infty(\Omega). \end{cases} \qquad (4.2)$$

For a given Banach space X, we are denoting by $\mathcal{M}_N^{\mathrm{sym}}(X)$ the space of the symmetric square matrices of order N with entries in X, and $W^{1,\infty}(\Omega)$ stands for the Sobolev space of all bounded and measurable functions in Ω with weak derivatives in $L^\infty(\Omega)$ (see Section 4.1 for its precise definition).

According to (4.2), the differential operator (4.1) fits within the abstract setting of Chapters 1 and 2. Actually, it can be expressed in the form

$$\mathfrak{L} = - \sum_{i,j=1}^{N} a_{ij}(x) \frac{\partial^2}{\partial x_i \partial x_j} + \sum_{j=1}^{N} \tilde{b}_j(x) \frac{\partial}{\partial x_j} + c(x), \quad x \in \Omega,$$

where

$$\tilde{b}_j := b_j - \sum_{i=1}^{N} \frac{\partial a_{ij}}{\partial x_i} \in L^\infty(\Omega), \qquad 1 \leq j \leq N.$$

Therefore, $\mathfrak{L}$ satisfies Assumption H2 of Chapter 2.

Throughout this chapter, we are imposing the following general assumptions:

B1 Ω is a bounded domain of $\mathbb{R}^N$, $N \geq 1$, whose boundary consists of two disjoint open and closed subsets, denoted by Γ_0 and Γ_1, of class $\mathcal{C}^1$

$$\partial\Omega := \Gamma_0 \cup \Gamma_1.$$

Necessarily, Γ_0 and Γ_1 must possess finitely many components. Either Γ_0, or Γ_1, might be empty.

By Theorem 1.9, Ω satisfies the uniform interior sphere property in the strong sense if, in addition, it is of class $\mathcal{C}^2$. In particular, Assumption H1 of Chapter 2 holds if Ω is of class $\mathcal{C}^2$.

B2 $\beta \in \mathcal{C}(\Gamma_1)$ is a continuous real function, $\mathbf{n}$ denotes the *outward unit normal* vector field of Ω, and

$$\nu := A\mathbf{n}$$

is the *conormal vector field*, i.e.,

$$\frac{\partial u}{\partial \nu} = \langle \nabla u, A\mathbf{n} \rangle = \langle A\nabla u, \mathbf{n} \rangle \quad \text{for all } u \in \mathcal{C}^1(\Gamma_1).$$

The vector field ν satisfies Assumption H3 of Chapter 2. Indeed,

$$\langle \nu, \mathbf{n} \rangle = \langle A\mathbf{n}, \mathbf{n} \rangle \geq \mu |\mathbf{n}|^2 = \mu > 0,$$

where μ is the ellipticity constant of $\mathfrak{L}$ in Ω. Therefore, (1.25) holds.

Summarizing, under conditions (4.2) and B1, B2, Assumptions H2 and H3 of Chapter 2 hold, as well as Assumption H1 if, in addition, Ω is of class $\mathcal{C}^2$.

Under Assumption B2, we denote by

$$\mathfrak{B} : \mathcal{C}(\Gamma_0) \otimes \mathcal{C}^1(\Gamma_1) \to \mathcal{C}(\partial\Omega)$$

the boundary operator

$$\mathfrak{B}\psi := \begin{cases} \psi & \text{on } \Gamma_0, \\ \frac{\partial\psi}{\partial\nu} + \beta\psi & \text{on } \Gamma_1, \end{cases} \quad \psi \in \mathcal{C}(\Gamma_0) \otimes \mathcal{C}^1(\Gamma_1). \tag{4.3}$$

Essentially, this chapter introduces the concept of weak solution for the linear boundary value problem

$$\begin{cases} (\mathfrak{L} + \omega)u = f & \text{in } \Omega, \\ \mathfrak{B}u = 0 & \text{on } \partial\Omega, \end{cases} \tag{4.4}$$

and it shows the existence of an $\omega_0 \in \mathbb{R}$ such that, for every $\omega > \omega_0$ and $f \in L^2(\Omega)$, (4.4) has a unique weak solution. Precisely, the distribution of this chapter is as follows. Section 4.1 begins by introducing the concepts of weak derivatives, Sobolev spaces, and Hölder spaces of continuous functions. Then, it collects the Sobolev imbedding theorems and the compactness result of F. Rellich and V. I. Kondrachov. As the contents of Section 4.1 can be thought of as a series of well-known properties of L^p-spaces, we are not giving the proofs of these results here. Section 4.2 constructs the *trace operators* and studies their main properties. Section 4.3 introduces the concept of *weak solution* for (4.4) and studies some of its properties through the associated bilinear form, denoted by $\mathfrak{a}(\cdot, \cdot)$. Section 4.4 shows the continuity of $\mathfrak{a}$, and Section 4.5 shows the coercivity of $\mathfrak{a}$ for sufficiently large ω when $\beta \geq 0$. The existence of a weak solution in this case is derived from the theorem of P. D. Lax and A. N. Milgram (see Theorem 3.5). Finally, Section 4.6 shows the existence of a weak solution for a general $\beta \in \mathcal{C}(\Gamma_1)$ and sufficiently large ω. This result might go back to [147].

Throughout the rest of this book, we will use the concept of classical solution introduced by the next definition.

Definition 4.1 (Classical solution). *A function $u \in \mathcal{C}^1(\bar{\Omega})$ is said to be a* **classical solution** *of* (4.4) *if it is twice classically differentiable almost everywhere in Ω,*

$$(\mathfrak{L} + \omega)u = f \qquad almost\ everywhere\ in\ \ \Omega,$$

and

$$\mathfrak{B}u = 0 \qquad on\ \ \partial\Omega.$$

4.1 Preliminaries. Sobolev spaces

4.1.1 *Test functions*

Extending the concept already introduced in the beginning of Section 2.6, we will subsequently denote by $\mathcal{C}_0^\infty(\Omega)$ the set of functions $\phi : \bar{\Omega} \to \mathbb{R}$ of class $\mathcal{C}^\infty$ with compact support

$$\operatorname{supp} \phi := \overline{\phi^{-1}(\mathbb{R} \setminus \{0\})} \subset \Omega.$$

As we are assuming that Ω is bounded, it is apparent that

$$\mathcal{C}_0^\infty(\Omega) = \{\phi \in \mathcal{C}^\infty(\bar{\Omega}) \ : \ \operatorname{supp} \phi \subset \Omega\}.$$

More generally, for every

$$\Gamma \in \{\Gamma_0, \Gamma_1, \partial\Omega\},$$

we will denote by $\mathcal{C}_\Gamma^\infty(\Omega)$ the set

$$\mathcal{C}_\Gamma^\infty(\Omega) := \{\phi \in \mathcal{C}^\infty(\bar\Omega) \;:\; \operatorname{supp}\phi \subset \bar\Omega \setminus \Gamma\}.$$

Consequently, a function $\phi \in \mathcal{C}^\infty(\bar\Omega)$ belongs to $\mathcal{C}_\Gamma^\infty(\Omega)$ if and only if it vanishes on some open neighborhood of Γ. According to these notations,

$$\mathcal{C}_{\partial\Omega}^\infty(\Omega) = \mathcal{C}_0^\infty(\Omega).$$

All those functions will be referred to as *test functions*.

4.1.2   Weak derivatives. Sobolev spaces

Besides $L^\infty(\Omega)$, which has already been introduced in the beginning of Chapter 1, we will subsequently consider, for every $p \in [1,\infty)$, the Banach space $L^p(\Omega)$ of all measurable real functions $u : \Omega \to \mathbb{R}$, as discussed by Lebesgue, such that

$$\int_\Omega |u|^p < \infty.$$

It is folklore that $L^p(\Omega)$ is a Banach space with norm

$$\|u\|_p = \|u\|_{L^p(\Omega)} := \left(\int_\Omega |u|^p\right)^{\frac{1}{p}}, \qquad u \in L^p(\Omega),$$

which is usually referred to as the L^p-norm. Actually, $L^2(\Omega)$ is a Hilbert space, as $\|\cdot\|_2$ is the norm induced by the scalar product

$$\langle u, v\rangle_{L^2} := \int_\Omega u(x)v(x)\,dx, \qquad u, v \in L^2(\Omega).$$

Similarly, for every $p \in [1,\infty)$, we will denote by $L_{\mathrm{loc}}^p(\Omega)$ the set of all measurable real functions $u : \Omega \to \mathbb{R}$ such that

$$\int_K |u|^p < \infty \qquad \text{for all compact subset } K \subset \Omega.$$

A fundamental property of these spaces establishes that, for every $p, q \in [1,\infty]$ with

$$\frac{1}{p} + \frac{1}{q} = 1,$$

and $u \in L^p(\Omega)$, $v \in L^q(\Omega)$, the product function uv belongs to $L^1(\Omega)$ and

$$\|uv\|_1 \le \|u\|_p \|v\|_q. \tag{4.5}$$

This estimate is usually referred to as the *Hölder inequality*.

Subsequently, for any multi-index

$$\alpha = (\alpha_1, ..., \alpha_N) \in \mathbb{N}^N$$

we denote

$$|\alpha| := \sum_{j=1}^{N} \alpha_j, \qquad D^\alpha := \frac{\partial^{|\alpha|}}{\partial x_1^{\alpha_1} \cdots \partial x_N^{\alpha_N}}.$$

Integrating by parts, it becomes apparent that

$$\int_\Omega u D^\alpha \phi = (-1)^{|\alpha|} \int_\Omega \phi D^\alpha u \qquad (4.6)$$

for all $u \in \mathcal{C}^{|\alpha|}(\Omega)$ and $\phi \in \mathcal{C}_0^\infty(\Omega)$. Actually, the left-hand side of (4.6) makes sense even if $u \in L_{\text{loc}}^1(\Omega)$, as $D^\alpha \phi$ has compact support in Ω. Consequently, the concept of *weak derivatives*, or *derivatives in the weak sense*, introduced by the next definition is rather natural.

Definition 4.2 (Weak derivatives). *Let u, $v \in L_{\text{loc}}^1(\Omega)$ and $\alpha \in \mathbb{N}^N$. Then, it is said that*

$$v = D^\alpha u \quad \text{in the weak sense}$$

or, equivalently, that v is the weak derivative of order α of u, if

$$\int_\Omega u D^\alpha \phi = (-1)^{|\alpha|} \int_\Omega \phi v \qquad \text{for all } \phi \in \mathcal{C}_0^\infty(\Omega).$$

This concept is consistent because the weak derivative $D^\alpha u$ is uniquely determined if it exists. Indeed, if v, $w \in L_{\text{loc}}^1(\Omega)$ satisfy

$$\int_\Omega v \phi = \int_\Omega w \phi \qquad \text{for all } \phi \in \mathcal{C}_0^\infty(\Omega),$$

then, $v = w$ almost everywhere in Ω.

For every $p \in [1, \infty]$ and $k \in \mathbb{N}$, the Sobolev space $W^{k,p}(\Omega)$ is defined as the set of functions $u \in L_{\text{loc}}^1(\Omega)$ such that the weak derivative $D^\alpha u$ exists for every $\alpha \in \mathbb{N}^N$ with $|\alpha| \le k$ and $D^\alpha u \in L^p(\Omega)$, endowed with the norm

$$\|u\|_{W^{k,p}(\Omega)} := \begin{cases} \left(\displaystyle\sum_{0 \le |\alpha| \le k} \|D^\alpha u\|_{L^p(\Omega)}^p \right)^{\frac{1}{p}} & \text{if } p < \infty, \\[6mm] \displaystyle\sum_{0 \le |\alpha| \le k} \|D^\alpha u\|_{L^\infty(\Omega)} & \text{if } p = \infty, \end{cases}$$

for all $u \in W^{k,p}(\Omega)$. Equivalently, one might define these norms through

$$\|u\|_{W^{k,p}(\Omega)} := \sum_{0 \leq |\alpha| \leq k} \|D^\alpha u\|_{L^p(\Omega)}$$

if $p < \infty$. It is folklore that all these normed spaces are Banach spaces. Actually, they are Hilbert spaces if $p = 2$ with inner product

$$\langle u, v \rangle_{W^{k,2}(\Omega)} := \sum_{0 \leq |\alpha| \leq k} \int_\Omega D^\alpha u \, D^\alpha v, \qquad u, v \in W^{k,2}(\Omega),$$

because $L^2(\Omega)$ is a Hilbert space. In particular,

$$\langle u, v \rangle_{W^{1,2}(\Omega)} := \int_\Omega \langle \nabla u, \nabla v \rangle + \int_\Omega uv, \qquad u, v \in W^{1,2}(\Omega).$$

Applying (4.5) with $p = q = 2$, it becomes apparent that all these inner products are well defined.

As usual, these features might induce us to adopt the notation

$$H^k(\Omega) := W^{k,2}(\Omega), \qquad k \in \mathbb{N}. \tag{4.7}$$

According to it, $H^0(\Omega) = L^2(\Omega)$. More generally, $W^{0,p}(\Omega) = L^p(\Omega)$ for all $p \in [1, \infty]$. Also, for every $k \in \mathbb{N}$ and $p \in [1, \infty]$, we will denote by $W^{k,p}_{\text{loc}}(\Omega)$ the set

$$W^{k,p}_{\text{loc}}(\Omega) := \bigcap_{\substack{D \text{ open} \\ \bar{D} \subset \Omega}} W^{k,p}(D).$$

Note that $\bar{D}$ is compact, since Ω is bounded. Naturally, the weak derivatives satisfy the algebraic properties collected in the next result.

Lemma 4.1. *Let $k \in \mathbb{N}$, $p \in [1, \infty]$, and $u, v \in W^{k,p}(\Omega)$. Then,*

i) *For every $\alpha, \beta \in \mathbb{N}^N$ with $|\alpha| + |\beta| \leq k$, $D^\alpha u \in W^{k-|\alpha|,p}(\Omega)$ and*

$$D^\beta(D^\alpha u) = D^\alpha(D^\beta u) = D^{\alpha+\beta} u.$$

ii) *For every $A, B \in \mathbb{R}$ and $\alpha \in \mathbb{N}^N$ with $|\alpha| \leq k$, $Au + Bv \in W^{k,p}(\Omega)$ and*

$$D^\alpha(Au + Bv) = AD^\alpha u + BD^\alpha v.$$

iii) *For every $\zeta \in \mathcal{C}_0^\infty(\Omega)$ and $\alpha \in \mathbb{N}^N$ with $|\alpha| \leq k$, $\zeta u \in W^{k,p}(\Omega)$ and*

$$D^\alpha(\zeta u) = \sum_{\beta \leq \alpha} \binom{\alpha}{\beta} D^\beta \zeta \, D^{\alpha-\beta} u, \qquad \binom{\alpha}{\beta} := \frac{|\alpha|!}{|\beta|! \, |\alpha - \beta|!}.$$

Based on these properties and the completeness of $L^p(\Omega)$, it is easy to see that the spaces $W^{k,p}(\Omega)$, $k \in \mathbb{N}$, $p \in [1, \infty]$, are indeed Banach spaces. The next result establishes that, for every $p \in [1, \infty)$ and $k \in \mathbb{N}$, $W^{k,p}(\Omega)$ is the completion of $\mathcal{C}^\infty(\bar{\Omega})$ with respect to the $W^{k,p}(\Omega)$-norm. It should not be forgotten that, according to Assumption B1, throughout this chapter we are assuming that Ω is a connected bounded open set of class $\mathcal{C}^1$.

Theorem 4.1 (Global approximation by smooth functions).
Let $k \in \mathbb{N}$, $p \in [1, \infty)$, and $u \in W^{k,p}(\Omega)$. Then, there exists a sequence $\phi_n \in \mathcal{C}^\infty(\bar{\Omega})$, $n \geq 1$, such that

$$\lim_{n \to \infty} \|\phi_n - u\|_{W^{k,p}(\Omega)} = 0.$$

When $u \in W^{k,p}(\Omega) \cap \mathcal{C}(\bar{\Omega})$, then $\{\phi_n\}_{n \geq 1}$ can be chosen so that

$$\lim_{n \to \infty} \left(\|\phi_n - u\|_{W^{k,p}(\Omega)} + \|\phi_n - u\|_{\mathcal{C}(\bar{\Omega})} \right) = 0.$$

4.1.3 *Hölder spaces of continuous functions*

The following definition introduces some important basic concepts.

Definition 4.3. *Let $u \in \mathcal{C}(\Omega)$ and $\nu \in (0, 1]$. Then,*

i) *u is said to be (globally) Lipschitz continuous in Ω if there exists a constant $C \geq 0$ such that*

$$|u(x) - u(y)| \leq C|x - y| \qquad \text{for all } x, y \in \Omega.$$

ii) *u is said to be locally Lipschitz continuous in Ω if for every $x \in \Omega$ there exists $R > 0$ such that $B_R(x) \subset \Omega$ and u is (globally) Lipschitz continuous in $B_R(x)$.*

iii) *u is said to be (globally) Hölder continuous in Ω, with exponent ν, if there exists a constant $C \geq 0$ such that*

$$|u(x) - u(y)| \leq C|x - y|^\nu \qquad \text{for all } x, y \in \Omega.$$

iv) *u is said to be locally Hölder continuous in Ω, with exponent ν, if for every $x \in \Omega$ there exists $R > 0$ such that $B_R(x) \subset \Omega$ and u is (globally) Hölder continuous in $B_R(x)$ with exponent ν.*

Obviously, u is Hölder continuous with exponent $\nu = 1$ if and only if it is Lipschitz continuous.

Throughout the rest of this book, for a given $\nu \in (0,1]$, we denote by $\mathcal{C}^{0,\nu}(\bar{\Omega})$ the Banach space of all functions $u \in \mathcal{C}(\bar{\Omega})$ that are Hölder continuous in Ω with exponent ν, endowed with the Hölder-norm

$$\|u\|_{\mathcal{C}^{0,\nu}(\bar{\Omega})} := \|u\|_{\mathcal{C}(\bar{\Omega})} + \sup_{\substack{x,y\in\bar{\Omega}\\x\neq y}} \frac{|u(x)-u(y)|}{|x-y|^{\nu}}, \qquad u \in \mathcal{C}^{0,\nu}(\bar{\Omega}).$$

More generally, for every $k \in \mathbb{N}$ and $\nu \in (0,1]$, we denote by $\mathcal{C}^{k,\nu}(\bar{\Omega})$ the Banach subspace of $\mathcal{C}^k(\bar{\Omega})$ consisting of all functions $u \in \mathcal{C}^k(\bar{\Omega})$ such that

$$D^{\alpha}u \in \mathcal{C}^{0,\nu}(\bar{\Omega}) \quad \text{for all} \quad \alpha \in \mathbb{N}^N \quad \text{with} \quad |\alpha| = k,$$

equipped with the norm

$$\|u\|_{\mathcal{C}^{k,\nu}(\bar{\Omega})} := \|u\|_{\mathcal{C}^k(\bar{\Omega})} + \sum_{|\alpha|=k} \sup_{\substack{x,y\in\bar{\Omega}\\x\neq y}} \frac{|D^{\alpha}u(x)-D^{\alpha}u(y)|}{|x-y|^{\nu}}, \quad u \in \mathcal{C}^{k,\nu}(\bar{\Omega}).$$

In other words, the space $\mathcal{C}^{k,\nu}(\bar{\Omega})$ consists of all functions u that are k times continuously differentiable and whose k^{th}-partial derivatives are Hölder continuous of exponent ν.

Eventually, for every $\nu \in (0,1]$, we will also consider the set of functions $\mathcal{C}^{k,\nu^-}(\bar{\Omega})$ defined through

$$\mathcal{C}^{k,\nu^-}(\bar{\Omega}) = \bigcap_{0<\theta<\nu} \mathcal{C}^{k,\theta}(\bar{\Omega}).$$

Also, for every $k \in \mathbb{N}$ and $\nu \in (0,1]$, we will consider the set of functions $\mathcal{C}^{k,\nu}(\Omega)$ defined by

$$\mathcal{C}^{k,\nu}(\Omega) := \bigcap_{\substack{D \text{ open}\\ \bar{D}\subset\Omega}} \mathcal{C}^{k,\nu}(\bar{D}).$$

The $W^{k,p}(\Omega)$ spaces are analogous in a certain sense to the $\mathcal{C}^{k,\nu}(\bar{\Omega})$ spaces. In the $W^{k,p}(\Omega)$ spaces, classical differentiability is replaced by weak differentiability and Hölder continuity by L^p-integrability. The results in the next sections make precise this idea.

4.1.4 *Sobolev's imbeddings*

In the next result we collect the main Sobolev imbedding theorems. Subsequently, for every real number $r \in \mathbb{R}$, $r \geq 0$, we denote by $[r]$ the entire part of r, i.e., the unique integer $[r] \in \mathbb{N}$ such that

$$[r] \leq r < [r] + 1.$$

Theorem 4.2 (Sobolev's imbeddings). *Let* $k \in \mathbb{N}$, $k \geq 1$, *and* $p \in (1, \infty)$.

i) *If* $kp < N$, *then*

$$W^{k,p}(\Omega) \subset L^{\frac{Np}{N-kp}}(\Omega)$$

and the associated injection is continuous, i.e., there exists a constant $C := C(k, p, N, \Omega)$ *such that*

$$\|u\|_{L^{\frac{Np}{N-kp}}(\Omega)} \leq C \|u\|_{W^{k,p}(\Omega)}$$

for all $u \in W^{k,p}(\Omega)$.

ii) *If* $kp > N$, *then*

$$W^{k,p}(\Omega) \subset C^{k-\left[\frac{N}{p}\right]-1,\nu}(\bar{\Omega}), \tag{4.8}$$

where

$$\nu = \begin{cases} \left[\frac{N}{p}\right] + 1 - \frac{N}{p} \in (0,1), & if \ \left[\frac{N}{p}\right] < \frac{N}{p}, \\ 1^-, & if \ \left[\frac{N}{p}\right] = \frac{N}{p}. \end{cases} \tag{4.9}$$

Moreover, the associated injection is continuous, i.e., there exists a constant $C := C(k, p, N, \Omega)$ *such that*

$$\|u\|_{C^{k-\left[\frac{N}{p}\right]-1,\nu}(\bar{\Omega})} \leq C \|u\|_{W^{k,p}(\Omega)}$$

for all $u \in W^{k,p}(\Omega)$. *When* $\nu = 1^-$, *this estimate should be understood in the sense that, for every* $\theta \in (0,1)$, *there exists a constant* $C := C(k, p, N, \Omega, \theta)$ *such that*

$$\|u\|_{C^{k-\left[\frac{N}{p}\right]-1,\theta}(\bar{\Omega})} \leq C \|u\|_{W^{k,p}(\Omega)}$$

for all $u \in W^{k,p}(\Omega)$.

In the special –but relevant– case when $k = 1$, Theorem 4.2 provides us with the next result.

Corollary 4.1. *Suppose* $1 < p < \infty$. *Then,*

$$W^{1,p}(\Omega) \subset \begin{cases} L^{\frac{Np}{N-p}}(\Omega), & if \ p < N, \\ C^{0,1-\frac{N}{p}}(\bar{\Omega}), & if \ p > N, \end{cases} \tag{4.10}$$

with continuous embeddings.

Note that $\left[\frac{N}{p}\right] = 0$ if $p > N$ and hence, (4.8) becomes

$$W^{1,p}(\Omega) \subset \mathcal{C}^{0,\nu}(\bar{\Omega}) \quad \text{with} \quad \nu = 1 - \frac{N}{p} \in (0,1).$$

According to (4.10), we also have that

$$W^{1,\infty}(\Omega) \subset \bigcap_{0<\theta<1} \mathcal{C}^{0,\theta}(\bar{\Omega}) = \mathcal{C}^{0,1^-}(\bar{\Omega}).$$

Actually, the following characterization holds.

Theorem 4.3. *$u \in W^{k,\infty}(\Omega)$ if and only if $u \in \mathcal{C}^{k-1,1}(\bar{\Omega})$, i.e., if $D^\alpha u$ is Lipschitz continuous in Ω for every $\alpha \in \mathbb{N}^N$ with $|\alpha| = k-1$. Consequently, $u \in W^{k,\infty}_{\mathrm{loc}}(\Omega)$ if and only if $D^\alpha u$ is locally Lipschitz continuous in Ω for all these α's. In other words,*

$$W^{k,\infty}(\Omega) = \mathcal{C}^{k-1,1}(\bar{\Omega}) \qquad and \qquad W^{k,\infty}_{\mathrm{loc}}(\Omega) = \mathcal{C}^{k-1,1}(\Omega).$$

In particular, $u \in W^{1,\infty}(\Omega)$ if and only if u is Lipschitz continuous in Ω, and $u \in W^{1,\infty}_{\mathrm{loc}}(\Omega)$ if and only if u is locally Lipschitz continuous in Ω.

According to Corollary 4.1, $u \in \mathcal{C}^{0,\nu}(\bar{\Omega})$ with $\nu = 1 - N/p \in (0,1)$ if $u \in W^{1,p}(\Omega)$ with $p > N$, but the converse fails to be true, as a function $u \in \mathcal{C}^{0,\nu}(\bar{\Omega})$ with $\nu \in (0,1)$ does not necessarily belong to

$$\bigcup_{p>N} W^{1,p}(\Omega).$$

The following regularity result clarifies this fact.

Theorem 4.4. *Suppose*

$$u \in \bigcup_{N<p\leq\infty} W^{1,p}_{\mathrm{loc}}(\Omega).$$

Then, u is differentiable almost everywhere (a.e.) in Ω, and its classical gradient equals its weak gradient a.e. in Ω. Consequently, if

$$u \in \bigcup_{N<p\leq\infty} W^{2,p}_{\mathrm{loc}}(\Omega),$$

then $u \in \mathcal{C}^1(\Omega)$ and it is twice classically differentiable a.e. in Ω. Moreover, the classical derivative $D^\alpha u$ equals the corresponding weak derivative a.e. in Ω for all multi-index $\alpha \in \mathbb{N}^N$ with $|\alpha| \leq 2$.

In particular, by Theorem 4.3, any locally Lipschitz continuous function must be differentiable almost everywhere. Note that the inclusion

$$\bigcup_{N < p \le \infty} W_{\text{loc}}^{2,p}(\Omega) \subset \mathcal{C}^1(\Omega)$$

follows from (4.8) and Theorem 4.3. Moreover, $u \in W_{\text{loc}}^{2,p}(\Omega)$ implies

$$\frac{\partial u}{\partial x_i} \in W_{\text{loc}}^{1,p}(\Omega), \qquad 1 \le i \le N,$$

and hence the second assertion of Theorem 4.4 is a direct consequence from the first one.

Letting $p \uparrow N$ in (4.10) one might expect the validity of the injection $W^{1,N}(\Omega) \subset L^\infty(\Omega)$, but this fails to be true if $N > 1$. Indeed, if $N \ge 2$ and $\Omega = B_1(0)$, then the function

$$u(x) := \text{Log Log} \left(1 + \frac{1}{|x|} \right), \qquad x \in \Omega,$$

belongs to $W^{1,N}(\Omega) \setminus L^\infty(\Omega)$. Obviously, the limiting inclusion obtained by letting $p \downarrow N$ in (4.10) cannot be true either, because $\mathcal{C}(\bar\Omega) \subset L^\infty(\Omega)$ and we have just seen that $W^{1,N}(\Omega) \nsubseteq L^\infty(\Omega)$.

4.1.5 *Compact imbeddings*

Given two Banach spaces X, Y and a linear continuous operator $T \in \mathcal{L}(X,Y)$, it is said that T is compact if $T(A)$ is a *pre-compact* subset of Y (i.e., the closure $\overline{T(A)}$ is a compact subset of Y) for every bounded subset A of X. In such case, it will simply said that

$$T \in \mathcal{K}(X,Y).$$

If $X \subset Y$ and $\bar A$ is compact in Y for all bounded set $A \subset X$, we will simply write that

$$X \hookrightarrow Y.$$

The main compactness result concerning Sobolev spaces is the next one.

Theorem 4.5 (of F. Rellich and V. I. Kondrachov). *Suppose* $k \in \mathbb{N}$, $k \ge 1$, *and* $1 \le p < \infty$.

i) *If* $kp < N$, *then, for every*

$$1 \le q < \frac{Np}{N - kp},$$

the imbedding of $W^{k,p}(\Omega)$ *into* $L^q(\Omega)$ *is a compact operator, i.e.,*

$$W^{k,p}(\Omega) \hookrightarrow L^q(\Omega).$$

ii) *If $kp > N$ and ν is given through (4.9), then,*

$$W^{k,p}(\Omega) \hookrightarrow \mathcal{C}^{k-\left[\frac{N}{p}\right]-1,\beta}(\bar{\Omega})$$

for all $\beta < \nu$.

Moreover,

$$W^{1,p}(\Omega) \hookrightarrow L^p(\Omega) \quad \text{for all } p \in [1,\infty]. \tag{4.11}$$

Certainly, (4.11) is reminiscent from the classical theorem of G. Ascoli [18] and C. Arzela [17].

4.2 Trace operators

Whenever $u \in \mathcal{C}(\bar{\Omega})$ the *trace restriction* of u to the boundary $\partial\Omega$, denoted by $u|_{\partial\Omega}$, is well defined in the classical sense. But a function $u \in W^{1,p}(\Omega)$ does not need to be continuous in Ω. Actually, it is only defined almost everywhere in Ω. Moreover, according to Assumption B1, $\partial\Omega$ is an $(N-1)$-dimensional surface of class $\mathcal{C}^1$ and, hence, it has zero N-dimensional Lebesgue measure. Therefore, it does not make sense to define the restriction of $u \in W^{1,p}(\Omega)$ to any portion of $\partial\Omega$. Nevertheless, the next *trace theorem* holds.

Theorem 4.6 (trace theorem). *Let $p \in [1,\infty)$ and $\Gamma \in \{\Gamma_0, \Gamma_1, \partial\Omega\}$. Then, there exists a unique linear continuous operator*

$$\mathcal{T}_\Gamma \in \mathcal{L}\left(W^{1,p}(\Omega), L^p(\Gamma)\right)$$

such that

$$\mathcal{T}_\Gamma u = u|_\Gamma \quad \text{for all } u \in W^{1,p}(\Omega) \cap \mathcal{C}(\bar{\Omega});$$

$\mathcal{T}_\Gamma$ *will be called the **trace operator of** $W^{1,p}(\Omega)$ **on** Γ, and, for every $u \in W^{1,p}(\Omega)$, $\mathcal{T}_\Gamma u \in L^p(\Gamma)$ will be referred to as the **trace of** u **on** Γ.*

By the uniqueness, we also have that

$$\mathcal{T}_{\partial\Omega} = \mathcal{T}_{\Gamma_0} \otimes \mathcal{T}_{\Gamma_1}.$$

Proof. First, we will show that there exists a constant $C > 0$ such that

$$\|u\|_{L^p(\Gamma)} \leq C\|u\|_{W^{1,p}(\Omega)} \tag{4.12}$$

for all $u \in \mathcal{C}^1(\bar{\Omega})$. Then, the theorem will follow through a density argument. As Γ is a compact subset of $\bar{\Omega}$, to prove (4.12) it suffices to show that for each $x_0 \in \Gamma$ there exist an open neighborhood $\tilde{\Gamma}$ of x_0 in Γ and a constant $C = C(x_0) > 0$ such that

$$\int_{\tilde{\Gamma}} |u|^p \, dS \leq C \left(\int_{\Omega} |u|^p + \int_{\Omega} |\nabla u|^p \right) \tag{4.13}$$

for all $u \in \mathcal{C}^1(\bar{\Omega})$. Indeed, fix $x_0 \in \Gamma$ and suppose, in addition, that Γ is flat in a neighborhood of $x_0 \in \Gamma$, in the sense that it lies on the plane

$$\Pi := \{ x \in \mathbb{R}^N \ : \ x_N = 0 \}.$$

As $\partial\Omega$ is of class $\mathcal{C}^1$, this can be reached after a local change of coordinates of class $\mathcal{C}^1$, but the details of the proof in the general case are postponed. Then, there exists $r = r(x_0) > 0$ such that

$$B_r^+ := B_r(x_0) \cap \{ x \in \mathbb{R}^N \ : \ x_N > 0 \} \subset \Omega,$$
$$B_r^- := B_r(x_0) \cap \{ x \in \mathbb{R}^N \ : \ x_N < 0 \} \subset \mathbb{R}^N \setminus \bar{\Omega}.$$

Necessarily, $B_r(x_0) \cap \Gamma \subset \Pi$ (see Figure 4.1).

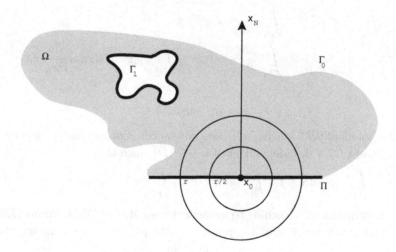

Fig. 4.1   The case when Γ ($= \Gamma_0$) is flat around $x_0 \in \Gamma$

Now, consider $B_{r/2}(x_0)$, the concentric open ball of radius $r/2$, and select $\zeta \in \mathcal{C}_0^\infty(B_r(x_0))$ such that $\zeta \geq 0$ in $B_r(x_0)$ and $\zeta = 1$ in $B_{r/2}(x_0)$. Obviously, ζ can be chosen to be radially symmetric. Figure 4.2 represents one of those ζ's. Subsequently, for any $\delta \in (0, r]$, we will denote by

$$\Gamma_\delta := \Gamma \cap B_\delta(x_0)$$

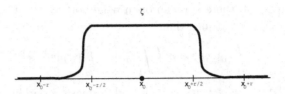

Fig. 4.2   A slice of an admissible ζ

the portion of Γ within $B_\delta(x_0)$, and, using the notations already introduced in Section 1.8, we set

$$x_{[N]} := (x_1, ..., x_{N-1}) \in \mathbb{R}^{N-1} \equiv \Pi.$$

Then, it is apparent that

$$\int_{\Gamma_{r/2}} |u|^p \, dS \leq \int_{\Gamma_r} \zeta |u|^p \, dS = \int_{\partial B_r^+} \zeta |u|^p \, dS.$$

Thus, by the divergence theorem, we find that

$$\int_{\Gamma_{r/2}} |u|^p \, dS \leq - \int_{B_r^+} \frac{\partial}{\partial x_N} \left(\zeta |u|^p \right) \, dx$$

$$= - \int_{B_r^+} \frac{\partial \zeta}{\partial x_N} |u|^p \, dx - \int_{B_r^+} \zeta p |u|^{p-1} \text{sign } u \frac{\partial u}{\partial x_N} \, dx$$

because

$$\mathbf{n} = (0, ..., 0, -1)$$

at the points of ∂B_r^+ where $\zeta |u|^p$ does not vanish. Consequently, there exist $C_1 > 0$ and $C_2 > 0$, independent of $u \in C^1(\bar{\Omega})$, such that

$$\int_{\Gamma_{r/2}} |u|^p \, dS \leq C_1 \int_{B_r^+} |u|^p \, dx + C_2 \int_{B_r^+} |u|^{p-1} |\nabla u| \, dx. \qquad (4.14)$$

On the other hand, according to a classical inequality of W. H. Young [228], for every $a > 0$, $b > 0$, and $1 < p, q < \infty$, with $p^{-1} + q^{-1} = 1$, one has that

$$ab \leq \frac{a^p}{p} + \frac{b^q}{q},$$

and hence,

$$|\nabla u| |u|^{p-1} \leq \frac{1}{p} |\nabla u|^p + \frac{p-1}{p} |u|^p \qquad \text{in } \Omega,$$

because $q = p/(p-1)$. Thus, setting

$$C := \max \left\{ C_1 + \frac{p-1}{p} C_2, \frac{C_2}{p} \right\},$$

we obtain from (4.14) that

$$\int_{\Gamma_{r/2}} |u|^p \, dS \leq C \left(\int_{B_r^+} |u|^p \, dx + \int_{B_r^+} |\nabla u|^p \, dx \right). \tag{4.15}$$

As $C > 0$ is independent of u, the estimate (4.13) is an easy consequence from (4.15).

When Γ is not flat around x_0, we can proceed as follows. By Definition 1.5, there exist $R := R(x_0) > 0$, an open neighborhood D of $0 \in \mathbb{R}^N$, and a bijection $\Phi : B_R(x_0) \to D$ such that $\Phi(x_0) = 0$, $D\Phi(x_0) \in \text{Iso}(\mathbb{R}^N)$, $\Phi \in \mathcal{C}^1(B_R(x_0); D)$, $\Phi^{-1} \in \mathcal{C}^1(D; B_R(x_0))$, and

$$\Phi(B_R(x_0) \cap \Omega) = \{x \in D \; : \; x_N > 0\},$$
$$\Phi(B_R(x_0) \cap \partial\Omega) = \{x \in D \; : \; x_N = 0\}.$$

In other words, Φ is a local diffeomorphism straightening $\partial\Omega$ at x_0 into the hyperplane $x_N = 0$. Now, let $r = r(x_0) > 0$ for which $\bar{B}_r(0) \subset D$ satisfies all the properties of the special case just dealt with before (in the present case $x_0 = 0$), and, for every $\delta \in (0, r]$, let $\mathcal{U}_\delta$ denote the open neighborhood of x_0 defined by

$$\mathcal{U}_\delta := \Phi^{-1}(B_\delta(0)).$$

Subsequently, for every $\delta \in (0, r]$, we also set

$$\Gamma_\delta := \mathcal{U}_\delta \cap \Gamma$$

and

$$v(x) := u(\Phi^{-1}(x)), \qquad x \in D = \Phi(B_R(x_0)).$$

By (4.15), there exists a constant $C_1 > 0$, independent of v, such that

$$\int_{\Phi(\Gamma_{r/2})} |v|^p \, dS \leq C_1 \left(\int_{\Phi(\mathcal{U}_r \cap \Omega)} |v|^p \, dx + \int_{\Phi(\mathcal{U}_r \cap \Omega)} |\nabla v|^p \, dx \right).$$

Consequently, performing the change of variable $x = \Phi(y)$, it becomes apparent that there exists $C_2 = C_2(x_0, \Omega) > 0$ such that

$$\int_{\Gamma_{r/2}} |u|^p \, dS \leq C_2 \left(\int_{\mathcal{U}_r \cap \Omega} |u|^p \, dy + \int_{\mathcal{U}_r \cap \Omega} |\nabla u|^p \, dy \right)$$
$$\leq C_2 \left(\int_\Omega |u|^p \, dy + \int_\Omega |\nabla u|^p \, dy \right)$$

for all $u \in \mathcal{C}^1(\bar{\Omega})$. This ends the proof of (4.13).

As Γ is compact, there exist $m \in \mathbb{N}$ and m points

$$x_{0,j} \in \Gamma, \qquad 1 \leq j \leq m,$$

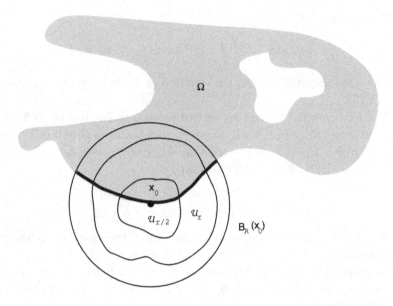

Fig. 4.3 $\mathcal{U}_{r/2} = \Phi^{-1}(B_{r/2}(0))$ and $\mathcal{U}_r = \Phi^{-1}(B_r(0))$

such that

$$\Gamma = \bigcup_{j=1}^{m} \Gamma_{r(x_{0,j})/2}.$$

Therefore,

$$
\begin{aligned}
\int_{\Gamma} |u|^p \, dS &\leq \sum_{j=1}^{m} \int_{\Gamma_{r(x_{0,j})/2}} |u|^p \, dS \\
&\leq \sum_{j=1}^{m} C_2(x_{0,j}, \Omega) \left(\int_{\Omega} |u|^p \, dx + \int_{\Omega} |\nabla u|^p \, dx \right) \\
&\leq C_3 \left(\int_{\Omega} |u|^p \, dx + \int_{\Omega} |\nabla u|^p \, dx \right),
\end{aligned}
$$

where

$$C_3 := m \max_{1 \leq j \leq m} C_2(x_{0,j}, \Omega).$$

As the constant C_3 only depends on the geometry of Ω, the proof of (4.12) is complete.

Now, we will construct the trace operator $\mathcal{T}_\Gamma$. Naturally, for every $u \in \mathcal{C}^1(\bar{\Omega})$, we define

$$\mathcal{T}_\Gamma u := u|_\Gamma$$

as the restriction of u to the component Γ of the boundary $\partial\Omega$.

Suppose $u \in W^{1,p}(\Omega)$. Then, according to Theorem 4.1, there exists a sequence $\phi_n \in \mathcal{C}^\infty(\bar{\Omega})$, $n \geq 1$, such that

$$\lim_{n\to\infty} \phi_n = u \quad \text{in } W^{1,p}(\Omega).$$

By (4.12), we have that

$$\|\mathcal{T}_\Gamma\phi_n - \mathcal{T}_\Gamma\phi_m\|_{L^p(\Gamma)} = \|\phi_n - \phi_m\|_{L^p(\Gamma)} \leq C\|\phi_n - \phi_m\|_{W^{1,p}(\Omega)}$$

and hence, the limit

$$\mathcal{T}_\Gamma u := \lim_{n\to\infty} \phi_n|_\Gamma$$

is well defined in $L^p(\Gamma)$. This definition is consistent. Indeed, if $\psi_n \in \mathcal{C}^\infty(\bar{\Omega})$, $n \geq 1$, is another sequence for which

$$\lim_{n\to\infty} \psi_n = u \quad \text{in } W^{1,p}(\Omega),$$

then, it follows from (4.12) that

$$\|\mathcal{T}_\Gamma\phi_n - \mathcal{T}_\Gamma\psi_n\|_{L^p(\Gamma)} \leq C\|\phi_n - \psi_n\|_{W^{1,p}(\Omega)}$$

for all $n \geq 1$, and, therefore,

$$\lim_{n\to\infty} \|\phi_n - \psi_n\|_{L^p(\Gamma)} = 0.$$

Consequently,

$$\mathcal{T}_\Gamma u := \lim_{n\to\infty} \phi_n|_\Gamma = \lim_{n\to\infty} \psi_n|_\Gamma \quad \text{in } L^p(\Gamma)$$

and the definition of $\mathcal{T}_\Gamma$ is indeed consistent. Obviously, $\mathcal{T}_\Gamma$ is linear, by definition. Moreover, owing to (4.12),

$$\|\phi_n\|_{L^p(\Gamma)} \leq C\|\phi_n\|_{W^{1,p}(\Omega)}, \qquad n \geq 1,$$

and hence, letting $n \to \infty$ provides us with the estimate

$$\|\mathcal{T}_\Gamma u\|_{L^p(\Gamma)} \leq C\|u\|_{W^{1,p}(\Omega)}$$

for all $u \in W^{1,p}(\Omega)$, which shows the continuity of $\mathcal{T}_\Gamma$.

Finally, let $u \in W^{1,p}(\Omega) \cap \mathcal{C}(\bar{\Omega})$. Then, according to Theorem 4.1, there exists a sequence $\phi_n \in \mathcal{C}^\infty(\bar{\Omega})$, $n \geq 1$, such that

$$\lim_{n\to\infty} \left(\|\phi_n - u\|_{W^{1,p}(\Omega)} + \|\phi_n - u\|_{\mathcal{C}(\bar{\Omega})} \right) = 0.$$

Consequently,

$$\mathcal{T}_\Gamma u := \lim_{n\to\infty} \mathcal{T}_\Gamma\phi_n = \lim_{n\to\infty} \phi_n|_\Gamma = u|_\Gamma.$$

The uniqueness of $\mathcal{T}_\Gamma$ is based upon the fact that

$$\mathcal{T}_\Gamma\phi = \phi|_\Gamma \quad \text{for all } \phi \in \mathcal{C}^\infty(\bar{\Omega}),$$

as $\mathcal{C}^\infty(\bar{\Omega})$ is dense in $W^{1,p}(\Omega)$. The proof is complete. $\qquad\square$

Subsequently, for every $\Gamma \in \{\Gamma_0, \Gamma_1, \partial\Omega\}$ and $p \in [1, \infty)$, we will consider the closed subspaces

$$W_\Gamma^{1,p}(\Omega) := \mathcal{T}_\Gamma^{-1}(0) = N[\mathcal{T}_\Gamma], \qquad (4.16)$$

where $\mathcal{T}_\Gamma$ is the trace operator of $W^{1,p}(\Omega)$ on Γ. Then, the following density result holds.

Theorem 4.7. *Let $\Gamma \in \{\Gamma_0, \Gamma_1, \partial\Omega\}$ and $p \in [1, \infty)$. Then,*

$$W_\Gamma^{1,p}(\Omega) = \overline{\mathcal{C}_\Gamma^\infty(\Omega)}^{W^{1,p}(\Omega)}. \qquad (4.17)$$

In other words, for any given $u \in W^{1,p}(\Omega)$, one has that $u \in W_\Gamma^{1,p}(\Omega)$, i.e., $\mathcal{T}_\Gamma u = 0$, if and only if, for some sequence $\phi_n \in \mathcal{C}_\Gamma^\infty(\Omega)$, $n \geq 1$,

$$\lim_{n\to\infty} \|\phi_n - u\|_{W^{1,p}(\Omega)} = 0. \qquad (4.18)$$

Remark 4.1. According to (4.17), we can extend the definition (4.16) up to cover the case $p = \infty$ by simply setting

$$W_\Gamma^{1,\infty}(\Omega) := \overline{\mathcal{C}_\Gamma^\infty(\Omega)}^{W^{1,\infty}(\Omega)}.$$

Proof. Let $u \in W^{1,p}(\Omega)$ for which there exists a sequence $\phi_n \in \mathcal{C}_\Gamma^\infty(\Omega)$, $n \geq 1$, satisfying (4.18). Then, by Theorem 4.6,

$$\mathcal{T}_\Gamma u = \lim_{n\to\infty} \mathcal{T}_\Gamma \phi_n = \lim_{n\to\infty} \phi_n|_\Gamma = 0$$

and, therefore, $u \in W_\Gamma^{1,p}(\Omega)$.

It remains to prove that if $u \in W^{1,p}(\Omega)$ satisfies $\mathcal{T}_\Gamma u = 0$, then, there exists a sequence $\phi_n \in \mathcal{C}_\Gamma^\infty(\Omega)$, $n \geq 1$, satisfying (4.18). This property is based on the following estimate

$$\int_{A_\epsilon} |u|^p dx \leq \frac{2^{p-1}}{p} \epsilon^p \int_{A_\epsilon} |\nabla u|^p, \qquad \epsilon > 0, \ \epsilon \sim 0, \qquad (4.19)$$

where we are denoting

$$A_\epsilon := \{x \in \Omega \ : \ \text{dist}\,(x, \Gamma) < \epsilon\}$$

for sufficiently small $\epsilon > 0$. For these ϵ's,

$$\partial A_\epsilon = \Gamma \cup \Gamma_\epsilon \ \text{ with } \ \Gamma_\epsilon := \{x \in \Omega \ : \ \text{dist}(x, \Gamma) = \epsilon\}.$$

To prove (4.19), let $\phi_n \in \mathcal{C}^\infty(\bar\Omega)$, $n \geq 1$, satisfying (4.18). Such a sequence exists by Theorem 4.1. Then, due to Theorem 4.6,

$$0 = \mathcal{T}_\Gamma u = \lim_{n\to\infty} \mathcal{T}_\Gamma \phi_n = \lim_{n\to\infty} \phi_n|_\Gamma. \qquad (4.20)$$

For sufficiently small $\epsilon > 0$, say $0 < \epsilon \leq \epsilon_0$, and every $x \in \Gamma_\epsilon$, there exists some point $y_x := \mathcal{P}_\Gamma x \in \Gamma$ such that

$$\epsilon = |x - y_x| = \text{dist}\,(x, \Gamma)$$

(see Figure 4.4). Fix $0 < \epsilon \leq \epsilon_0$. Then, for every $n \geq 1$, $0 < \delta < \epsilon$, and $x \in \Gamma_\delta$, the following identities hold

$$\phi_n(x) = \phi_n(y_x) + \int_0^\delta \frac{d}{dt}\phi_n\left(y_x + \frac{x - y_x}{\delta}t\right)\,dt$$

$$= \phi_n(y_x) + \int_0^\delta \left\langle \nabla\phi_n\left(y_x + \frac{x - y_x}{\delta}t\right), \frac{x - y_x}{\delta}\right\rangle\,dt$$

and hence,

$$|\phi_n(x)| \leq |\phi_n(y_x)| + \int_0^\delta \left|\nabla\phi_n\left(y_x + \frac{x - y_x}{\delta}t\right)\right|\,dt. \tag{4.21}$$

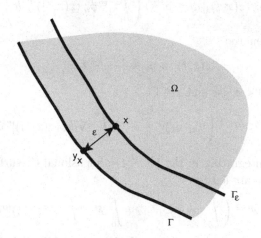

Fig. 4.4 The projection of Γ_ϵ into Γ, $0 < \epsilon \leq \epsilon_0$

On the other hand, for every $a \geq 0$, $b \geq 0$, and $p \geq 1$,

$$(a + b)^p \leq 2^{p-1}(a^p + b^p). \tag{4.22}$$

Indeed, (4.22) is obvious if $ab = 0$, for as $1 \leq 2^{p-1}$. So, suppose $ab > 0$. Then, dividing by a^p, it becomes apparent that (4.22) holds if and only if

$$(1 + x)^p \leq 2^{p-1}(1 + x^p)$$

for all $x \geq 0$. This estimate holds true because the auxiliary function

$$g(x) := \frac{(1+x)^p}{1+x^p}, \qquad x \geq 0,$$

satisfies

$$g \geq 0, \qquad g(0) = 1, \qquad \lim_{x \to \infty} g(x) = 1, \qquad g(1) = 2^{p-1} \geq 1,$$

and $g'(x) = 0$ if and only if $x = 1$. Consequently,

$$g(x) \leq g(1) = 2^{p-1} \qquad \text{for all } x \geq 0.$$

Applying (4.22), we find from (4.21) that

$$|\phi_n(x)|^p \leq 2^{p-1}\left[|\phi_n(y_x)|^p + \left(\int_0^\delta \left|\nabla \phi_n\left(y_x + \frac{x - y_x}{\delta} t\right)\right| dt\right)^p\right]$$

and, according to Hölder inequality,

$$\int_0^\delta |\nabla \phi_n(z(x,t))|\, dt \leq \delta^{\frac{p-1}{p}} \left(\int_0^\delta |\nabla \phi_n(z(x,t))|^p\, dt\right)^{\frac{1}{p}},$$

where we have denoted

$$z(x,t) := y_x + \frac{x - y_x}{\delta} t.$$

Thus, for every $0 < \delta < \epsilon$ and $x \in \Gamma_\delta$,

$$|\phi_n(x)|^p \leq 2^{p-1}\left(|\phi_n(y_x)|^p + \delta^{p-1} \int_0^\delta |\nabla \phi_n(z(x,t))|^p\, dt\right)$$

and, therefore, integrating in the $(N-1)$-dimensional $\mathcal{C}^1$-surface Γ_δ and, later, in $(0,\epsilon)$, we are led to

$$\int_{A_\epsilon} |\phi_n|^p \leq 2^{p-1}\left(\int_{A_\epsilon} |\phi_n(y_x)|^p dx + \int_0^\epsilon \delta^{p-1} \int_{A_\delta} |\nabla \phi_n(z)|^p dz\, d\delta\right)$$

because A_ϵ can be parametrized through the union of the Γ_δ's for δ ranging in $(0,\epsilon)$. By the theorem of dominated convergence of Lebesgue, it follows from (4.20) that

$$\lim_{n \to \infty} \int_{A_\epsilon} |\phi_n(y_x)|^p dx = 0,$$

since $y_x \in \Gamma$ for all $x \in A_\epsilon$. Consequently, letting $n \to \infty$ in the previous inequality, (4.18) implies that

$$\int_{A_\epsilon} |u|^p \leq 2^{p-1} \int_0^\epsilon \delta^{p-1}\left(\int_{A_\delta} |\nabla u|^p\right) d\delta \leq 2^{p-1}\frac{\epsilon^p}{p} \int_{A_\epsilon} |\nabla u|^p,$$

which concludes the proof of (4.19).

Let $n_0 \in \mathbb{N}$ such that $1/n < \epsilon_0$ if $n \geq n_0$, and, for every $n \geq n_0$, pick $\zeta_n \in C_\Gamma^\infty(\Omega)$ such that $0 \leq \zeta_n \leq 1$,

$$\zeta_n = \begin{cases} 1 & \text{in } \bar{\Omega} \setminus A_{\frac{1}{n}}, \\ 0 & \text{in } A_{\frac{1}{2n}}, \end{cases} \quad \text{and} \quad \|\nabla \zeta_n\|_{C(\bar{\Omega})} \leq C\, n,$$

for some constant $C > 0$ independent of n. By using an appropriate partition of the unity and straightening $\partial\Omega$ locally, it is easy to realize the existence of such a sequence of test functions. Subsequently, we consider the auxiliary sequence

$$u_n := \zeta_n u, \qquad n \geq n_0.$$

Obviously,

$$u_n \in W^{1,p}(\Omega), \qquad \text{and} \quad u_n = \begin{cases} u & \text{in } \bar{\Omega} \setminus A_{\frac{1}{n}}, \\ 0 & \text{in } A_{\frac{1}{2n}}, \end{cases} \tag{4.23}$$

for all $n \geq n_0$, by construction. Moreover,

$$\|u_n - u\|_{W^{1,p}(\Omega)}^p = \|u(\zeta_n - 1)\|_{W^{1,p}(\Omega)}^p$$

$$= \int_{A_{\frac{1}{n}}} |u(\zeta_n - 1)|^p + \int_{A_{\frac{1}{n}}} |\nabla [u(\zeta_n - 1)]|^p$$

$$\leq \|u\|_{L^p(A_{1/n})}^p + \int_{A_{\frac{1}{n}}} |\nabla [u(\zeta_n - 1)]|^p.$$

Thus, letting $n \to \infty$ shows that

$$\limsup_{n \to \infty} \|u_n - u\|_{W^{1,p}(\Omega)}^p \leq \limsup_{n \to \infty} \int_{A_{\frac{1}{n}}} |\nabla [u(\zeta_n - 1)]|^p. \tag{4.24}$$

On the other hand, by Lemma 4.1 and (4.22), we find that

$$\int_{A_{\frac{1}{n}}} |\nabla [u(\zeta_n - 1)]|^p = \int_{A_{\frac{1}{n}}} |(\zeta_n - 1)\nabla u + u\nabla \zeta_n|^p$$

$$\leq 2^{p-1} \int_{A_{\frac{1}{n}}} (|(\zeta_n - 1)\nabla u|^p + |u\nabla \zeta_n|^p)$$

$$\leq 2^{p-1} \|\nabla u\|_{L^p(A_{1/n})}^p + 2^{p-1} \int_{A_{\frac{1}{n}}} |u\nabla \zeta_n|^p$$

for all $n \geq n_0$. Thus, since

$$\lim_{n \to \infty} \|\nabla u\|_{L^p(A_{1/n})}^p = 0,$$

it follows from (4.24) that

$$\limsup_{n\to\infty} \|u_n - u\|^p_{W^{1,p}(\Omega)} \leq 2^{p-1} \limsup_{n\to\infty} \int_{A_{\frac{1}{n}}} |u\nabla\zeta_n|^p. \qquad (4.25)$$

Finally, by the choice of ζ_n, (4.19) implies that

$$\int_{A_{\frac{1}{n}}} |u\nabla\zeta_n|^p \leq C^p n^p \int_{A_{\frac{1}{n}}} |u|^p$$

$$\leq C^p n^p \frac{2^{p-1}}{p\,n^p} \int_{A_{\frac{1}{n}}} |\nabla u|^p$$

$$= \frac{C^p}{p} 2^{p-1} \|\nabla u\|^p_{L^p\left(A_{1/n}\right)} \to 0$$

as $n \to \infty$. Therefore, (4.25) shows that

$$\lim_{n\to\infty} \|u_n - u\|_{W^{1,p}(\Omega)} = 0. \qquad (4.26)$$

It remains to prove that the u_n's can be approximated in $W^{1,p}(\Omega)$ by test functions $\phi_n \in C^\infty_\Gamma(\Omega)$, $n \geq 1$. This can be accomplished because, according to (4.23), $u_n = 0$ in $A_{\frac{1}{2n}}$ for all $n \geq n_0$. Indeed, consider the auxiliary function

$$\eta(x) := \begin{cases} Ce^{\frac{1}{|x|^2-1}} & \text{if } |x| \leq 1, \\ 0 & \text{if } |x| \geq 1, \end{cases}$$

where the constant C is given through

$$C := \left(\int_{B_1(0)} e^{\frac{1}{|x|^2-1}} \, dx \right)^{-1}.$$

Then,

$$\eta \in C^\infty_0(\mathbb{R}^N), \quad \eta \geq 0, \quad \text{supp } \eta = B_1(0), \quad \int_{\mathbb{R}^N} \eta = 1. \qquad (4.27)$$

Now, for every $\epsilon > 0$, we introduce

$$\eta_\epsilon(x) := \epsilon^{-N} \eta\left(\frac{x}{\epsilon}\right), \qquad x \in \mathbb{R}^N.$$

According to (4.27), it becomes apparent that

$$\eta_\epsilon \in C^\infty_0(\mathbb{R}^N), \quad \eta_\epsilon \geq 0, \quad \text{supp } \eta = B_\epsilon(0), \quad \int_{\mathbb{R}^N} \eta_\epsilon = 1, \qquad (4.28)$$

for all $\epsilon > 0$. Further, for every $n \geq n_0$ and $0 < \epsilon < \frac{1}{4n}$, we consider the function

$$
\begin{aligned}
\psi_{n,\epsilon}(x) &= \int_\Omega \eta_\epsilon(x-y) u_n(y)\, dy \\
&= \int_{\Omega \backslash A_{\frac{1}{4n}}} \eta_\epsilon(x-y) u_n(y)\, dy, \qquad x \in \Omega.
\end{aligned}
$$

By (4.28), it readily follows that

$$
\psi_{n,\epsilon} \in \mathcal{C}^\infty_\Gamma(\Omega), \qquad \lim_{\epsilon \downarrow 0} \|\psi_{n,\epsilon} - u_n\|_{W^{1,p}(\Omega)} = 0
$$

for all $n \geq n_0$. Fix $n \geq n_0$. Then, there exists $\epsilon := \epsilon(n)$ such that the test function

$$
\phi_n := \psi_{n,\epsilon(n)} \in \mathcal{C}^\infty_\Gamma(\Omega)
$$

satisfies

$$
\|\phi_n - u_n\|_{W^{1,p}(\Omega)} \leq \frac{1}{n}.
$$

Therefore, for every $n \geq n_0$, we obtain that

$$
\begin{aligned}
\|\phi_n - u\|_{W^{1,p}(\Omega)} &\leq \|\phi_n - u_n\|_{W^{1,p}(\Omega)} + \|u - u_n\|_{W^{1,p}(\Omega)} \\
&\leq \frac{1}{n} + \|u - u_n\|_{W^{1,p}(\Omega)}.
\end{aligned}
$$

Consequently, (4.26) implies that

$$
\lim_{n \to \infty} \|\phi_n - u\|_{W^{1,p}(\Omega)} = 0.
$$

The proof is complete. $\qquad\qquad\qquad\qquad\qquad\qquad\qquad\qquad\qquad\square$

This section concludes with another important property of the trace operator.

Proposition 4.1. *Let* $\Gamma \in \{\Gamma_0, \Gamma_1, \partial\Omega\}$, $p \in [1, \infty)$, *and* $g \in \mathcal{C}^1(\bar{\Omega})$. *Then, for every* $u \in W^{1,p}(\Omega)$,

$$
\mathcal{T}_\Gamma(gu) = g\mathcal{T}_\Gamma u.
$$

Proof. Let $u \in W^{1,p}(\Omega)$. According to Theorem 4.1, there exists a sequence $\phi_n \in \mathcal{C}^\infty(\bar{\Omega})$, $n \geq 1$, such that

$$
\lim_{n \to \infty} \|\phi_n - u\|_{W^{1,p}(\Omega)} = 0.
$$

As $g \in \mathcal{C}^1(\bar{\Omega})$, it is easy to see that

$$
\lim_{n \to \infty} \|g\phi_n - gu\|_{W^{1,p}(\Omega)} = 0.
$$

Hence, by Theorem 4.6, it becomes apparent that

$$
\mathcal{T}_\Gamma(gu) = \lim_{n \to \infty} \mathcal{T}_\Gamma(g\phi_n) = \lim_{n \to \infty} (g\phi_n) = g\mathcal{T}_\Gamma u
$$

in $L^p(\Gamma)$. $\qquad\qquad\qquad\qquad\qquad\qquad\qquad\qquad\qquad\qquad\qquad\qquad\square$

Remark 4.2. To simplify the notation, throughout the rest of this book we will denote

$$u|_\Gamma := \mathcal{T}_\Gamma u \in L^p(\Gamma)$$

for all $p \geq 1$, $u \in W^{1,p}(\Omega)$, and $\Gamma \in \{\Gamma_0, \Gamma_1, \partial\Omega\}$, if there is no ambiguity.

4.3   Weak solutions

This section considers the linear boundary value problem (4.4) for an arbitrary $\omega \in \mathbb{R}$.

Definition 4.4. *Let* $1 \leq p < \infty$, $f \in L^p(\Omega)$, *and* $u \in W^{2,p}(\Omega)$. *Then,* u *is said to be a solution of* (4.4) *if* $(\mathfrak{L} + \omega)u = f$ *in* Ω *and* $\mathfrak{B}u = 0$ *on* $\partial\Omega$ *in the sense of traces, i.e.,* $\mathcal{T}_{\Gamma_0} u = 0$, *or, equivalently,* $u \in W^{1,p}_{\Gamma_0}(\Omega)$ *and*

$$\langle \mathcal{T}_{\Gamma_1} \nabla u, \nu \rangle + \beta \mathcal{T}_{\Gamma_1} u = 0 \qquad \text{on } \Gamma_1.$$

Let $p \in [2, \infty)$, $f \in L^p(\Omega)$, and suppose $u \in W^{2,p}(\Omega) \cap W^{1,p}_{\Gamma_0}(\Omega)$ is a solution of problem (4.4). Then, according to Definition 4.2 and Lemma 4.1, for every $\phi \in \mathcal{C}^\infty_{\Gamma_0}(\Omega)$, we have that

$$
\begin{aligned}
\int_\Omega f\phi &= \int_\Omega \phi \left(\mathfrak{L} + \omega\right)u \\
&= -\int_\Omega \phi \operatorname{div}(A\nabla u) + \int_\Omega \phi\langle b, \nabla u \rangle + \int_\Omega (c + \omega)u\phi \\
&= -\int_\Omega \operatorname{div}(\phi A\nabla u) + \int_\Omega \langle \nabla\phi, A\nabla u \rangle + \int_\Omega \phi\langle b, \nabla u \rangle + \int_\Omega (c + \omega)u\phi \\
&= -\int_{\partial\Omega} \langle \phi A\nabla u, n \rangle \, dS + \int_\Omega \langle A\nabla u, \nabla\phi \rangle + \int_\Omega \left(\langle b, \nabla u \rangle + cu + \omega u\right)\phi.
\end{aligned}
$$

Moreover,

$$\int_{\partial\Omega} \langle \phi A\nabla u, n \rangle \, dS = \int_{\Gamma_1} \langle \phi A\nabla u, n \rangle \, dS,$$

since ϕ vanishes in a neighborhood of Γ_0, and hence,

$$\int_{\partial\Omega} \langle \phi A\nabla u, n \rangle \, dS = \int_{\Gamma_1} \phi \frac{\partial u}{\partial\nu} \, dS = -\int_{\Gamma_1} \beta u\phi \, dS,$$

because

$$0 = \mathfrak{B}u = \frac{\partial u}{\partial\nu} + \beta u \qquad \text{on } \Gamma_1$$

(see Remark 4.2). Consequently,

$$\int_\Omega f\phi = \int_\Omega \langle A\nabla u, \nabla\phi\rangle + \int_\Omega (\langle b, \nabla u\rangle + cu + \omega u)\phi + \int_{\Gamma_1} \beta u\phi\, dS$$

for all $\phi \in C^\infty_{\Gamma_0}(\Omega)$, and, therefore,

$$\mathfrak{a}(u, \phi) = \langle f, \phi\rangle_{L^2(\Omega)} \qquad \forall\, \phi \in C^\infty_{\Gamma_0}(\Omega), \tag{4.29}$$

where

$$\mathfrak{a} \;:\; W^{1,2}_{\Gamma_0}(\Omega) \times W^{1,2}_{\Gamma_0}(\Omega) \longrightarrow \mathbb{R}$$

stands for the bilinear form defined by

$$\mathfrak{a}(u, v) := \int_\Omega \langle A\nabla u, \nabla v\rangle + \int_\Omega (\langle b, \nabla u\rangle + cu + \omega u)\, v + \int_{\Gamma_1} \beta uv\, dS$$

for all $u, v \in W^{1,2}_{\Gamma_0}(\Omega)$ (see Remark 4.2). Throughout the rest of this book, $\mathfrak{a}(\cdot, \cdot)$ will be referred to as the *bilinear form associated to* (4.4), or, equivalently, to the tern $(\mathfrak{L} + \omega, \mathfrak{B}, \Omega)$.

The following concept plays a fundamental role in the theory of partial differential equations.

Definition 4.5 (Weak solution). *For any* $f \in L^2(\Omega)$, *a function* $u \in W^{1,2}_{\Gamma_0}(\Omega)$ *is said to be a* **weak solution** *of* (4.4) *if* (4.29) *is satisfied.*

We have just seen that every solution $u \in W^{2,p}(\Omega)$, $p \geq 2$, of (4.4) provides us with a weak solution. Actually, the following characterization holds.

Proposition 4.2. *Let* $f \in L^2(\Omega)$ *and* $u \in W^{2,p}(\Omega) \cap W^{1,p}_{\Gamma_0}(\Omega)$ *for some* $p \geq 2$. *Then,* u *solves* (4.4) *in the sense of Definition 4.4 if and only if* u *is a weak solution of* (4.4).

Proof. We already know that u is a weak solution if it solves (4.4). So, suppose u is a weak solution of (4.4). Then, it satisfies (4.29) and, consequently, for every $\phi \in C^\infty_{\Gamma_0}(\Omega)$, we have that

$$\int_\Omega f\phi = \mathfrak{a}(u, \phi)$$

$$= \int_\Omega \langle A\nabla u, \nabla\phi\rangle + \int_\Omega (\langle b, \nabla u\rangle + cu + \omega u)\phi + \int_{\Gamma_1} \beta u\phi\, dS$$

$$= \int_\Omega \phi(\mathfrak{L} + \omega)u + \int_\Omega \operatorname{div}(\phi A\nabla u) + \int_{\Gamma_1} \beta u\phi\, dS$$

$$= \int_\Omega \phi(\mathfrak{L} + \omega)u + \int_{\partial\Omega} \langle \phi A\nabla u, n\rangle\, dS + \int_{\Gamma_1} \beta u\phi\, dS,$$

whence

$$\int_\Omega f\phi = \int_\Omega \phi(\mathfrak{L}+\omega)u + \int_{\Gamma_1} \phi\left(\frac{\partial u}{\partial \nu} + \beta u\right) dS. \qquad (4.30)$$

Now, fix $\psi \in \mathcal{C}^\infty(\Gamma_1)$ and let $\phi_n \in \mathcal{C}^\infty_{\Gamma_0}(\Omega)$, $n \geq 1$, be a sequence of test functions such that

$$\phi_n|_{\Gamma_1} = \psi \quad \text{and} \quad \operatorname{supp}\phi_n \subset \left\{ x \in \Omega \; : \; \operatorname{dist}(x,\Gamma_1) \leq \frac{1}{n} \right\}$$

for sufficiently large n, and

$$\|\phi_n\|_{L^\infty(\Omega)} \leq C, \qquad n \geq 1,$$

for some positive constant C. By (4.30),

$$\int_\Omega f\phi_n = \int_\Omega \phi_n(\mathfrak{L}+\omega)u + \int_{\Gamma_1} \phi_n\left(\frac{\partial u}{\partial \nu} + \beta u\right) dS$$

for all $n \geq 1$. Hence, letting $n \to \infty$ in this identity, the theorem of dominated convergence of Lebesgue shows that

$$\int_{\Gamma_1} \psi\left(\frac{\partial u}{\partial \nu} + \beta u\right) dS = 0.$$

As this identity holds for every $\psi \in \mathcal{C}^\infty(\Gamma_1)$, it becomes apparent that

$$\frac{\partial u}{\partial \nu} + \beta u = 0 \qquad \text{on} \quad \Gamma_1 \qquad (4.31)$$

and, therefore,

$$\mathfrak{B}u = 0 \qquad \text{on} \quad \partial\Omega.$$

Moreover, substituting (4.31) into (4.30), yields

$$\int_\Omega f\phi = \int_\Omega \phi(\mathfrak{L}+\omega)u \qquad \forall\, \phi \in \mathcal{C}^\infty_{\Gamma_0}(\Omega)$$

and, consequently,

$$(\mathfrak{L}+\omega)u = f \qquad \text{in} \quad \Omega.$$

The proof is complete. □

4.4 Continuity of the associated bilinear form

This section shows the continuity of the bilinear form $\mathfrak{a}(\cdot, \cdot)$ associated to (4.4). The main result reads as follows.

Proposition 4.3. *For every $\omega \in \mathbb{R}$ there exists a constant $C := C(\omega) > 0$ such that*

$$|\mathfrak{a}(u,v)| \leq C \|u\|_{W_{\Gamma_0}^{1,2}(\Omega)} \|v\|_{W_{\Gamma_0}^{1,2}(\Omega)} \tag{4.32}$$

for all $u, v \in W_{\Gamma_0}^{1,2}(\Omega)$.

Proof. Let $u, v \in W_{\Gamma_0}^{1,2}(\Omega)$. Then, by the Cauchy–Schwarz inequality, and according to Remark 4.2, we find that

$$
|\mathfrak{a}(u,v)| \leq \int_\Omega |\langle A\nabla u, \nabla v\rangle| + \int_\Omega |\langle b, \nabla u\rangle + cu + \omega u| \, |v| + \int_{\Gamma_1} |\beta| \, |u| \, |v| \, dS
$$

$$
\leq \int_\Omega |A\nabla u| \, |\nabla v| + \int_\Omega \left(|b| \, |\nabla u| + \|c + \omega\|_{L^\infty(\Omega)} \, |u| \right) |v|
$$

$$
+ \|\beta\|_{L^\infty(\Gamma_1)} \int_{\Gamma_1} |u| \, |v| \, dS.
$$

Thus, setting

$$
\begin{aligned}
\|A\|_\infty &:= \max_{1 \leq i,j \leq N} \left\{ \|a_{ij}\|_{L^\infty(\Omega)} \right\}, \\
\|b\|_\infty &:= \max_{1 \leq j \leq N} \left\{ \|b_j\|_{L^\infty(\Omega)} \right\},
\end{aligned} \tag{4.33}
$$

and applying the Hölder inequality, we are led to

$$
\begin{aligned}
|\mathfrak{a}(u,v)| \leq\ & \|A\|_\infty \|\nabla u\|_{L^2(\Omega)} \|\nabla v\|_{L^2(\Omega)} + \|b\|_\infty \|\nabla u\|_{L^2(\Omega)} \|v\|_{L^2(\Omega)} \\
& + \|c + \omega\|_{L^\infty(\Omega)} \|u\|_{L^2(\Omega)} \|v\|_{L^2(\Omega)} \\
& + \|\beta\|_{L^\infty(\Gamma_1)} \|\mathcal{T}_{\Gamma_1} u\|_{L^2(\Gamma_1)} \|\mathcal{T}_{\Gamma_1} v\|_{L^2(\Gamma_1)}.
\end{aligned}
$$

On the other hand, owing to Theorem 4.6, we have that

$$
\|\mathcal{T}_{\Gamma_1} w\|_{L^2(\Gamma_1)} \leq \|\mathcal{T}_{\Gamma_1}\|_{\mathcal{L}(W^{1,2}(\Omega), L^2(\Gamma_1))} \|w\|_{W^{1,2}(\Omega)}
$$

for all $w \in W^{1,2}(\Omega)$. Therefore, (4.32) holds with

$$
\begin{aligned}
C(\omega) = 2\big(&\|A\|_\infty + \|b\|_\infty + \|c\|_{L^\infty(\Omega)} + |\omega| \\
& + \|\beta\|_{L^\infty(\Gamma_1)} \|\mathcal{T}_{\Gamma_1}\|_{\mathcal{L}(W^{1,2}(\Omega), L^2(\Gamma_1))}^2 \big).
\end{aligned}
$$

The proof is complete. $\qquad\square$

4.5   Invertibility of (4.4) when $\beta \geq 0$

Throughout this section we suppose that $\beta \geq 0$.

4.5.1   *Coercivity of the associated bilinear form*

The next result shows the coercivity of $\mathfrak{a}(\cdot, \cdot)$ for sufficiently large ω.

Theorem 4.8 (of Gårding). *Suppose $\beta \geq 0$. Then, there exists $\omega_0 \in \mathbb{R}$ such that for every $\omega \geq \omega_0$ there is a constant $\alpha := \alpha(\omega) > 0$ for which*

$$|\mathfrak{a}(u, u)| \geq \alpha \|u\|^2_{W^{1,2}_{\Gamma_0}(\Omega)} \qquad \forall \, u \in W^{1,2}_{\Gamma_0}(\Omega). \tag{4.34}$$

Consequently, $\mathfrak{a}(\cdot, \cdot)$ is coercive in the Hilbert space

$$H := W^{1,2}_{\Gamma_0}(\Omega) \tag{4.35}$$

for all $\omega \geq \omega_0$.

Proof.   Remember that the scalar product of H is given by

$$\langle u, v \rangle_H := \int_\Omega \langle \nabla u, \nabla v \rangle + \int_\Omega uv, \qquad u, v \in H.$$

As $\beta \geq 0$ implies

$$\int_{\Gamma_1} \beta u^2 \, dS \geq 0, \qquad u \in H,$$

it becomes apparent that

$$\mathfrak{a}(u, u) \geq \int_\Omega \langle A \nabla u, \nabla u \rangle + \int_\Omega \left(\langle b, \nabla u \rangle + cu + \omega u \right) u \tag{4.36}$$

for all $u \in H$. Let $\mu > 0$ denote the ellipticity constant of $\mathfrak{L}$ in Ω. Then,

$$\langle A \nabla u, \nabla u \rangle \geq \mu |\nabla u|^2 \qquad \text{in } \Omega,$$

and hence,

$$\int_\Omega \langle A \nabla u, \nabla u \rangle \geq \mu \|\nabla u\|^2_{L^2(\Omega)}. \tag{4.37}$$

Moreover,

$$\left| \int_\Omega \langle b, \nabla u \rangle u \right| \leq \|b\|_\infty \int_\Omega |u| \, |\nabla u|,$$

where $\|b\|_\infty$ is given by (4.33), and

$$|u| \, |\nabla u| = \epsilon |u| \frac{|\nabla u|}{\epsilon} \leq \frac{\epsilon^2}{2} |u|^2 + \frac{|\nabla u|^2}{2\epsilon^2} \qquad \text{in } \Omega$$

for all $\epsilon > 0$. Thus, setting $\eta := \epsilon^2$, shows that

$$|\int_\Omega \langle b, \nabla u \rangle u| \leq \frac{\|b\|_\infty}{2} \left(\eta \|u\|_{L^2(\Omega)}^2 + \frac{1}{\eta} \|\nabla u\|_{L^2(\Omega)}^2 \right)$$

for all $\eta > 0$. In particular,

$$\int_\Omega \langle b, \nabla u \rangle u \geq -\frac{\|b\|_\infty}{2} \left(\eta \|u\|_{L^2(\Omega)}^2 + \frac{1}{\eta} \|\nabla u\|_{L^2(\Omega)}^2 \right). \tag{4.38}$$

Also,

$$\int_\Omega (cu + \omega u) u \geq \left(\omega + \inf_\Omega c \right) \|u\|_{L^2(\Omega)}^2. \tag{4.39}$$

Consequently, substituting (4.37), (4.38) and (4.39) into (4.36), we find that, for every $\eta > 0$ and $u \in H$,

$$\begin{aligned}
\mathfrak{a}(u, u) &\geq \mu \|\nabla u\|_{L^2(\Omega)}^2 - \frac{\|b\|_\infty}{2} \left(\eta \|u\|_{L^2(\Omega)}^2 + \frac{1}{\eta} \|\nabla u\|_{L^2(\Omega)}^2 \right) \\
&\qquad + \left(\omega + \inf_\Omega c \right) \|u\|_{L^2(\Omega)}^2 \\
&= \left(\mu - \frac{\|b\|_\infty}{2\eta} \right) \|\nabla u\|_{L^2(\Omega)}^2 + \left(\omega + \inf_\Omega c - \frac{\|b\|_\infty}{2} \eta \right) \|u\|_{L^2(\Omega)}^2.
\end{aligned}$$

Pick $\delta \in (0, \mu)$, let $\eta > 0$ sufficiently large so that

$$\mu - \frac{\|b\|_\infty}{2\eta} > \mu - \delta,$$

and set

$$\omega_0 := -\inf_\Omega c + \frac{\|b\|_\infty}{2} \eta + \mu - \delta.$$

Then, for every $\omega \geq \omega_0$ and $u \in H$, we have that

$$\begin{aligned}
\mathfrak{a}(u, u) &\geq \left(\mu - \frac{\|b\|_\infty}{2\eta} \right) \|\nabla u\|_{L^2(\Omega)}^2 + \left(\omega + \inf_\Omega c - \frac{\|b\|_\infty}{2} \eta \right) \|u\|_{L^2(\Omega)}^2 \\
&\geq (\mu - \delta) \left(\|\nabla u\|_{L^2(\Omega)}^2 + \|u\|_{L^2(\Omega)}^2 \right) \\
&= (\mu - \delta) \|u\|_{W_{\Gamma_0}^{1,2}(\Omega)}^2.
\end{aligned}$$

This concludes the proof. $\qquad\qquad\qquad\qquad\qquad\qquad\qquad\qquad\quad \square$

4.5.2   *Existence of weak solutions. The resolvent operator*

This section applies Theorem 3.5 to obtain the next fundamental result.

Theorem 4.9 (of invertibility of $(\mathfrak{L}, \mathfrak{B}, \Omega)$). *Suppose the bilinear form $\mathfrak{a}(\cdot, \cdot)$ associated to the problem* (4.4) *is coercive for some $\omega \in \mathbb{R}$. Then,* (4.4) *possesses a unique weak solution $u \in W^{1,2}_{\Gamma_0}(\Omega)$. Moreover, if we denote it by*

$$u := (\mathfrak{L} + \omega)^{-1} f \in W^{1,2}_{\Gamma_0}(\Omega),$$

*then, the **resolvent operator***

$$L^2(\Omega) \xrightarrow{\ (\mathfrak{L}+\omega)^{-1}\ } W^{1,2}_{\Gamma_0}(\Omega)$$

$$f \quad \mapsto \quad (\mathfrak{L}+\omega)^{-1} f$$

is linear and continuous, i.e.,

$$(\mathfrak{L} + \omega)^{-1} \in \mathcal{L}\left(L^2(\Omega), W^{1,2}_{\Gamma_0}(\Omega)\right). \tag{4.40}$$

Therefore, if we denote by J the compact imbedding

$$J \ : \ W^{1,2}_{\Gamma_0}(\Omega) \hookrightarrow L^2(\Omega),$$

then, the composition

$$J(\mathfrak{L} + \omega)^{-1} \ : \ L^2(\Omega) \longrightarrow L^2(\Omega)$$

is a linear and compact operator.

Remark 4.3. Owing to Theorem 4.8, when $\beta \geq 0$, there exists $\omega_0 > 0$ such that $\mathfrak{a}(\cdot, \cdot)$ is coercive for all $\omega \geq \omega_0$. Therefore, the conclusions of Theorem 4.9 hold true for all $\omega \geq \omega_0$.

Proof.   Consider the Hilbert space $H := W^{1,2}_{\Gamma_0}(\Omega)$ and the bilinear form $\mathfrak{a} : H^2 \to \mathbb{R}$ associated to (4.4). By Proposition 4.3, $\mathfrak{a}(\cdot, \cdot)$ is continuous, and we are assuming that it is coercive for some $\omega \in \mathbb{R}$. Moreover, for every $f \in L^2(\Omega)$, the map

$$\langle f, \cdot \rangle_{L^2(\Omega)} \ : \ W^{1,2}_{\Gamma_0}(\Omega) \longrightarrow \mathbb{R}, \qquad u \mapsto \langle f, u \rangle_{L^2(\Omega)} = \int_\Omega fu,$$

is linear and continuous, and hence it belongs to the topological dual H' of H. Therefore, according to Theorem 3.5, for every $f \in L^2(\Omega)$ there exists a unique

$$u := (\mathfrak{L} + \omega)^{-1} f \in H$$

such that

$$\mathfrak{a}(u, v) = \langle f, v \rangle_{L^2(\Omega)} \qquad \forall\, v \in H.$$

Obviously, u is the unique weak solution of (4.4).

Subsequently, we consider arbitrary $A, B \in \mathbb{R}$, $f, g \in L^2(\Omega)$, and $v \in H$. Then,

$$\begin{aligned}
\langle Af + Bg, v \rangle_{L^2(\Omega)} &= A\langle f, v \rangle_{L^2(\Omega)} + B\langle g, v \rangle_{L^2(\Omega)} \\
&= A\,\mathfrak{a}\left((\mathfrak{L} + \omega)^{-1} f, v\right) + B\,\mathfrak{a}\left((\mathfrak{L} + \omega)^{-1} g, v\right) \\
&= \mathfrak{a}\left(A(\mathfrak{L} + \omega)^{-1} f + B(\mathfrak{L} + \omega)^{-1} g, v\right).
\end{aligned}$$

Consequently, by the uniqueness of the weak solution,

$$(\mathfrak{L} + \omega)^{-1}(Af + Bg) = A(\mathfrak{L} + \omega)^{-1} f + B(\mathfrak{L} + \omega)^{-1} g$$

and, therefore, $(\mathfrak{L} + \omega)^{-1}$ is a linear operator.

Now, let $\{f_n\}_{n \geq 1}$ be a sequence in $L^2(\Omega)$ such that

$$\lim_{n \to \infty} \|f_n - f\|_{L^2(\Omega)} = 0 \tag{4.41}$$

and set

$$u := (\mathfrak{L} + \omega)^{-1} f, \qquad u_n := (\mathfrak{L} + \omega)^{-1} f_n, \qquad n \geq 1.$$

By definition, we have that

$$\mathfrak{a}(u_n - u, u_n - u) = \langle f_n - f, u_n - u \rangle_{L^2(\Omega)}$$

for all $n \geq 1$. Consequently, as $\mathfrak{a}$ is coercive, there exists a constant $C > 0$ such that

$$\begin{aligned}
\|u_n - u\|_H^2 &\leq C\,|\mathfrak{a}(u_n - u, u_n - u)| \\
&= C\,|\langle f_n - f, u_n - u \rangle_{L^2(\Omega)}| \\
&\leq C\,\|f_n - f\|_{L^2(\Omega)} \|u_n - u\|_H
\end{aligned}$$

for all $n \geq 1$. Therefore, (4.41) implies

$$\lim_{n \to \infty} \|u_n - u\|_H = 0$$

and shows (4.40).

The compactness of the imbedding J was already established by Theorem 4.5. The proof is complete. □

4.6 Invertibility of (4.4) for arbitrary β

The main goal of this section is to extend the result of Remark 4.3 up to cover the general case when β changes sign on Γ_1. To accomplish this task, besides Assumptions B1 and B2, we will impose

B3. There exist $\psi \in C^2(\bar{\Omega})$ and $\gamma > 0$ such that

$$\frac{\partial \psi}{\partial \nu}(x) \geq \gamma \qquad \text{for all} \quad x \in \Gamma_1.$$

By Lemma 2.1, B3 holds if $\partial\Omega$ is of class C^2. According to Proposition 2.1, under these assumptions, there exist $\omega_0 \in \mathbb{R}$ and $h \in C^2(\bar{\Omega})$, with $h(x) > 0$ for all $x \in \bar{\Omega}$, such that h is a strict supersolution of $(\mathfrak{L} + \omega, \mathfrak{B}, \Omega)$ for all $\omega \geq \omega_0$. Actually, going back to the proof of Proposition 2.1, it becomes apparent that we can make the choice

$$h(x) = e^{M\psi(x)}, \qquad x \in \bar{\Omega}, \tag{4.42}$$

for any $M > 0$ satisfying

$$M\gamma + \beta > 0 \qquad \text{on } \Gamma_1. \tag{4.43}$$

Subsequently, we set

$$b_h := b - \frac{2}{h}A\nabla h, \qquad c_h := \frac{\mathfrak{L}h}{h}, \tag{4.44}$$

and consider the operator

$$\mathfrak{L}_h := -\operatorname{div}(A\nabla\,\cdot) + \langle b_h, \nabla\,\cdot\rangle + c_h. \tag{4.45}$$

Then,

$$b_h \in (L^\infty(\Omega))^N, \qquad c_h \in L^\infty(\Omega),$$

because $h \in C^2(\bar{\Omega})$, and, much like in the proof of Theorem 1.7, a straightforward computation shows that

$$\mathfrak{L}u = h\,\mathfrak{L}_h\frac{u}{h} \qquad \text{in } \Omega \tag{4.46}$$

for all function u almost everywhere twice classically differentiable in Ω.

Similarly, we introduce

$$\beta_h := \frac{\mathfrak{B}h}{h} \tag{4.47}$$

and the boundary operator

$$\mathfrak{B}_h := \begin{cases} \mathfrak{D} & \text{on } \Gamma_0, \\ \frac{\partial}{\partial \nu} + \beta_h & \text{on } \Gamma_1. \end{cases} \tag{4.48}$$

By Assumption B3, (4.42) and (4.43) imply that

$$\beta_h = \frac{\mathfrak{B}h}{h} = M\frac{\partial\psi}{\partial\nu} + \beta \geq M\gamma + \beta > 0 \quad \text{on} \quad \Gamma_1. \tag{4.49}$$

Moreover, $\beta_h \in \mathcal{C}(\Gamma_1)$ and

$$\mathfrak{B}u = h\,\mathfrak{B}_h\frac{u}{h} \quad \text{on} \quad \partial\Omega \tag{4.50}$$

for all $u \in \mathcal{C}^1(\bar\Omega)$. Therefore, if u is a classical solution of (4.4) in Ω, as discussed in Definition 4.1, and we perform the change of variable

$$v := \frac{u}{h} \tag{4.51}$$

then, v is twice classically differentiable a.e. in Ω, and, thanks to (4.46) and (4.50), it is a classical solution of

$$\begin{cases} (\mathfrak{L}_h + \omega)v = \frac{f}{h} & \text{in } \Omega, \\ \mathfrak{B}_h v = 0 & \text{on } \partial\Omega. \end{cases} \tag{4.52}$$

As the converse is as well true, (4.51) establishes a bijection between the classical solutions of (4.4) and (4.52). According to (4.48) and (4.49), the transformed problem (4.52) fits within the framework of Section 4.5, as $\beta_h \geq 0$. Not surprisingly, (4.51) also establishes a bijection between the corresponding weak solutions. Indeed, let

$$\mathfrak{a}, \quad \mathfrak{a}_h \; : \; W^{1,2}_{\Gamma_0}(\Omega) \times W^{1,2}_{\Gamma_0}(\Omega) \longrightarrow \mathbb{R}$$

denote the bilinear forms associated to (4.4) and (4.52), respectively, i.e.,

$$\mathfrak{a}(u,w) := \int_\Omega \langle A\nabla u, \nabla w\rangle + \int_\Omega (\langle b, \nabla u\rangle + cu + \omega u)\,w + \int_{\Gamma_1} \beta uw\,dS$$

$$\mathfrak{a}_h(u,w) := \int_\Omega \langle A\nabla u, \nabla w\rangle + \int_\Omega (\langle b_h, \nabla u\rangle + c_h u + \omega u)\,w + \int_{\Gamma_1} \beta_h uw\,dS,$$

for all $u, w \in W^{1,2}_{\Gamma_0}(\Omega)$, where the integrals on Γ_1 should be understood in the sense of traces. The following result establishes a fundamental relationship between these bilinear forms.

Lemma 4.2. *For every* $v, w \in W^{1,2}_{\Gamma_0}(\Omega)$,

$$\mathfrak{a}\left(hv, \frac{w}{h}\right) = \mathfrak{a}_h(v, w).$$

Proof.   Let $v, w \in W^{1,2}_{\Gamma_0}(\Omega)$. Then, by definition,

$$\mathfrak{a}\left(hv, \frac{w}{h}\right) = \int_\Omega \langle A\nabla(hv), \nabla\frac{w}{h}\rangle + \int_\Omega \langle b, \nabla(hv)\rangle\frac{w}{h} + \int_\Omega (c+\omega)vw$$
$$+ \int_{\Gamma_1} \beta\mathcal{T}_{\Gamma_1}(hv)\mathcal{T}_{\Gamma_1}\frac{w}{h}\, dS,$$

where $\mathcal{T}_{\Gamma_1}$ is the trace operator of $W^{1,p}(\Omega)$ on Γ_1. Moreover, by Proposition 4.1 and Remark 4.2,

$$\int_{\Gamma_1} \beta\mathcal{T}_{\Gamma_1}(hv)\mathcal{T}_{\Gamma_1}\frac{w}{h}\, dS = \int_{\Gamma_1} \beta vw\, dS.$$

Thus, expanding the underlying derivatives we are led to

$$\mathfrak{a}\left(hv, \frac{w}{h}\right) = \int_\Omega \langle hA\nabla v + vA\nabla h, \frac{h\nabla w - w\nabla h}{h^2}\rangle$$
$$+ \int_\Omega \langle b, h\nabla v + v\nabla h\rangle\frac{w}{h} + \int_\Omega (c+\omega)vw + \int_{\Gamma_1} \beta vw\, dS$$
$$= \int_\Omega \langle A\nabla v, \nabla w\rangle - \int_\Omega \langle A\nabla v, \nabla h\rangle\frac{w}{h} + \int_\Omega \langle A\nabla h, \nabla w\rangle\frac{v}{h}$$
$$- \int_\Omega \langle A\nabla h, \nabla h\rangle\frac{vw}{h^2} + \int_\Omega \langle b, \nabla v\rangle w + \int_\Omega \langle b, \nabla h\rangle\frac{vw}{h}$$
$$+ \int_\Omega (c+\omega)vw + \int_{\Gamma_1} \beta vw\, dS.$$

On the other hand, it follows from (4.44) that

$$\int_\Omega \langle b, \nabla v\rangle w - \int_\Omega \langle A\nabla v, \nabla h\rangle\frac{w}{h} = \int_\Omega \langle b - \frac{A\nabla h}{h}, \nabla v\rangle w$$
$$= \int_\Omega \langle b_h, \nabla v\rangle w + \int_\Omega \langle A\nabla h, \nabla v\rangle\frac{w}{h}$$

and hence,

$$\mathfrak{a}\left(hv, \frac{w}{h}\right) = \int_\Omega \langle A\nabla v, \nabla w\rangle + \int_\Omega \langle b_h, \nabla v\rangle w + \int_\Omega \langle A\nabla h, \nabla v\rangle\frac{w}{h}$$
$$+ \int_\Omega \langle A\nabla h, \nabla w\rangle\frac{v}{h} - \int_\Omega \langle A\nabla h, \nabla h\rangle\frac{vw}{h^2} + \int_\Omega \langle b, \nabla h\rangle\frac{vw}{h}$$
$$+ \int_\Omega (c+\omega)vw + \int_{\Gamma_1} \beta vw\, dS.$$

Thus, taking into account that

$$\int_\Omega \langle A\nabla h, \nabla v\rangle\frac{w}{h} + \int_\Omega \langle A\nabla h, \nabla w\rangle\frac{v}{h} = \int_\Omega \langle A\frac{\nabla h}{h}, \nabla(vw)\rangle,$$

$$\int_\Omega \langle A\frac{\nabla h}{h}, \nabla(vw)\rangle - \int_\Omega \langle A\nabla h, \nabla h\rangle\frac{vw}{h^2} = \int_\Omega \langle A\nabla h, \nabla\frac{vw}{h}\rangle,$$

and

$$\int_\Omega \langle A\nabla h, \nabla\frac{vw}{h}\rangle = \int_\Omega \text{div}\left(\frac{vw}{h}A\nabla h\right) - \int_\Omega \text{div}\,(A\nabla h)\frac{vw}{h}$$

$$= \int_{\Gamma_1} \frac{vw}{h}\frac{\partial h}{\partial\nu}\,dS - \int_\Omega \text{div}\,(A\nabla h)\frac{vw}{h},$$

we find that

$$\mathfrak{a}\left(hv,\frac{w}{h}\right) = \int_\Omega \langle A\nabla v, \nabla w\rangle + \int_\Omega \langle b_h, \nabla v\rangle w - \int_\Omega \text{div}\,(A\nabla h)\frac{vw}{h}$$

$$+ \int_\Omega \langle b, \nabla h\rangle\frac{vw}{h} + \int_\Omega (c+\omega)vw + \int_{\Gamma_1}\left(\beta+\frac{1}{h}\frac{\partial h}{\partial\nu}\right)vw\,dS$$

and, consequently,

$$\mathfrak{a}\left(hv,\frac{w}{h}\right)$$

$$= \int_\Omega \langle A\nabla v, \nabla w\rangle + \int_\Omega \langle b_h, \nabla v\rangle w + \int_\Omega \frac{(\mathfrak{L}+\omega)h}{h}vw + \int_{\Gamma_1}\frac{\mathfrak{B}h}{h}vw\,dS$$

$$= \int_\Omega \langle A\nabla v, \nabla w\rangle + \int_\Omega \langle b_h, \nabla v\rangle w + \int_\Omega (c_h+\omega)vw + \int_{\Gamma_1}\beta_h vw\,dS$$

$$= \mathfrak{a}_h(v,w),$$

which concludes the proof. $\qquad\square$

The next result is a corollary from Lemma 4.2.

Theorem 4.10. *Suppose B1–B3, and let h be the function (4.42), with M satisfying (4.43). Then, $v \in W_{\Gamma_0}^{1,2}(\Omega)$ is a weak solution of (4.52) if and only if $u := hv \in W_{\Gamma_0}^{1,2}(\Omega)$ is a weak solution of (4.4).*

Proof. By definition, v is a weak solution of (4.52) if and only if

$$\mathfrak{a}_h(v,\phi) = \langle\frac{f}{h},\phi\rangle_{L^2(\Omega)} \tag{4.53}$$

for all $\phi \in \mathcal{C}_{\Gamma_0}^\infty(\Omega)$. By Lemma 4.2, (4.53) can be expressed in the form

$$\mathfrak{a}(hv,\frac{\phi}{h}) = \langle f,\frac{\phi}{h}\rangle_{L^2(\Omega)},$$

or, equivalently,

$$\mathfrak{a}(u,\frac{\phi}{h}) = \langle f,\frac{\phi}{h}\rangle_{L^2(\Omega)} \qquad\text{for all}\quad \phi \in \mathcal{C}_{\Gamma_0}^\infty(\Omega). \tag{4.54}$$

As $\mathcal{C}_{\Gamma_0}^\infty(\Omega)$ is dense in the subspace of $\mathcal{C}^2(\bar\Omega)$ defined by

$$\mathcal{C}_{\Gamma_0}^2(\Omega) := \{\, u \in \mathcal{C}^2(\bar\Omega) \;:\; \operatorname{supp} \phi \subset \Omega \cup \Gamma_1 \,\},$$

it follows from Proposition 4.3 that (4.54) holds if and only if

$$\mathfrak{a}(u,\psi) = \langle f, \psi \rangle_{L^2(\Omega)} \qquad \text{for all} \quad \psi \in \mathcal{C}_{\Gamma_0}^\infty(\Omega),$$

i.e., if and only if u is a weak solution of (4.4). □

Theorem 4.11. *Suppose* B1–B3. *Then, there exists* $\omega_0 \in \mathbb{R}$ *such that, for every* $\omega \geq \omega_0$, *the problem* (4.4) *possesses a unique weak solution*

$$u := (\mathfrak{L} + \omega)^{-1} f \in W_{\Gamma_0}^{1,2}(\Omega).$$

Moreover, all the conclusions of Theorem 4.9 hold true.

Proof. Let h be the function (4.42), with M satisfying (4.43). Then, by the choice of h, $\beta_h \geq 0$. Thus, according to Theorem 4.9 and Remark 4.3, there exists $\omega_0 > 0$ such that, for each $\omega \geq \omega_0$, the problem (4.52) possesses a unique weak solution

$$v := (\mathfrak{L}_h + \omega)^{-1} \frac{f}{h} \in W_{\Gamma_0}^{1,2}(\Omega).$$

By Theorem 4.10, the function

$$u := hv = h(\mathfrak{L}_h + \omega)^{-1} \frac{f}{h} \in W_{\Gamma_0}^{1,2}(\Omega)$$

provides us with the unique weak solution of (4.4). The remaining assertions of the theorem can be easily obtained from the identity

$$(\mathfrak{L} + \omega)^{-1} = h(\mathfrak{L}_h + \omega)^{-1} \frac{\cdot}{h}$$

by applying Theorem 4.9 to the transformed problem (4.52). □

4.7 Comments on Chapter 4

The exposition of Section 4.1 is based on Chapter 7 of D. Gilbarg and N. Trudinger [79], and on Chapter 5 of L. C. Evans [60]. Some monographs and classical textbooks about Sobolev spaces are R. A. Adams [1], A. Friedman [67], J. L. Lions and E. Magenes [130], P. Malliavin [156], V. G. Maz'ja [159], C. B. Morrey [167], J. Nečas [170], H. Triebel [220], and J. Wloka [226], among others.

Although the basic concepts of measure and integral used in this book go back to H. Lebesgue [124], it was F. Riesz [189] who first introduced and studied the spaces $L^p(I)$, I being an interval, for $1 < p < \infty$. These contributions have shown to be a milestone for the development of mathematical analysis. Though inequality (4.5) is attributed to O. Hölder [101], it might be attributed to L. J. Rogers [193] too.

As in most of the specialized literature, we have attributed the spaces $W^{k,p}(\Omega)$ to S. L. Sobolev [210], though a number of spaces of weakly differentiable functions had been previously introduced by C. B. Morrey [166, 167]. Later, L. Schwartz [199] established the foundations of the abstract theory of distributions.

The most pioneering approximation results in the sprit of Theorem 4.1 go back to K. O. Friedrichs [69]. Theorem 4.1 goes back to Theorem 2.3 of J. Deny and J. L. Lions [52] for $k = 1$, and it was later extended by N. G. Meyers and J. Serrin [161] to cover the general case when $k \geq 1$.

In Definition 4.3, the concept of Lipschitz function goes back to R. Lipschitz [133], and the concept of Hölder function seem to go back to O. Hölder [101], though the Hölder spaces introduced in Section 4.1.3 might go back to A. Korn [116] and J. Schauder [195, 196]. The interested reader is advised to refer to Chapters 4 and 6 of D. Gilbarg and N. Trudinger [79] for further details.

Except for the Hölder condition on the maximal order derivative in (4.8), which goes back to C. B. Morrey [166], Theorem 4.2 is attributed to S. L. Sobolev [210, 211]. Theorem 4.3 goes back, at least, to Problem 7.7 of D. Gilbarg and N. Trudinger [79] (see Section 5.8.2 of L. C. Evans [60] for a proof). Naturally, Theorem 4.4 is based upon the Lebesgue differentiation theorem. It goes back to Theorem 12 of A. P. Calderón and A. Zygmund [37] and Theorem VIII.1 of E. M. Stein [215] (cf. Section 5.8.3 of L. C. Evans [60]). The particular feature that any locally Lipschitz continuous function must be differentiable almost everywhere is attributed to H. Rademacher [90]. The compactness results collected by Theorem 4.5 are attributed to F. Rellich [188] in case $p = 2$ and to V. I. Kondrachov [115] in the general case $p > 1$. Theorem 4.5 was later extended by E. Gagliardo [74] to cover the general case $p \geq 1$.

The exposition of Section 4.2 is based on Section 5.5 of L. C. Evans [60]. It includes the detailed proofs of the trace theorems because the results of L. C. Evans [60] were proven in the very special case when $\Gamma = \partial\Omega$, as in most of the available literature. Although the proof of Theorem 4.6 is a detailed re-elaboration of the proof of Theorem 1 in Section 5.5 of L. C. Evans [60], our proof of Theorem 4.7 differs substantially from the proof of Theorem 2 in Section 5.5 of L. C. Evans [60], which was suggested by W.

Schlag [197] (see p. 292 of [60]). Sharper results about the exact range of the trace operator can be found on p. 196 of the classical textbook of H. Brézis [29], and in J. L. Lions and E. Magenes [130].

According to Chapter 9 of D. R. Adams and L. I. Hedberg [2], the results of Section 4.2 go back at least to S. L. Sobolev's fundamental paper [209], which appeared in 1937, but was translated to English only in 1963. In his celebrated paper, S. L. Sobolev generalized some earlier work by K. O. Friedrichs [68].

The variational approach to problem (4.4) studied in Sections 4.3-4.5 can be traced back as far as the works of D. Hilbert [98] and H. Lebesgue [123], at least in the very special case when $\mathfrak{L} = -\Delta$ and $\Gamma_1 = \emptyset$. Theorem 4.8 goes back to L. Gårding [77]. In its full generality, Theorem 4.9 should be attributed to P. D. Lax and A. N. Milgram [122].

Although H. Amann [9] found some "inverse positivity" results for the classical solutions of (4.4) when Ω is of class $\mathcal{C}^2$ and $\beta \in \mathcal{C}^1(\Gamma_1)$, the existence results of Section 4.6 seem to go back to J. López-Gómez [147] for arbitrary β.

In a number of circumstances along this chapter, the condition that $\partial\Omega$ is of class $\mathcal{C}^1$ is far from necessary. For instance, for introducing the Hölder spaces of continuous functions. While, in many others, it is a crucial hypothesis for the validity of the results. For instance, Theorem 4.1 can fail when $\partial\Omega$ is not of class $\mathcal{C}^1$.

A careful reading of this chapter reveals that $\beta \in \mathcal{C}(\Gamma_1)$ can be relaxed up to consider $\beta \in L^\infty(\Gamma_1)$. Moreover, for the validity of the last assertion of Theorem 4.6 it suffices to impose $u \in W^{1,2}(\Omega) \cap \mathcal{C}(\Omega \cup \Gamma)$. But the 'principle of greatest generality' remains outside the scope of this book.

Chapter 5

Regularity of weak solutions

Throughout this chapter, besides the general assumptions of Chapter 4, we assume that $\partial\Omega$ is of class $\mathcal{C}^2$. Then, according to Lemma 2.1, Assumptions B1, B2 and B3 of Chapter 4 hold, as well as H1, H2 and H3 of Chapter 2. It should be remembered that B3 was stated at the beginning of Section 4.6. Consequently, thanks to Theorem 4.11, there exists $\omega_0 \in \mathbb{R}$ such that (4.4) possesses a unique weak solution

$$u := (\mathfrak{L} + \omega)^{-1} f \in W^{1,2}_{\Gamma_0}(\Omega)$$

for every $\omega \geq \omega_0$. The main goal of this chapter is to sketch a proof of the fact that the weak solution u must be a classical solution of (4.4), in the sense of Definition 4.1, if $f \in L^\infty(\Omega)$. By the weak counterpart of the classical theory developed in Chapter 2 that will be studied in Chapter 7, this regularity result entails the strong positivity of the resolvent operators

$$(\mathfrak{L} + \omega)^{-1}, \qquad \omega \geq \omega_0,$$

associated to $(\mathfrak{L}, \mathfrak{B}, \Omega)$. More precisely, our main goal in this chapter is to provide the reader with a first informal approach to the problem of the regularity of the weak solutions of (4.4) from a classical very fruitful perspective. Consequently, this chapter has an expository nature.

Essentially, the main goal is accomplished through the elliptic L^p theory. According to it, every weak solution u of (4.4) lies in the Sobolev space $W^{2,p}(\Omega)$ if $f \in L^p(\Omega)$, $p \geq 2$. Consequently,

$$u \in \bigcap_{p \geq 2} W^{2,p}(\Omega) \qquad \text{if} \quad f \in L^\infty(\Omega)$$

and hence, by Theorems 4.2(ii) and 4.4, $u \in \mathcal{C}^{1,1^-}(\bar{\Omega})$ and it is twice classically differentiable almost everywhere in Ω. By Proposition 4.2, u must be a solution of (4.4) in the sense of Definition 4.4. Moreover, owing to

Theorem 4.4, for any $\alpha \in \mathbb{N}^N$ with $|\alpha| \leq 2$, the classical derivative $D^\alpha u$ equals the corresponding weak derivative of u and hence

$$(\mathfrak{L} + \omega)u = f \qquad \text{a.e. in } \Omega,$$

in the classical sense. Also, as $u \in W^{1,2}_{\Gamma_0}(\Omega)$, we find from (4.16) that $\mathcal{T}_{\Gamma_0} u = 0$ and, consequently, thanks to Theorem 4.6,

$$u = 0 \qquad \text{on } \Gamma_0,$$

because $u \in W^{1,p}(\Omega) \cap C^1(\bar{\Omega})$. Similarly,

$$\frac{\partial u}{\partial \nu} + \beta u = 0 \qquad \text{on } \Gamma_1,$$

and, therefore, any weak solution u of (4.4) must be a classical solution satisfying $u \in C^{1,1^-}(\bar{\Omega})$.

The fact that any weak solution of (4.4) belongs to the Sobolev space $W^{2,p}(\Omega)$ if $f \in L^p(\Omega)$, $p \geq 2$, will be proven through the *method of continuity*, from the classical *elliptic L^p-estimates* (see Theorem 9.14 of D. Gilbarg and N. Trudinger [79]) and the corresponding result for the Laplace operator $-\Delta$, which is a very classical result in potential theory. Basically, the L^p regularity is a consequence from the theory of singular integrals of A. P. Calderón and A. Zygmund [36], attributable to K. O. Friedrichs [70] and to S. Agmon, A. Douglis, and L. Nirenberg [4]. The pioneering results of K. O. Friedrichs [70] extended a classical result of H. Weyl [225], usually referred to as the *Weyl lemma*, establishing that any weak solution of the Laplace equation

$$-\Delta u = f \in L^2$$

must be of class C^∞ almost everywhere in any domain D where $f \in C^\infty(D)$. The *method of continuity* was introduced by J. Schauder [195, 196] to obtain maximal Hölder regularity for general second order elliptic operators with Hölder continuous coefficients. Essentially, it consists in establishing a continuous deformation, or homotopy, between the original elliptic differential operator and the classical Laplace operator for transferring all the available regularity results from classical potential theory to the original setting. The underlying ideas have been a milestone for the development of modern nonlinear analysis and its applications to the theory of nonlinear partial differential equations (see [140] and Chapter 12 of [153] for some very recent advances in this direction).

This chapter pays special attention to the case of Dirichlet boundary conditions ($\Gamma_1 = \emptyset$). Consequently, as we shall use the method of continuity to derive the general L^p regularity result, this chapter might actually be regarded as the resolution of Problem 9.8 of [79]. Getting general L^p regularity results under the general boundary operator $\mathfrak{B}$ dealt with in this book is much harder technically and it remains outside the general scope of this book. Some recent general L^p a priori estimates are now available with "modern" proofs in R. Denk, M. Hiebber and J. Prüss [50, 51].

This chapter will provide with a general regularity scheme which might be substantially expanded, completed, and polished up to generate a more specialized monograph on L^p regularity theory. It is distributed as follows. Section 5.1 collects a fundamental theorem of A. P. Calderón [35] establishing the L^p regularity in $\mathbb{R}^N$. Section 5.2 discusses the validity of the underlying regularity result in bounded domains. Section 5.3 collects the L^p a priori estimates established by Theorem 9.14 of D. Gilbarg and N. Trudinger [79]. Section 5.4 studies the fundamental theorem supporting the method of continuity. Although it is Theorem 5.2 on p. 75 of D. Gilbarg and N. Trudinger [79], the reader should be aware of the fact that the proof of Theorem 5.2 of [79] contains a gap, as the operator T introduced there might not be a contraction. Section 5.5 contains the resolution of Problem 9.8 of [79]. Finally, Section 5.6 provides a scheme for extending the previous regularity results to the general case when $\Gamma_1 \neq \emptyset$. As, in our approach here, the underlying Banach spaces do actually change when the deformation parameter of the differential operator varies, one must overcome an additional highly technical difficulty which did not arise in the previous analysis of the Dirichlet problem.

5.1 $L^p(\mathbb{R}^N)$-estimates for the Laplacian

For every $f \in L^1(\mathbb{R}^N)$, the *Fourier transform* of f is the bounded and continuous function

$$\mathfrak{F}f(\xi) := \int_{\mathbb{R}^N} e^{-i\langle x,\xi\rangle} f(x)\,dx, \qquad \xi \in \mathbb{R}^N,$$

while the *inverse Fourier transform* of f is defined by

$$\mathfrak{F}^{-1}f(x) := \frac{1}{(2\pi)^N} \int_{\mathbb{R}^N} e^{i\langle x,\xi\rangle} f(\xi)\,d\xi, \qquad x \in \mathbb{R}^N.$$

If $f \in L^1(\mathbb{R}^N)$ and $\mathfrak{F}f \in L^1(\mathbb{R}^N)$, then the *Fourier inversion formula*

$$f = \mathfrak{F}^{-1}\mathfrak{F}f$$

holds. A fundamental property of the Fourier transform is the *product formula*, establishing that if $f, g \in L^1(\mathbb{R}^N)$, then the *convolution* of f and

g, denoted by $f * g$ and defined as

$$f * g(x) := \int_{\mathbb{R}^N} f(x - y)g(y)\, dy = \int_{\mathbb{R}^N} f(y)g(x - y)\, dy$$

satisfies $f * g \in L^1(\mathbb{R}^N)$ and

$$\mathfrak{F}(f * g) = \mathfrak{F}(f)\mathfrak{F}(g). \qquad (5.1)$$

Another important property of the Fourier transform establishes that

$$\mathfrak{F}\frac{\partial f}{\partial x_j}(\xi) = i\xi_j\mathfrak{F}f(\xi), \qquad 1 \le j \le N,$$

for sufficiently regular f, and hence,

$$\mathfrak{F}\Delta f(\xi) = -|\xi|^2\mathfrak{F}f(\xi), \qquad \xi \in \mathbb{R}^N. \qquad (5.2)$$

According to (5.2), the problem of solving the Laplace equation

$$(-\Delta + 1)u = f \qquad \text{in } \mathbb{R}^N \qquad (5.3)$$

can be transformed into the problem of finding some sufficiently regular function u such that

$$\mathfrak{F}u = \frac{\mathfrak{F}f}{1 + |\xi|^2} \qquad \text{in } \mathbb{R}^N. \qquad (5.4)$$

The function

$$G(x) := \mathfrak{F}^{-1}\left(\frac{1}{1 + |\xi|^2}\right)(x) = \frac{1}{(2\pi)^N}\int_{\mathbb{R}^N}\frac{e^{i\langle x,\xi\rangle}}{1 + |\xi|^2}\, d\xi \qquad (5.5)$$

is known as the *Bessel kernel*. According to Section V.3.1 of E. M. Stein [215],

$$G(x) = \frac{1}{4\pi}\int_0^\infty t^{-\frac{N}{2}}e^{-\pi\frac{|x|^2}{t} - \frac{t}{4\pi}}\, dt, \qquad x \in \mathbb{R}^N.$$

Thus, combining the Fourier inversion formula with (5.1), (5.4) and (5.5), it becomes apparent that the best candidate to solve (5.3) is

$$u(x) = G * f(x) = \int_{\mathbb{R}^N} G(x - y)f(y)\, dy, \qquad x \in \mathbb{R}^N.$$

The next fundamental result of A. P. Calderón [35] confirms it.

Theorem 5.1 (of $L^p(\mathbb{R}^N)$-regularity). *Let $1 < p < \infty$. Then, for every $f \in L^p(\mathbb{R}^N)$, the function*

$$u := G * f$$

satisfies $u \in W^{2,p}(\mathbb{R}^N)$ and it solves (5.3) a.e. in $\mathbb{R}^N$. Moreover, the map

$$
\begin{array}{ccc}
L^p(\mathbb{R}^N) & \xrightarrow{(-\Delta+1)^{-1}} & W^{2,p}(\mathbb{R}^N) \\
f & \mapsto & G * f
\end{array}
\tag{5.6}
$$

is a linear topological isomorphism. In particular, there exists a constant $C > 0$ such that

$$
C^{-1}\|f\|_{L^p(\mathbb{R}^N)} \leq \|G * f\|_{W^{2,p}(\mathbb{R}^N)} \leq C\|f\|_{L^p(\mathbb{R}^N)}
\tag{5.7}
$$

for all $f \in L^p(\mathbb{R}^N)$.

Theorem 5.1 is a consequence from the fact that the *Riesz transform*

$$
\mathfrak{R}_j f := \mathfrak{F}^{-1}\left(-i\frac{\xi_j}{|\xi|}\mathfrak{F}(f)\right), \qquad 1 \leq j \leq N,
$$

can be extended from $L^2(\mathbb{R}^N)$ to a bounded operator on $L^p(\mathbb{R}^N)$, for all $1 < p < \infty$, which is a fundamental consequence of the theory of singular integrals of A. P. Calderón and A. Zygmund [36]. Different proofs of Theorem 5.1 are given in A. P. Calderón [35] and in Theorem V.3 of E. M. Stein [215].

As a byproduct from (5.7), we find that

$$
\|u\|_{W^{2,p}(\mathbb{R}^N)} \leq C\left(\| -\Delta u\|_{L^p(\mathbb{R}^N)} + \|u\|_{L^p(\mathbb{R}^N)}\right)
\tag{5.8}
$$

for all $u \in W^{2,p}(\mathbb{R}^N)$. Indeed, for every $u \in W^{2,p}(\mathbb{R}^N)$, there exists $f \in L^p(\mathbb{R}^N)$ such that $u = G * f$ and hence,

$$
\begin{aligned}
\|u\|_{W^{2,p}(\mathbb{R}^N)} &= \|G * f\|_{W^{2,p}(\mathbb{R}^N)} \leq C\|f\|_{L^p(\mathbb{R}^N)} \\
&= C\| -\Delta u + u\|_{L^p(\mathbb{R}^N)} \\
&\leq C\left(\| -\Delta u\|_{L^p(\mathbb{R}^N)} + \|u\|_{L^p(\mathbb{R}^N)}\right).
\end{aligned}
$$

Theorem 5.1 can be generalized to cover the more general case when $-\Delta$ is substituted by a second order elliptic operator with constant coefficients of the form

$$
\mathfrak{L} = -\operatorname{div}(A\nabla \cdot), \qquad A \in \mathcal{M}_N^{\mathrm{sym}}(\mathbb{R}).
$$

In such case, the *generalized Bessel kernel*

$$
G_A := \mathfrak{F}^{-1}\left(\frac{1}{\xi^T A\xi + 1}\right)
$$

plays a similar role as the former $G = G_I$.

More generally, for every $s > 0$, one can consider the Bessel kernel

$$G_s(x) := \mathfrak{F}^{-1}\left(\frac{1}{(1+|\xi|^2)^{s/2}}\right) = \frac{1}{(2\pi)^N} \int_{\mathbb{R}^N} \frac{e^{i\,\langle x,\xi\rangle}}{(1+|\xi|^2)^{s/2}}\, d\xi$$

$$= \frac{1}{(4\pi)^{s/2}\Gamma(s/2)} \int_0^\infty t^{\frac{s-N}{2}-1} e^{-\pi\frac{|x|^2}{t} - \frac{t}{4\pi}}\, dt, \qquad x \in \mathbb{R}^N,$$

where Γ stands for gamma of L. Euler

$$\Gamma(\alpha) := \int_0^\infty t^{\alpha-1} e^{-t}\, dt, \qquad \alpha > 0,$$

as well as the associated *Bessel potential space*

$$L^{s,p}(\mathbb{R}^N) := \{\, G_s * f \;:\; f \in L^p(\mathbb{R}^N)\,\}, \qquad 1 < p < \infty,$$

equipped with the norm

$$\|G_s * f\|_{L^{s,p}(\mathbb{R}^N)} := \|f\|_{L^p(\mathbb{R}^N)}.$$

Adopting this perspective, A. P. Calderón [35] actually proved that

$$L^{k,p}(\mathbb{R}^N) = W^{k,p}(\mathbb{R}^N), \qquad 1 < p < \infty, \tag{5.9}$$

with equivalent norms, for all integer $k \geq 0$. Theorem 5.1 has already established (5.9) for $k = 2$. By (5.9), the Bessel potential spaces can be regarded as generalized Sobolev spaces incorporating fractional order derivatives of arbitrary order.

Another fruitful approach to the problem of the construction of generalized Sobolev spaces incorporating fractional order derivatives was proposed by E. N. Slobodeckii [205]. According to it, for every $s \in (0,1)$ and $p \in [1,\infty)$, the Slobodeckii space $W^{s,p}(\Omega)$ stands for the Banach space of all functions $u \in L^p(\Omega)$ such that

$$I_p(u) := \int_{\Omega \times \Omega} \frac{|u(x) - u(y)|^p}{|x-y|^{N+ps}}\, dx\, dy < \infty$$

equipped with the Slobodeckii norm

$$\|u\|_{W^{s,p}(\Omega)} = [I_p(u)]^{1/p}.$$

When $s \in \mathbb{R} \setminus \mathbb{Z}$ satisfies $s > 1$, the Sobolev–Slobodeckii space $W^{s,p}(\Omega)$ is defined as the set of functions $u \in W^{[s],p}(\Omega)$ such that $D^\alpha u \in W^{s-[s],p}(\Omega)$ for all α with $|\alpha| = [s]$. Naturally, in this case the Sobolev–Slobodeckii norm is defined by

$$\|u\|_{W^{s,p}(\Omega)} = \|u\|_{W^{[s],p}(\Omega)} + \sum_{|\alpha|=[s]} \|D^\alpha u\|_{W^{s-[s],p}(\Omega)}.$$

By using local coordinate charts, the Sobolev–Slobodeskii spaces can be extended to any sufficiently smooth manifold. In particular, it makes sense to consider the spaces $W^{s,p}(\Gamma)$ for all $s > 0$, $p \in [1, \infty)$, and $\Gamma \in \{\Gamma_0, \Gamma_1, \partial\Omega\}$. The next remark collects some useful properties of these spaces.

Remark 5.1.

(a) For every $s > 0$ and $\Gamma \in \{\Gamma_0, \Gamma_1, \partial\Omega\}$, the imbedding $W^{s,2}(\Gamma) \hookrightarrow L^2(\Gamma)$ is compact (see, e.g., Theorem 7.10 of J. Wloka [226]).

(b) For every $\Gamma \in \{\Gamma_0, \Gamma_1, \partial\Omega\}$, $k \in \{1, 2\}$ and $p > 1$, the trace operator $\mathcal{T}_\Gamma$ of $W^{1,p}(\Omega)$ on Γ satisfies

$$\mathcal{T}_\Gamma \in \mathcal{L}\left(W^{k,p}(\Omega), W^{k-\frac{1}{p},p}(\Gamma)\right)$$

and hence

$$\mathcal{B} \in \mathcal{L}\left(W^{2,p}(\Omega), W^{2-\frac{1}{p},p}(\Gamma_0) \times W^{1-\frac{1}{p},p}(\Gamma_1)\right) \qquad \text{for all } p > 1$$

(see, e.g., Theorem 8.7 of J. Wloka [226]).

5.2 $L^p(\Omega)$-estimates for the Laplacian

Subsequently, we denote by $\psi(r)$, $r > 0$, the function defined by

$$\psi(r) := \begin{cases} \dfrac{r^{2-N}}{(2-N)\omega_N}, & N > 2, \\[2mm] \dfrac{\log r}{2\pi}, & N = 2, \end{cases}$$

where ω_N stands for the surface "area" of the unit sphere in $\mathbb{R}^N$

$$\omega_N := \frac{2\pi^{N/2}}{\Gamma(N/2)}.$$

Then, the function

$$K(x, y) := \psi(|x - y|), \qquad (x, y) \in \mathbb{R}^N \times \mathbb{R}^N, \quad x \neq y,$$

provides us with the normalized *fundamental solution* of the Laplace equation, and the *Green representation formula* establishes that

$$u(y) = \int_\Omega K(x, y)\Delta u(x)\, dx + \int_{\partial\Omega}\left[u(x)\frac{\partial K}{\partial \mathbf{n}_x}(x, y) - K(x, y)\frac{\partial u}{\partial \mathbf{n}_x}\right] dS_x$$

for all $u \in \mathcal{C}^2(\bar{\Omega})$ and $y \in \Omega$ (see, e.g., Chapter 4 of F. John [108], or Section 2.4 of D. Gilbarg and N. Trudinger [79]). For a measurable function f, the integral

$$\mathcal{N}(f) := \int_\Omega K(x, \cdot)f(x)\, dx$$

is called the *Newtonian potential with density* f, and, independently of the values of u and $\partial u/\partial \mathbf{n}_x$ on $\partial \Omega$, the function

$$\mathcal{H}(u) := \int_{\partial \Omega} \left[u(x) \frac{\partial K}{\partial \mathbf{n}_x}(x, \cdot) - K(x, \cdot) \frac{\partial u}{\partial \mathbf{n}_x}(x) \right] dS_x$$

is harmonic in Ω. Thus, the Green representation formula actually establishes that any function $u \in \mathcal{C}^2(\bar{\Omega})$ differs from the Newtonian potential with density Δu in a harmonic function. Precisely,

$$u = \mathcal{N}(\Delta u) + \mathcal{H}(u). \tag{5.10}$$

As $\mathcal{H}(u)$ is real analytic in Ω, because it inherits the regularity of K outside its singularity, (5.10) entails that u and $\mathcal{N}(\Delta u)$ must have the same regularity. This is an extremely important observation, because of the next consequence of the theory of A. P. Calderón and A. Zygmund [36].

Theorem 5.2. *For every $1 < p < \infty$ and $f \in L^p(\Omega)$, the function*

$$w := - \int_{\Omega} K(x, \cdot) f(x)\, dx$$

belongs to $W^{2,p}(\Omega)$ and it satisfies

$$-\Delta w = f \qquad a.e.\ in\ \Omega.$$

According to Theorem 5.2, the Green representation formula makes sense as soon as $u \in W^{2,p}(\Omega)$ for some $1 < p < \infty$, because, in such case, the traces of u and $\partial u/\partial \mathbf{n}_x$ on $\partial \Omega$ are well defined in $L^p(\partial \Omega)$. Consequently, by (5.10) and Theorem 5.2, the most natural space for the solutions of the Laplace equation

$$-\Delta u = f \in L^p(\Omega)$$

is $W^{2,p}(\Omega)$, much like in the context of Theorem 5.1. The reader is advised to read the proof of Theorem 9.9 of D. Gilbarg and N. Trudinger [79] for a proof of Theorem 5.2.

Now, suppose $h \in \mathcal{C}^2(\bar{\Omega})$ is harmonic in Ω. Then, by the second Green identity, we obtain that

$$0 = \int_{\Omega} h \Delta u\, dx + \int_{\partial \Omega} \left(u \frac{\partial h}{\partial \mathbf{n}} - h \frac{\partial u}{\partial \mathbf{n}} \right) dS \tag{5.11}$$

for all $u \in \mathcal{C}^2(\bar{\Omega})$. Thus, setting

$$G(x, y) := K(x, y) + h(x), \qquad x, y \in \bar{\Omega}, \quad x \neq y,$$

and adding up (5.10) and (5.11) yields

$$u(y) = \int_\Omega G(\cdot, y)\Delta u \, dx + \int_{\partial\Omega} \left(u(x)\frac{\partial G}{\partial \mathbf{n}_x}(x, y) - G(x, y)\frac{\partial u}{\partial \mathbf{n}_x} \right) dS_x \quad (5.12)$$

for all $u \in C^2(\bar{\Omega})$ and $y \in \Omega$. Note that $G(x, y)$ again is a fundamental solution of the Laplace equation and, therefore, (5.12) provides us with a generalized version of the Green representation formula (5.10). Now, for a given $y \in \Omega$, let choose the harmonic function $h := h(\cdot, y)$ to satisfy

$$h(x, y) = -K(x, y) \qquad \forall \, x \in \partial\Omega.$$

Then, $G(x, y) = 0$ for all $x \in \partial\Omega$, and hence it follows from (5.12) that

$$u(y) = \int_\Omega G(x, y)\Delta u(x) \, dx + \int_{\partial\Omega} u(x)\frac{\partial G}{\partial \mathbf{n}_x}(x, y) \, dS_x$$

for all $y \in \Omega$ and $u \in C^2(\bar{\Omega})$. For such choice of h, G is called the *Green function* of the Dirichlet problem in the domain Ω, and the function

$$P(x, y) := \frac{\partial G}{\partial \mathbf{n}_x}(x, y), \qquad x \in \partial\Omega, \quad y \in \Omega,$$

is usually referred to as the *Poisson kernel* of the problem. Naturally,

$$u(y) = \int_\Omega G(x, y)f(x) \, dx + \int_{\partial\Omega} P(x, y)g(x) \, dS_x$$

provides us with the unique solution of the Dirichlet problem

$$\begin{cases} \Delta u = f & \text{in } \Omega, \\ u = g & \text{on } \partial\Omega. \end{cases}$$

In particular,

$$u(y) := -\int_\Omega G(x, y)f(x) \, dx, \qquad y \in \Omega, \quad (5.13)$$

is the unique solution of

$$\begin{cases} -\Delta u = f & \text{in } \Omega, \\ u = 0 & \text{on } \partial\Omega. \end{cases} \quad (5.14)$$

The following result of S. Agmon, A. Douglis, and L. Nirenberg [4] ascertains the regularity of u. It should be weighted against Theorem 5.1.

Theorem 5.3. *Let $1 < p < \infty$ and $f \in L^p(\Omega)$. Then, the function u defined by (5.13) satisfies $u \in W^{2,p}(\Omega) \cap W_0^{1,p}(\Omega)$ and it is the unique solution of (5.14) in the sense of Definition 4.4. Moreover, the map*

$$L^p(\Omega) \xrightarrow{(-\Delta)^{-1}} W^{2,p}(\Omega) \cap W_0^{1,p}(\Omega)$$

$$\quad (5.15)$$

$$f \quad \mapsto \quad (-\Delta)^{-1}f := u$$

*is a linear topological isomorphism. Therefore, there exists a constant $C > 0$
such that*

$$C^{-1}\|f\|_{L^p(\Omega)} \leq \|u\|_{W^{2,p}(\Omega)} \leq C\|f\|_{L^p(\Omega)} \tag{5.16}$$

for all $f \in L^p(\Omega)$.

Suppose $p \geq 2$ and $u \in W^{2,p}(\Omega) \cap W_0^{1,p}(\Omega)$ satisfies (5.14) in the sense of Definition 4.4. Then, by Proposition 4.2, u provides us with a weak solution of (5.14). Consequently, the uniqueness assertion in Theorem 5.3 is a by-product of the uniqueness of the weak solution obtained from Theorem 4.9.

Suppose $u \in W^{2,p}(\Omega) \cap W_0^{1,p}(\Omega)$. Then,

$$f := -\Delta u \in L^p(\Omega)$$

and hence, thanks to Theorem 5.3,

$$\|u\|_{W^{2,p}(\Omega)} \leq C \left(\| - \Delta u \|_{L^p(\Omega)} + \|u\|_{L^p(\Omega)} \right). \tag{5.17}$$

This inequality extends the validity of the global *elliptic estimate* (5.8) to any bounded domain Ω of class $\mathcal{C}^2$.

The proof of Theorem 5.3 is based upon the representation formula (5.13). Unfortunately, though (5.13) should imply

$$\frac{\partial^2 u}{\partial y_i \partial y_j} := -\int_\Omega \frac{\partial^2 G}{\partial y_i \partial y_j}(x, \cdot) f(x)\, dx \qquad \text{in } \Omega$$

for all $i, j \in \{1, ..., N\}$, it turns out that

$$\frac{\partial^2 G}{\partial y_i \partial y_j}(x, \cdot) \notin L^1(\Omega),$$

as a consequence of the singularity of G at $y = x$. Consequently, once again, the theory of singular integrals of A. P. Calderón and A. Zygmund [36] is imperative to overcome such a technical difficulty in the proof of the theorem, whose technical details are omitted here.

5.3   General elliptic $L^p(\Omega)$-estimates when $\Gamma_1 = \emptyset$

This section collects a general version of the global elliptic estimate (5.17). Such estimate will show the solvability of (4.4) in $L^p(\Omega)$, under homogeneous Dirichlet boundary conditions, when $\mathfrak{L} + \omega$ is coercive. To state it with the appropriate generality, consider a family $\mathcal{A}$ of symmetric matrices

$$A = (a_{ij})_{1 \leq i,j \leq N} \in \mathcal{M}_N^{\text{sym}}(W^{1,\infty}(\Omega))$$

with

$$\sup_{A \in \mathcal{A}} \|A\|_{L^\infty(\Omega)} < \infty,$$

a family $\mathcal{B}$ of vectors fields $b \in (L^\infty(\Omega))^N$ and a family $\mathcal{C}$ of functions $c \in L^\infty(\Omega)$ such that

$$\sup_{b \in \mathcal{B}} \|b\|_{L^\infty(\Omega)} < \infty, \qquad \sup_{c \in \mathcal{C}} \|c\|_{L^\infty(\Omega)} < \infty.$$

Then, the next result holds.

Theorem 5.4. *Let $\mathcal{A}, \mathcal{B}, \mathcal{C}$ as above and suppose that there is a constant $\mu > 0$ such that*

$$\xi^T A(x) \xi \geq \mu |\xi|^2 \tag{5.18}$$

for all $A \in \mathcal{A}$, $x \in \Omega$, and $\xi \in \mathbb{R}^N$. Then, for every $1 < p < \infty$, there exists a constant $C := C(p) > 0$ such that

$$\|u\|_{W^{2,p}(\Omega)} \leq C \left(\|\mathfrak{L}u\|_{L^p(\Omega)} + \|u\|_{L^p(\Omega)} \right) \tag{5.19}$$

for all $u \in W^{2,p}(\Omega) \cap W_0^{1,p}(\Omega)$, $a \in \mathcal{A}$, $b \in \mathcal{B}$ and $c \in \mathcal{C}$, where

$$\mathfrak{L} := - \operatorname{div} (A\nabla \cdot) + \langle b, \nabla \cdot \rangle + c.$$

Theorem 5.4 is a direct consequence from Theorem 9.14 of D. Gilbarg and N. Trudinger [79], where we refer for the proof.

5.4 The method of continuity

It relies upon the following theorem.

Theorem 5.5 (of continuity). *Let E be a Banach space, F a normed linear space, and $\mathfrak{L}_0, \mathfrak{L}_1 \in \mathcal{L}(E, F)$ two linear continuous operators from E into F. For every $t \in [0,1]$, set*

$$\mathfrak{L}_t := (1 - t)\mathfrak{L}_0 + t\mathfrak{L}_1 \tag{5.20}$$

and suppose that there is a constant $C > 0$ such that

$$\|x\|_E \leq C \|\mathfrak{L}_t x\|_F \tag{5.21}$$

for all $t \in [0,1]$ and $x \in E$. Then, the following conditions are equivalent:

(a) $R[\mathfrak{L}_0] = F$,
(b) $R[\mathfrak{L}_t] = F$ for all $t \in [0,1]$.

In other words, $\mathfrak{L}_0$ is onto if and only if $\mathfrak{L}_t$ is onto for all $t \in [0,1]$.

Equivalently, by the open mapping theorem, $\mathfrak{L}_0 \in \mathrm{Iso}(E,F)$ if and only if $\mathfrak{L}_t \in \mathrm{Iso}(E,F)$ for all $t \in [0,1]$, because (5.21) entails the injectivity of $\mathfrak{L}_t$ for all $t \in [0,1]$.

Proof.   Obviously, (b) implies (a). Now, suppose
$$R[\mathfrak{L}_s] = F$$
for some $s \in [0,1]$. According to (5.21), $\mathfrak{L}_s$ is injective and, hence, by the open mapping theorem,
$$\mathfrak{L}_s \in \mathrm{Iso}(E,F).$$
According to (5.20), we have that
$$\mathfrak{L}_s x = \mathfrak{L}_t x + (\mathfrak{L}_s - \mathfrak{L}_t)\, x = \mathfrak{L}_t x + (t-s)\,(\mathfrak{L}_0 - \mathfrak{L}_1)\, x$$
for all $x \in E$ and $t \in [0,1]$. Thus,
$$x = \mathfrak{L}_s^{-1}\mathfrak{L}_t x + (t-s)\mathfrak{L}_s^{-1}(\mathfrak{L}_0 - \mathfrak{L}_1)x$$
for all $x \in E$ and $t \in [0,1]$. Therefore, for any given $y \in F$, there exists $x \in E$ such that
$$y = \mathfrak{L}_t x$$
if and only if
$$x = \mathfrak{L}_s^{-1}y + (t-s)\mathfrak{L}_s^{-1}(\mathfrak{L}_0 - \mathfrak{L}_1)x,$$
which can be equivalently expressed as
$$\left[I_E - (t-s)\mathfrak{L}_s^{-1}(\mathfrak{L}_0 - \mathfrak{L}_1)\right] x = \mathfrak{L}_s^{-1}y. \qquad (5.22)$$
Owing to (5.21), we have that
$$\|\mathfrak{L}_s^{-1}z\|_E \le C\|z\|_F$$
for all $z \in F$, and hence,
$$\|(t-s)\mathfrak{L}_s^{-1}(\mathfrak{L}_0 - \mathfrak{L}_1)\|_{\mathcal{L}(E,F)} \le |t-s|C\left(\|\mathfrak{L}_0\|_{\mathcal{L}(E,F)} + \|\mathfrak{L}_1\|_{\mathcal{L}(E,F)}\right).$$
Consequently, if
$$|t-s| < \delta := \frac{1}{C\left(\|\mathfrak{L}_0\|_{\mathcal{L}(E,F)} + \|\mathfrak{L}_1\|_{\mathcal{L}(E,F)}\right)}$$
then,
$$I_E - (t-s)\mathfrak{L}_s^{-1}(\mathfrak{L}_0 - \mathfrak{L}_1) \in \mathrm{Iso}(E,E)$$
and we find from (5.22) that
$$x = \left[I_E - (t-s)\mathfrak{L}_s^{-1}(\mathfrak{L}_0 - \mathfrak{L}_1)\right]^{-1}\mathfrak{L}_s^{-1}y$$
satisfies $\mathfrak{L}_t x = y$. Therefore,
$$\mathfrak{L}_t \in \mathrm{Iso}(E,F) \qquad \text{if } |t-s| < \delta.$$
As δ does not depend on s and we can divide the interval $[0,1]$ in subintervals with length lesser than δ, the proof is complete. $\qquad\square$

5.5 Regularity of weak solutions when $\Gamma_1 = \emptyset$

Throughout this section we suppose that $p \in [2, \infty)$ and consider the Banach spaces

$$E := W^{2,p}(\Omega) \cap W_0^{1,p}(\Omega), \qquad F := L^p(\Omega),$$

the operators

$$\mathfrak{L}_0 := -\Delta + \omega \ : \ E \to F, \tag{5.23}$$

$$\mathfrak{L}_1 := -\operatorname{div}(A\nabla \cdot) + \langle b, \nabla \cdot \rangle + c + \omega \ : \ E \to F, \tag{5.24}$$

and the homotopy

$$\mathfrak{L}_t := (1-t)\mathfrak{L}_0 + t\mathfrak{L}_1 \ : \ E \to F, \qquad t \in [0,1], \tag{5.25}$$

where $\omega > 0$ is sufficiently large to guarantee the coercivity of the bilinear form associated to $(\mathfrak{L}_1, \mathfrak{D}, \Omega)$. According to (5.23) and (5.24), it is apparent that

$$\mathfrak{L}_t := -\operatorname{div}([tA + (1-t)I_{\mathbb{R}^N}]\nabla \cdot) + \langle tb, \nabla \cdot \rangle + tc + \omega$$

for all $t \in [0,1]$. Subsequently, for every $t \in [0,1]$, we consider the problem

$$\begin{cases} \mathfrak{L}_t u = f & \text{in } \Omega, \\ u = 0 & \text{on } \partial\Omega, \end{cases} \tag{5.26}$$

as well as its associated bilinear form

$$\mathfrak{a}_t(u,v) := \int_\Omega \langle [tA + (1-t)I_{\mathbb{R}^N}]\nabla u, \nabla v \rangle + \int_\Omega (\langle tb, \nabla u \rangle + tcu + \omega u)\, v.$$

The next result collects some important properties of this homotopy.

Theorem 5.6. *Let $\omega \in \mathbb{R}$ be such that $\mathfrak{a}_1$ is coercive and consider the family $\{\mathfrak{L}_t\}_{t \in [0,1]}$ defined through (5.25). Let $\mu > 0$ denote the ellipticity constant of $\mathfrak{L}_1$. Then:*

(a) *For every $t \in [0,1]$, $\mathfrak{L}_t$ is uniformly elliptic in Ω with constant $\min\{1, \mu\} > 0$.*

(b) *There exists a constant $\alpha > 0$ such that*

$$|\mathfrak{a}_t(u,u)| \geq \alpha \|u\|_{W_0^{1,2}(\Omega)}^2$$

for all $u \in W_0^{1,2}(\Omega)$ and $t \in [0,1]$. In particular, the bilinear form $\mathfrak{a}_t$ is coercive for all $t \in [0,1]$.

(c) *There exists a constant $C > 0$ such that*

$$\|u\|_{W^{2,p}(\Omega)} \leq C \left(\|\mathfrak{L}_t u\|_{L^p(\Omega)} + \|u\|_{L^p(\Omega)} \right)$$

for all $u \in W^{2,p}(\Omega) \cap W_0^{1,p}(\Omega)$ and $t \in [0,1]$.

Proof. For every $t \in [0,1]$, the matrix of the coefficients of the principal part of $\mathfrak{L}_t$ is given by

$$A_t := tA + (1-t)I_{\mathbb{R}^N}.$$

Obviously, for every $\xi \in \mathbb{R}^N$ and $t \in [0,1]$, we have that

$$\begin{aligned}
\xi^T A_t \xi &= t\xi^T A\xi + (1-t)|\xi|^2 \\
&\geq (1-t+t\mu)|\xi|^2 \\
&\geq \min\{1,\mu\}|\xi|^2.
\end{aligned}$$

This ends the proof of Part (a).

Subsequently, we denote by α_j the coercivity constant of $\mathfrak{L}_j$, $j \in \{0,1\}$. Obviously,

$$\mathfrak{a}_t = (1-t)\mathfrak{a}_0(u,v) + t\mathfrak{a}_1 \qquad \text{for all } t \in [0,1].$$

Consequently,

$$\begin{aligned}
\mathfrak{a}_t(u,u) &= (1-t)\mathfrak{a}_0(u,u) + t\mathfrak{a}_1(u,u) \\
&\geq (1-t)\alpha_0\|u\|_{W_0^{1,2}(\Omega)}^2 + t\alpha_1\|u\|_{W_0^{1,2}(\Omega)}^2 \\
&\geq \min\{\alpha_0,\alpha_1\}\|u\|_{W_0^{1,2}(\Omega)}^2
\end{aligned}$$

for all $u \in W_0^{1,2}(\Omega)$ and $t \in [0,1]$. The proof of Part (b) is complete.

Finally, thanks to Theorem 5.4, Part (c) is a direct consequence from Part (a). This ends the proof. $\square$

Obviously, the operators $\mathfrak{L}_t : E \to F$ defined by (5.25) are linear and continuous, as there is a constant $C > 0$ such that

$$\|\mathfrak{L}_t u\|_{L^p(\Omega)} \leq C\|u\|_{W^{2,p}(\Omega)}$$

for all $t \in [0,1]$ and $u \in W^{2,p}(\Omega)$. The next result establishes that they are injective.

Proposition 5.1. $\mathfrak{L}_t \in \mathcal{L}(E,F)$ *is injective for all $t \in [0,1]$.*

Proof. Let $u, v \in E$ such that

$$\mathfrak{L}_t u = \mathfrak{L}_t v$$

for some $t \in [0, 1]$. Then, $w := u - v \in E$ satisfies $\mathfrak{L}_t w = 0$ in Ω and, hence, by Proposition 4.2, w is a weak solution of

$$\begin{cases} \mathfrak{L}_t w = 0 & \text{in } \Omega, \\ w = 0 & \text{on } \partial\Omega. \end{cases}$$

On the other hand, according to Theorems 5.6 and 4.9, for every $f \in L^p(\Omega) \left(\subset L^2(\Omega) \right)$ and $t \in [0, 1]$, the problem (5.26) has a unique weak solution. Therefore, $w = 0$, and hence, $u = v$. $\qquad\square$

Thanks to Proposition 5.1,

$$\mathfrak{L}_t \in \mathrm{Iso}(E, \mathfrak{L}_t(E)) \qquad \text{for all} \quad t \in [0, 1].$$

Therefore, for every $t \in [0, 1]$, there exists a constant $C_t > 0$ such that

$$\|u\|_{W^{2,p}(\Omega)} \leq C_t \|\mathfrak{L}_t u\|_{L^p(\Omega)} \qquad \text{for all} \quad u \in E.$$

The next result shows that C_t can be chosen so that

$$\sup_{t \in [0,1]} C_t < \infty$$

and infers a fundamental property from this estimate.

Theorem 5.7. *There exists a constant $C > 0$ such that*

$$\|u\|_{W^{2,p}(\Omega)} \leq C \|\mathfrak{L}_t u\|_{L^p(\Omega)} \tag{5.27}$$

for all $u \in E$ and $t \in [0, 1]$. Therefore, according to Theorem 5.5,

$$R[\mathfrak{L}_t] = R[\mathfrak{L}_0] \qquad \text{for all} \quad t \in [0, 1]. \tag{5.28}$$

Proof. The proof of (5.27) proceeds by contradiction. Suppose that there exists a sequence

$$(t_n, u_n) \in [0, 1] \times \left(W^{2,p}(\Omega) \cap W_0^{1,p}(\Omega) \setminus \{0\} \right), \qquad n \geq 1,$$

such that

$$\lim_{n \to \infty} \frac{\|u_n\|_{W^{2,p}(\Omega)}}{\|\mathfrak{L}_{t_n} u_n\|_{L^p(\Omega)}} = \infty. \tag{5.29}$$

As $[0, 1]$ is compact, we can assume, without loss of generality, that

$$\lim_{n \to \infty} t_n = t^* \in [0, 1].$$

Subsequently, we set

$$v_n := \frac{u_n}{\|u_n\|_{W^{2,p}(\Omega)}}, \qquad n \geq 1. \tag{5.30}$$

Then,

$$v_n \in E, \qquad \|v_n\|_{W^{2,p}(\Omega)} = 1, \qquad n \geq 1, \tag{5.31}$$

and (5.29) can be expressed as

$$\lim_{n \to \infty} \|\mathfrak{L}_{t_n} v_n\|_{L^p(\Omega)} = 0. \tag{5.32}$$

According to (5.31) and (4.11), we can suppose, without loss of generality, that there is $v^* \in L^p(\Omega)$ such that

$$\lim_{n \to \infty} \|v_n - v^*\|_{L^p(\Omega)} = 0. \tag{5.33}$$

Owing to Theorem 5.6(c), there exists a constant $C > 0$ such that

$$\|v_n - v_m\|_{W^{2,p}(\Omega)} \leq C \left(\|\mathfrak{L}_{t_n}(v_n - v_m)\|_{L^p(\Omega)} + \|v_n - v_m\|_{L^p(\Omega)} \right) \tag{5.34}$$

for all $n, m \geq 1$. By (5.33),

$$\lim_{n,m \to \infty} \|v_n - v_m\|_{L^p(\Omega)} = 0.$$

Moreover, for every $n, m \geq 1$,

$$\mathfrak{L}_{t_n}(v_n - v_m) = \mathfrak{L}_{t_n} v_n - \mathfrak{L}_{t_m} v_m + (\mathfrak{L}_{t_m} - \mathfrak{L}_{t_n}) v_m, \tag{5.35}$$

and, due to (5.25),

$$\mathfrak{L}_{t_m} - \mathfrak{L}_{t_n} = (t_m - t_n)(\mathfrak{L}_1 - \mathfrak{L}_0).$$

Thus, we find from (5.31) that

$$\| (\mathfrak{L}_{t_m} - \mathfrak{L}_{t_n}) v_m \|_F \leq |t_m - t_n| \|\mathfrak{L}_0 - \mathfrak{L}_1\|_{\mathcal{L}(E,F)}$$

and, therefore, it follows from (5.32) and (5.35) that

$$\lim_{n,m \to \infty} \|\mathfrak{L}_{t_n}(v_n - v_m)\|_{L^p(\Omega)} = 0.$$

Consequently, by (5.34), $\{v_n\}_{n \geq 1}$ is a Cauchy sequence in $W^{2,p}(\Omega)$. Thus, according to (5.33), v^* does actually belong to E and

$$\lim_{n \to \infty} \|v_n - v^*\|_E = 0. \tag{5.36}$$

In particular, by (5.31),

$$\|v^*\|_{W^{2,p}(\Omega)} = 1. \tag{5.37}$$

Obviously, (5.36) implies that

$$\lim_{n\to\infty} \|\mathfrak{L}_j v_n - \mathfrak{L}_j v^*\|_{L^p(\Omega)} = 0, \qquad j \in \{0,1\},$$

and, consequently,

$$\lim_{n\to\infty} \|\mathfrak{L}_{t_n} v_n - \mathfrak{L}_{t^*} v^*\|_{L^p(\Omega)} = 0. \tag{5.38}$$

Indeed, we have that

$$\|\mathfrak{L}_{t_n} v_n - \mathfrak{L}_{t^*} v^*\|_{L^p(\Omega)} \leq \|\mathfrak{L}_{t_n} v_n - \mathfrak{L}_{t^*} v_n\|_{L^p(\Omega)} + \|\mathfrak{L}_{t^*} v_n - \mathfrak{L}_{t^*} v^*\|_{L^p(\Omega)}$$

for all $n \geq 1$. Moreover, by (5.25) and (5.31),

$$\lim_{n\to\infty} \|\mathfrak{L}_{t_n} v_n - \mathfrak{L}_{t^*} v_n\|_{L^p(\Omega)} = 0,$$

and, due to (5.36),

$$\lim_{n\to\infty} \|\mathfrak{L}_{t^*} v_n - \mathfrak{L}_{t^*} v^*\|_{L^p(\Omega)} = 0.$$

Therefore, (5.38) holds, and we conclude from (5.32) and (5.38) that

$$\mathfrak{L}_{t^*} v^* = 0.$$

Consequently, thanks to Proposition 5.1, $v^* = 0$. This contradicts (5.37) and concludes the proof. $\square$

A further application of Theorem 5.5 to the operators

$$\mathfrak{M}_0 := -\Delta : \; E \to F,$$
$$\mathfrak{M}_1 := -\Delta + \omega : \; E \to F,$$

through the homotopy

$$\mathfrak{M}_t := (1-t)\mathfrak{M}_0 + t\mathfrak{M}_1, \qquad t \in [0,1],$$

reveals that

$$R[\mathfrak{L}_0] = R[-\Delta + \omega] = R[-\Delta].$$

Therefore, according to Theorems 5.3 and 5.7, we find that

$$R[\mathfrak{L}_t] = L^p(\Omega) \qquad \text{for all } \; t \in [0,1], \tag{5.39}$$

and, consequently, the next fundamental result holds.

Theorem 5.8 (of regularity of weak solutions). *Let $\omega \in \mathbb{R}$ be such that $\mathfrak{a}_1$ is coercive. Then, for every $p \in [2, \infty)$ and $f \in L^p(\Omega)$, the unique weak solution of*

$$\begin{cases} (\mathfrak{L} + \omega)u = f & in \ \Omega, \\ u = 0 & on \ \partial\Omega, \end{cases} \tag{5.40}$$

satisfies

$$u \in W^{2,p}(\Omega) \cap W_0^{1,p}(\Omega). \tag{5.41}$$

Therefore, by Proposition 4.2, u solves (5.40) in the sense of Definition 4.4. Furthermore, if $f \in L^\infty(\Omega)$, then $u \in W^{2,\infty^-}(\Omega)$, where

$$W^{2,\infty^-}(\Omega) := \bigcap_{p>1} W^{2,p}(\Omega)$$

and, consequently,

i) $u \in \mathcal{C}_0^1(\bar{\Omega}) \cap \mathcal{C}^{1,1^-}(\bar{\Omega})$.
ii) u *is twice classically differentiable almost everywhere in* Ω.
iii) u *is a classical solution of (5.40), in the sense of Definition 4.1.*

Proof. By Theorem 4.9, the problem (5.40) has a unique weak solution. According to (5.39), there exists $u \in W^{2,p}(\Omega) \cap W_0^{1,p}(\Omega)$ satisfying (5.40) in the sense of Definition 4.4. By Proposition 4.2, u must be a weak solution of (5.40). Therefore, the unique weak solution of (5.40) satisfies (5.41) and it solves (5.40) in the sense of Definition 4.4.

The remaining assertions of the theorem are direct consequences from Theorems 4.2(ii), 4.4 and 4.6. Indeed, let $f \in L^\infty(\Omega)$. Then, the unique weak solution of (5.40) satisfies

$$u \in \bigcap_{p=2}^{\infty} \left[W^{2,p}(\Omega) \cap W_0^{1,p}(\Omega) \right] \subset W^{2,\infty^-}(\Omega).$$

On the other hand, by Theorem 4.2(ii),

$$W^{2,p}(\Omega) \subset \mathcal{C}^{1,1-\frac{N}{p}}(\bar{\Omega})$$

for all $p > N$. Thus,

$$u \in W^{2,\infty^-}(\Omega) \subset \bigcap_{p>N} \mathcal{C}^{1,1-\frac{N}{p}}(\bar{\Omega}) = \mathcal{C}^{1,1^-}(\bar{\Omega}).$$

Moreover, by Theorem 4.4, u is twice classically differentiable a.e. in Ω and the classical derivative $D^\alpha u$ equals the corresponding weak derivative a.e. in Ω for all multi-index $\alpha \in \mathbb{N}^N$ with $|\alpha| \le 2$. Consequently,

$$(\mathfrak{L} + \omega)u = f \qquad \text{almost everywhere in} \ \Omega$$

in the classical sense, as u solves (5.40) in the sense of Definition 4.4.

Finally, since $u \in \mathcal{C}(\bar{\Omega})$, Theorem 4.6 implies that

$$0 = \mathcal{T}_{\partial\Omega} u = u|_{\partial\Omega}$$

and, hence, u is a classical solution of (5.40), as discussed by Definition 4.1. This ends the proof. $\qquad\qquad\qquad\qquad\qquad\qquad\qquad\qquad\square$

5.6 A first glance to the general case when $\Gamma_1 \neq \emptyset$

As for the existence of weak solutions in Section 4.6, the regularity of the weak solutions of (4.4) in the general case when β changes of sign can be easily derived from Theorem 4.10 through the regularity result for the special case when

$$\beta \geq 0. \tag{5.42}$$

Therefore, (5.42) will be assumed throughout this section.

When $\Gamma_1 \neq \emptyset$, Theorem 5.5 cannot be applied straightaway to get the $W^{2,p}(\Omega)$-regularity of the weak solutions of (4.4) for $f \in L^p(\Omega)$, $p \geq 2$, as we have done in the previous section for the special case when $\mathfrak{B} = \mathfrak{D}$ ($\Gamma_1 = \emptyset$), because the method of continuity now involves a homotopy of both differential operators and boundary operators on Γ_1 and, as a result, the Banach space E where the operators $\mathfrak{L}_t$, $t \in [0,1]$, are defined depends on t. This technical difficulty makes the analysis of the underlying regularity problem extraordinarily more involved than in case $\Gamma_1 = \emptyset$, however Theorem 5.5 also applies to get the regularity result in the general case. As giving all the technical details of the proofs of the necessary results to get the $W^{2,p}(\Omega)$-regularity of the weak solutions of (4.4) in the general case when $\Gamma_1 \neq \emptyset$ lies outside the scope of this book, in this section we will restrict ourselves to sketch an approach to this problem and to state the main result which will be used later.

Subsequently, we fix $p \in [2, \infty)$ and consider, for every $t \in [0,1]$, the boundary operator

$$\mathfrak{B}_t : W^{2,p}(\Omega) \to L^p(\Gamma_1)$$

defined through

$$\mathfrak{B}_t u := \langle \nabla u, tA\mathbf{n} + (1-t)\mathbf{n} \rangle + \beta u, \qquad u \in W^{2,p}(\Omega), \tag{5.43}$$

in the sense of traces. According to Theorem 4.6,

$$\mathfrak{B}_t \in \mathcal{L}\left(W^{2,p}(\Omega), L^p(\Gamma_1)\right) \qquad \text{for all } t \in [0,1].$$

Thus,

$$N[\mathfrak{B}_t] = \mathfrak{B}_t^{-1}(0)$$

is a closed linear subspace of $W^{2,p}(\Omega)$ and hence, we can consider the curve of Banach subspaces of $W^{2,p}(\Omega)$ defined by

$$E_t := W^{2,p}(\Omega) \cap W_{\Gamma_0}^{1,p}(\Omega) \cap N[\mathfrak{B}_t], \qquad t \in [0,1], \qquad (5.44)$$

as well as the associated homotopy

$$\mathfrak{L}_t : E_t \to F := L^p(\Omega)$$

defined by

$$\mathfrak{L}_t u := -\mathrm{div}\,([tA + (1-t)I_{\mathbb{R}^N}]\nabla u) + t\langle b, \nabla u\rangle + tcu + \omega u, \qquad (5.45)$$

for all $u \in E_t$ and $t \in [0,1]$; $\mathfrak{L}_t$ is well defined in $W^{2,p}(\Omega)$ and

$$\mathfrak{L}_t u = t\mathfrak{L}_1 u + (1-t)\mathfrak{L}_0 u \qquad (5.46)$$

for all $u \in W^{2,p}(\Omega)$ and $t \in [0,1]$. The homotopy $\mathfrak{L}_t : E_t \to F$ between $\mathfrak{L}_0 : E_0 \to F$ and $\mathfrak{L}_1 : E_1 \to F$ establishes a natural deformation between the solutions of (4.4) and the solutions of

$$\begin{cases} (-\Delta + \omega)u = f & \text{in } \Omega, \\ u = 0 & \text{on } \Gamma_0, \\ \frac{\partial u}{\partial \mathbf{n}} + \beta u = 0 & \text{on } \Gamma_1. \end{cases} \qquad (5.47)$$

By (5.42), it follows from Theorem 4.8 that there exists $\omega_0 > 0$ such that the bilinear forms associated to $\mathfrak{L}_0$ and $\mathfrak{L}_1$ are coercive for all $\omega \geq \omega_0$. Throughout the rest of this section we will assume that $\omega \geq \omega_0$.

According to L. N. Slobodeckii [205], F. E. Browder [30–32], S. Agmon, A. Douglis, and L. Nirenberg [4] (see Section 7 of H. Amann [9] too), it is well known that

$$R[\mathfrak{L}_0] = L^p(\Omega). \qquad (5.48)$$

Moreover, the next result holds. Essentially, it establishes the existence of a uniformly bounded family of isomorphisms between every pair of E_t's.

Theorem 5.9. *For every* $t \in [0,1]$, *there exists* $\Phi_t \in \mathrm{Iso}(E_0, E_t)$ *such that* $\Phi_0 = I$, *the map*

$$[0,1] \longrightarrow \mathcal{L}(E_0, W^{2,p}(\Omega))$$
$$t \longmapsto \Phi_t$$

is continuous, and there is a constant $C > 0$ *such that*

$$\|\Phi_t^{-1}\|_{\mathcal{L}(E_t, E_0)} \leq C \qquad \text{for all } t \in [0,1]. \qquad (5.49)$$

Proof. By (5.43), we have that

$$\mathfrak{B}_0 = \mathfrak{B}_t - t\left(\mathfrak{B}_1 - \mathfrak{B}_0\right) \qquad \text{in} \quad W^{2,p}(\Omega) \cap W^{1,p}_{\Gamma_0}(\Omega) \tag{5.50}$$

for all $t \in [0,1]$. By some advanced results in trace theory (e.g., R. A. Adams [1], P. L. Butzer and H. Berens [34], J. Nečas [170], H. Triebel [220] and J. Wloka [226]), we have that

$$R[\mathfrak{B}_t] = W^{1-1/p,p}(\Gamma_1) \qquad \forall\, t \in [0,1],$$

and, in particular,

$$R[\mathfrak{B}_t] = R[\mathfrak{B}_0] \qquad \text{for all } t \in [0,1]$$

(see Remark 5.1). By some sophisticated inverse trace theorems (see, e.g., Lemma 5.1 of H. Amann [9], Lemma II.5.8 of J. Nečas [170], and J. L. Lions [129]), it is apparent that $N[\mathfrak{B}_t]$ possesses a topological complement in $W^{2,p}(\Omega) \cap W^{1,p}_{\Gamma_0}(\Omega)$ for all $t \in [0,1]$. Equivalently, $\mathfrak{B}_t$ admits a right inverse for all $t \in [0,1]$. In other words, there exists a linear continuous operator

$$\mathfrak{R}_t : R[\mathfrak{B}_t] \to W^{2,p}(\Omega) \cap W^{1,p}_{\Gamma_0}(\Omega)$$

such that

$$\mathfrak{B}_t \mathfrak{R}_t q = q \qquad \text{for all } q \in R[\mathfrak{B}_t].$$

Consequently, (5.50) can be written in the form

$$\mathfrak{B}_0 = \mathfrak{B}_t \left[I - t\mathfrak{R}_t \left(\mathfrak{B}_1 - \mathfrak{B}_0\right) \right] \qquad \text{in} \quad W^{2,p}(\Omega) \cap W^{1,p}_{\Gamma_0}(\Omega)$$

and, hence, for every $t \in [0,1]$, we find that

$$u \in E_0 \iff u - t\mathfrak{R}_t \left(\mathfrak{B}_1 - \mathfrak{B}_0\right) u \in E_t.$$

Therefore, the most reasonable choice for Φ_t is

$$\Phi_t u := u - t\mathfrak{R}_t \left(\mathfrak{B}_1 - \mathfrak{B}_0\right) u, \qquad u \in E_0, \ t \in [0,1]. \tag{5.51}$$

The most delicate part of the proof of Theorem 5.9 is the construction of the right inverses $\mathfrak{R}_t$ so that the operators (5.51) satisfy (5.49). The technical details are left outside of this book. $\qquad \square$

As $W^{2,p}(\Omega) \cap W^{1,p}_{\Gamma_0}(\Omega)$ is a u.c. B-space, according to Theorem 3.10, there exists a continuous projection $\mathcal{P}_{N[\mathfrak{B}_t]}$ of $W^{2,p}(\Omega) \cap W^{1,p}_{\Gamma_0}(\Omega)$ on $N[\mathfrak{B}_t]$. Unfortunately, it is unknown whether $\mathcal{P}_{N[\mathfrak{B}_t]}$ is a linear operator and, consequently, in the proof of Theorem 5.9, one must invoke to an inverse trace theorem to get the existence of a linear continuous projection on $N[\mathfrak{B}_t]$.

Next, we will use Theorem 5.9 to show how (5.48) implies that

$$R[\mathfrak{L}_t] = L^p(\Omega) \qquad \text{for all } t \in [0,1]. \tag{5.52}$$

Adapting the proof of Proposition 5.1, it becomes apparent that

$$\mathfrak{L}_t : E_t \to L^p(\Omega)$$

is injective for all $t \in [0,1]$. Thus, there exists a constant $C_t > 0$ such that

$$\|x_t\|_{W^{2,p}(\Omega)} \leq C_t \|\mathfrak{L}_t x_t\|_{L^p(\Omega)} \qquad \text{for all } x_t \in E_t, \quad 0 \leq t \leq 1.$$

Much within the spirit of Theorems 5.4 and 5.7, the constant C_t can be chosen so that

$$\sup_{t \in [0,1]} C_t < \infty.$$

Indeed, the next result holds.

Theorem 5.10. *There is a constant $C > 0$ such that*

$$\|x_t\|_{W^{2,p}(\Omega)} \leq C \|\mathfrak{L}_t x_t\|_{L^p(\Omega)} \tag{5.53}$$

for all $t \in [0,1]$ and $x_t \in E_t$.

The proof of Theorem 5.10 combines the global interior elliptic estimates of the Appendix of H. Amann and J. López-Gómez [13] with the boundary estimates of H. Amann [12] (see Remark A3.2 of [13]). The details of the proof are not included.

Subsequently, for every $t \in [0,1]$, we denote by $\mathfrak{M}_t$ the operator

$$\mathfrak{M}_t = \mathfrak{L}_t \Phi_t : E_0 \to L^p(\Omega).$$

By construction, $\mathfrak{M}_t \in \mathcal{L}(E_0, L^p(\Omega))$ and it is injective for all $t \in [0,1]$. Also, since $\Phi_0 = I$, we find from (5.48) that

$$R[\mathfrak{M}_0] = R[\mathfrak{L}_0] = L^p(\Omega),$$

and hence, by the open mapping theorem,

$$\mathfrak{M}_0 \in \text{Iso}(E_0, L^p(\Omega)).$$

Moreover, according to Theorem 5.9 and (5.53), there exists a constant $C > 0$ such that

$$\|x\|_{E_0} \leq C \|\mathfrak{M}_t x\|_{L^p(\Omega)} \tag{5.54}$$

for all $t \in [0,1]$ and $x \in E_0$. More generally, suppose that

$$\mathfrak{M}_s \in \text{Iso}(E_0, L^p(\Omega))$$

for some $s \in [0,1]$. Then,

$$\mathfrak{M}_s x = \mathfrak{M}_t x + (\mathfrak{M}_s - \mathfrak{M}_t)\, x$$

for all $t \in [0,1]$ and $x \in E_0$. On the other hand,

$$\begin{aligned}
\mathfrak{M}_s - \mathfrak{M}_t &= \mathfrak{L}_s \Phi_s - \mathfrak{L}_t \Phi_t \\
&= (\mathfrak{L}_s - \mathfrak{L}_t)\, \Phi_s + \mathfrak{L}_t\, (\Phi_s - \Phi_t) \\
&= (s-t)(\mathfrak{L}_1 - \mathfrak{L}_0)\Phi_s + \mathfrak{L}_t\, (\Phi_s - \Phi_t).
\end{aligned}$$

Thus, for a given $y \in L^p(\Omega)$, there exists $x \in E_0$ such that $\mathfrak{M}_t x = y$ if and only if

$$[\mathfrak{M}_s - (s-t)(\mathfrak{L}_1 - \mathfrak{L}_0)\Phi_s - \mathfrak{L}_t\, (\Phi_s - \Phi_t)]\, x = y. \tag{5.55}$$

As the map $t \mapsto \Phi_t$ is uniformly continuous, there exists $\delta > 0$ such that $|t-s| \le \delta$ implies

$$\left\| (s-t)(\mathfrak{L}_1 - \mathfrak{L}_0)\Phi_s + \mathfrak{L}_t\, (\Phi_s - \Phi_t)\, \right\| \le \frac{1}{2C},$$

where C is the constant of (5.54). As, due to (5.54), $\|\mathfrak{M}_s^{-1}\| \le C$, for such a choice of δ, the operator on the left-hand side of (5.55) is invertible and, therefore, there is a unique x satisfying (5.55) whenever $|s-t| \le \delta$. As $[0,1]$ can be divided into a finite number of subintervals with length less than δ, it becomes apparent that

$$R[\mathfrak{M}_t] = L^p(\Omega) \qquad \text{for all } t \in [0,1].$$

Consequently, (5.52) holds. As a by-product, the next counterpart of Theorem 5.8 holds. Note that, thanks to (5.42) and Theorem 4.8, there exists ω_0 such that the bilinear form $\mathfrak{a}_1$ associated to $\mathfrak{L}_1$ is coercive for all $\omega \ge \omega_0$.

Theorem 5.11 (of regularity of weak solutions). *Let $\omega \in \mathbb{R}$ be such that $\mathfrak{a}_1$ is coercive. Then, for every $p \ge 2$ and $f \in L^p(\Omega)$, the unique weak solution of*

$$\begin{cases} (\mathfrak{L} + \omega)u = f & \text{in } \Omega, \\ \mathfrak{B}u = 0 & \text{on } \partial\Omega, \end{cases} \tag{5.56}$$

satisfies

$$u \in W^{2,p}(\Omega) \cap W^{1,p}_{\Gamma_0}(\Omega) \cap N[\mathfrak{B}_1].$$

Therefore, by Proposition 4.2, u solves (5.56) in the sense of Definition 4.4. Furthermore, if $f \in L^\infty(\Omega)$, then $u \in W^{2,\infty^-}(\Omega)$ and, consequently,

i) $u \in \mathcal{C}^1_{\Gamma_0}(\bar{\Omega}) \cap \mathcal{C}^{1,1^-}(\bar{\Omega})$.

ii) u *is twice classically differentiable almost everywhere in* Ω.

iii) u *is a classical solution of* (5.56), *in the sense of Definition 4.1.*

In the statement of Theorem 5.11, we are denoting by $C^1_{\Gamma_0}(\bar{\Omega})$ the closure of $C^\infty_{\Gamma_0}(\Omega)$ in $C^1(\bar{\Omega})$ (see Section 4.1.1, if necessary).

It should be noted that Theorem 5.11 also follows from the abstract theory of R. Denk, M. Hieber and J. Prüss [51].

5.7   Comments on Chapter 5

Far reaching results about the Weyl lemma were given by L. Hörmander [105].

Section 5.1 was inspired by Chapter 1 of D. R. Adams and L. I. Hedberg [2]. The Bessel potential spaces were systematically studied by N. Aronszajn and coworkers (see N. Aronszajn and K. T. Smith [15], N. Aronszajn F. Mulla and P. Szeptycki [16], and N. Aronszajn [14]). As already commented in Chapter 4, specialized monographs and textbooks discussing the Sobolev–Slobodeskii spaces are those of R. A. Adams [1], A. Friedman [67], J. L. Lions and E. Magenes [130], P. Malliavin [156], V. G. Maz'ja [159], C. B. Morrey [167], J. Nečas [170], H. Triebel [220], and J. Wloka [226], among others.

Section 5.2 is based on Chapter 4 of F. John [108], Theorem 9.9 of D. Gilbarg and N. Trudinger [79], and Theorem IX.32 of H. Brézis [29]. Section 5.3 provides a statement of Theorem 9.14 of D. Gilbarg and N Trudinger [79].

Section 5.4 is based on Section 5.2 of D. Gilbarg and N Trudinger [79], however the proof of Theorem 5.2 of [79] is wrong, because the map

$$Tx := L_s^{-1}y + (t-s)L_s^{-1}(L_0 - L_1)x$$

defined on p. 75 of [79] is far from being contractive if

$$|s-t| < \delta = [C\|L_0\| + \|L_1\|]^{-1},$$

as claimed by D. Gilbarg and N Trudinger therein. Theorem 5.4 does actually ascertain whether the interval $[\mathfrak{L}_0, \mathfrak{L}_1]$ of $\mathcal{L}(E,F)$ satisfies

$$[\mathfrak{L}_0, \mathfrak{L}_1] \subset \mathrm{Iso}(E,F).$$

This is a very classical result, which might be attributed to S. Banach [22]. As already mentioned in the presentation of this chapter, Section 5.5 consists of the resolution of Problem 9.8 of [79].

The underlying ideas outlined in Section 5.6 might be considerably expanded up to span a more specialized monograph on regularity of weak solutions. Within the context of Hölder regularity theory, the method of continuity for rather general classes of mixed boundary value problems had already been used by D. Gilbarg and N Trudinger [79] in the proof of Theorem 6.31 of [79].

Besides D. Gilbarg and N Trudinger [79], other classical texts covering the most fundamental features about the regularity of the weak solutions of elliptic partial differential equations are those of L. Hörmander [106] and Morrey [167]. Some optimal L^p a priori estimates are available in R. Denk, M. Hiebber and J. Prüss [50, 51], where the reader should refer to for further details.

Chapter 6

The Krein–Rutman theorem

Essentially, the theorem of M. G. Krein and M. A. Rutman culminated a series of generalizations of the celebrated results of O. Perron [175] and G. Frobenius [71, 73] about the spectrum of matrices with non-negative entries, establishing that the spectral radius of an $n \times n$ matrix with positive entries is an eigenvalue with a component-wise positive eigenvector in $\mathbb{R}^n$. Essentially, the Krein–Rutman theorem extended the Perron–Frobenius theorem to deal with strongly positive compact endomorphisms in ordered Banach spaces.

For a historical account and a bibliography of earlier work on compact positive operators, the reader should refer to the original monograph of M. G. Krein and M. A. Rutman [120].

Not surprisingly, compactness and positivity are crucial in all these extensions. Basically, the compactness localizes the spectrum and provides with the Fredholm alternative, while the positivity entails the spectral radius of the operator to be the unique eigenvalue on the spectral circle. Combining both, compactness and positivity, provides with the convergence of the monotone schemes approximating the spectral radius and its associated positive eigenvector.

6.1 Orderings. Ordered Banach spaces

Given a nonempty set X, an *ordering* in X is a relation in X which is reflexive, transitive and anti-symmetric. In this book, these relations are denoted by $\leq$. A nonempty set X together with an ordering, $(X, \leq)$, will be called an *ordered set*. Given an ordered set $(X, \leq)$ and a point $(x, y) \in X^2$, it is said that $y \geq x$ if $x \leq y$, while we write $x < y$ (or, equivalently, $y > x$)

if $x \leq y$ and $x \neq y$, The set

$$[x, y] := \{ z \in X \ : \ x \leq z \leq y \}$$

is called the *order interval* defined by x and y. Note that $[x, y] \neq \emptyset$ if, and only if, $x \leq y$.

Let B be a subset of X. Then,

- B is said to be *bounded* if $B \subset [x, y]$ for some $x, y \in X$.
- B is said to be *convex* if $[x, y] \subset B$ for every $x, y \in B$.

Given two arbitrary ordered sets $(X, \leq)$ and $(Y, \leq)$ (both ordering are denoted by $\leq$) and a map $f : X \to Y$, then

- f is said to be *increasing* if $f(x) \leq f(y)$ for every $x, y \in X$ with $x \leq y$.
- f is said to be *strictly increasing* if $f(x) < f(y)$ for every $x, y \in X$ with $x < y$.
- f is said to be *decreasing* if $f(x) \geq f(y)$ for every $x, y \in X$ with $x \leq y$.
- f is said to be *strictly decreasing* if $f(x) > f(y)$ for every $x, y \in X$ with $x < y$.

Given a real vector space V, an ordering $\leq$ in V is said to be *linear* if

$$x, y \in V, \quad x \leq y \quad \Longrightarrow \quad x + z \leq y + z \text{ and } \lambda x \leq \lambda y$$

for all $z \in V$ and

$$\lambda \in \mathbb{R}_+ := [0, \infty) \subset \mathbb{R}.$$

A real vector space V together with a linear ordering, $(V, \leq)$, will be called an *ordered vector space* (OVS).

Given a real vector space V, a subset $P \subset V$ is said to be a *cone* if

$$P + P \subset P, \quad \mathbb{R}_+ P \subset P, \quad \text{and} \quad P \cap (-P) = \{0\}. \tag{6.1}$$

According to (6.1), every cone is convex. When $(V, \leq)$ is an OVS, then it is easy to see that the set P defined through

$$P := \{ x \in V \ : \ x \geq 0 \}$$

is a cone, which is referred to as the *positive cone* associated to the ordering $\leq$. Conversely, given a real vector space V together with a cone $P \subset V$, the relation defined by

$$x \leq y \quad \text{if and only if} \quad y - x \in P \tag{6.2}$$

is a linear ordering in V whose associated positive cone is P. Actually, this correspondence establishes a bijection between the set of linear orderings of a real vector space and its set of cones, in such a way that each cone is the positive cone of its induced linear ordering through (6.2). The vectors of the punctate cone

$$\dot{P} := P \setminus \{0\} = \{ x \in V \ : \ x > 0 \}$$

are called *positive*.

Let $E = (E, \| \cdot \|)$ be a Banach space ordered by a cone P through (6.2). Then,

$$E := (E, \| \cdot \|, P)$$

is said to be an *ordered Banach space* (OBS) if P is closed, i.e., if

$$\bar{P} = P.$$

In such case, we will denote by

$$\overset{\circ}{P} := \operatorname{int} P$$

the interior of the cone P, which might be empty.

Throughout the rest of this book, given an OBS, $(E, \| \cdot \|, P)$, with $\operatorname{int} P \neq \emptyset$, and $x, y \in E$, it is said that $x \ll y$, or, equivalently, $y \gg x$, when

$$y - x \in \overset{\circ}{P}.$$

In particular,

$$x \gg 0 \quad \text{if and only if} \quad x \in \overset{\circ}{P}.$$

The following results will be extremely useful later.

Lemma 6.1. *Let* $(E, \| \cdot \|, P)$ *be an OBS with* $\overset{\circ}{P} \neq \emptyset$. *Then,*

$$x \gg 0, \quad \rho > 0 \implies \rho x \gg 0.$$

Proof. Suppose $x \gg 0$ and $\rho > 0$. Then, there exists $\epsilon > 0$ such that

$$x + y \in P \qquad \text{if} \quad \|y\| \leq \epsilon.$$

Thus, owing to (6.1),

$$\rho(x + y) \in P \qquad \text{if} \quad \|y\| \leq \epsilon,$$

and hence,

$$\rho x + z \in P \qquad \text{if} \quad \|z\| \leq \rho\epsilon.$$

Therefore, $\rho x \gg 0$, because $\rho\epsilon > 0$. $\qquad\square$

Lemma 6.2. *Let* $(E, \| \cdot \|, P)$ *be an OBS with* $\overset{\circ}{P} \neq \emptyset$. *Then,*

$$E = P - P := \{\, x - y \ : \ x, y \in P \,\}. \tag{6.3}$$

Proof. Fix $y \in \overset{\circ}{P}$, and let $x \in E$ arbitrary. Then, by Lemma 6.1,

$$z := \lambda y - x = \lambda \left(y - \lambda^{-1} x \right) \in P$$

for sufficiently large $\lambda > 0$. Therefore,

$$x = \lambda y - z \quad \text{with} \quad z \geq 0 \quad \text{and} \quad \lambda y \gg 0,$$

which concludes the proof.   □

Definition 6.1 (generating and total cones). *Let* $(E, \| \cdot \|, P)$ *be an OBS. Then:*

- *P is said to be generating if* $E = P - P$.
- *P is said to be total if the set of all finite linear combinations of its elements is dense in* E.

Obviously, P is total if it is generating.

According to Definition 6.1, Lemma 6.2 can be equivalently stated by simply saying that *P is generating if it has nonempty interior.*

Total cones induce a canonical dual ordering in the dual space

$$E' := \mathcal{L}(E, \mathbb{R}).$$

Indeed, given an OBS $(E, \| \cdot \|, P)$, let P^* denote the subset of E' defined through

$$P^* := \{\, x' \in E' \ : \ x'(x) \geq 0 \quad \forall\, x \in P \,\}. \tag{6.4}$$

By definition,

$$P^* + P^* \subset P^* \quad \text{and} \quad \mathbb{R}_+ P^* \subset P^*. \tag{6.5}$$

Moreover, the following result holds.

Lemma 6.3. P^* *is a cone of* E' *if and only if* P *is total. In such case,* P^* *is said to be the dual cone of* P *and*

$$E' := (E', \| \cdot \|_{E'}, P^*)$$

is an OBS — the dual OBS of E.

Proof. According to (6.5), to prove the first assertion it suffices to show that

$$P^* \cap (-P^*) = \{0\} \quad \text{if and only if } P \text{ is total.}$$

Indeed, since

$$P^* \cap (-P^*) = \{x' \in E' \; : \; x'(x) = 0 \quad \forall \, x \in P\},$$

it becomes apparent that

$$P^* \cap (-P^*) = \{0\}$$

if and only if

$$x' \in E' \quad \text{and} \quad x'(x) = 0 \quad \forall \; x \in P \quad \text{imply} \quad x' = 0.$$

As x' is linear and continuous, it is obvious that this occurs if the set of all finite linear combinations of elements of P is dense in E. Consequently, P^* is a cone if P is total.

Now, suppose that P is not total. Then, the linear manifold F generated by all finite linear combinations of vectors of P satisfies

$$\bar{F} \neq E.$$

Thus, according to the Hahn–Banach theorem (see, e.g., Corollary 1.8 of H. Brézis [29]), there exists $x' \in E' \setminus \{0\}$ such that $x'(x) = 0$ for all $x \in F$. In particular, $x'(x) = 0$ for all $x \in P$, and hence,

$$x' \in P^* \cap (-P^*).$$

Consequently, P^* is not a cone. Therefore, P^* is a cone if and only if P is total.

To conclude the proof of the lemma, it remains to show that P^* is closed in E'. Indeed, let $\{x'_n\}_{n \geq 1}$ be a sequence in $P^* \subset E'$ such that

$$\lim_{n \to \infty} \|x'_n - x'\|_{E'} = 0$$

for some $x' \in E'$. Then,

$$\lim_{n \to \infty} x'_n(x) = x'(x)$$

for all $x \in E$. Moreover, as $x'_n \in P^*$ for all $n \geq 1$, we have that $x'_n(x) \geq 0$ for all $x \in P$ and $n \geq 1$. Consequently, $x'(x) \geq 0$ for all $x \in P$ and, therefore, $x' \in P^*$. The proof is complete. $\square$

As a consequence from Lemma 6.3, P^* is a cone if P is generating. According to Lemma 6.2, this occurs if P has non-empty interior. The following concept will play an important role later.

Definition 6.2 (normal cone). *Let* $(E, \|\cdot\|, P)$ *be an* OBS *with total positive cone* P. *Then,* P *is said to be normal if the cone* P^* *is generating.*

Thanks to Lemma 6.3, the set P^* is a cone if P is total. Due to Definition 6.2, P is normal if it is total and

$$E' = P^* - P^*.$$

According to the first result of Section V.3.4 and Corollary V.3.5 of H. H. Schaefer [194], within the context of ordered Banach spaces the concept of normal cone introduced by Definition 6.2 coincides with the concept introduced on p. 215 of H. H. Schaefer [194]. Thus, by the first theorem of Section V.3.1 of H. H. Schaefer [194], the positive cone P is normal if and only if the norm $\|\cdot\|$ of E is equivalent to some *monotone* norm $\|\|\cdot\|\|$. By monotone, it means that

$$0 \leq x \leq y \quad \text{implies} \quad \|\|x\|\| \leq \|\|y\|\|.$$

More generally, the following characterization holds.

Theorem 6.1 (of characterization of normal cones). *Let* $(E, \|\cdot\|, P)$ *be an ordered Banach space. Then, the following properties are mutually equivalent:*

(a) *P is a normal cone.*

(b) *The norm $\|\cdot\|$ is* semi-monotone, *i.e., there exists a constant $\delta > 0$ such that*

$$0 \leq x \leq y \quad \text{implies} \quad \|x\| \leq \delta \|y\|.$$

(c) *There exists an equivalent monotone norm for E.*

(d) *For every $x, y \in P$, the interval $[x, y]$ is bounded.*

(e) *The order convex hull of every bounded set is bounded.*

(f) *There exists a positive real number $\alpha > 0$ such that*

$$\left. \begin{array}{c} x, y \in P \\ \|x\| = \|y\| = 1 \end{array} \right\} \implies \|x + y\| \geq \alpha.$$

Geometrically, when E is a Hilbert space, Theorem 6.1(f) entails that the angle between two arbitrary positive unit vectors must be bounded away from π. So, normal cones cannot be too large.

The proof of Theorem 6.1 can be accomplished from Theorem 5.1 of H. Amann [8], M. A. Krasnoselskij [118], H. H. Schaefer [194], and G. Jameson [107]. The technical details of the proof are omitted here.

6.2 Spectral theory of linear compact operators

This section collects a series of fundamental results about linear compact operators that will be used in the proof of the Krein–Rutman theorem. The proofs of these results will not be given here, as they can be easily reconstructed from available monographs on linear operator theory.

Let $(E, \|\cdot\|)$ be a real Banach space. Then, for every operator $T \in \mathcal{L}(E)$, the following limit exists:

$$\mathrm{spr}\,(T) := \lim_{n \to \infty} \|T^n\|_{\mathcal{L}(E)}^{1/n} \tag{6.6}$$

and it is called the *spectral radius* of T. Clearly,

$$\mathrm{spr}\,T \leq \|T\|_{\mathcal{L}(E)}$$

because

$$\|T^n\|_{\mathcal{L}(E)} \leq \|T\|_{\mathcal{L}(E)}^n, \qquad n \geq 1.$$

Moreover, for every $\zeta \in \mathbb{C}$ with $|\zeta| > \mathrm{spr}\,T$, we have that

$$\mathrm{spr}\,\frac{T}{\zeta} < 1$$

and hence the *resolvent operator*

$$\mathcal{R}(\zeta; T) := (\zeta I - T)^{-1} = \zeta^{-1}\left(I - \zeta^{-1}T\right)^{-1} \tag{6.7}$$

is given through the series

$$\mathcal{R}(\zeta; T) := \sum_{n=0}^{\infty} \zeta^{-(n+1)} T^n \tag{6.8}$$

which converges uniformly on compact subsets of $|\zeta| > \mathrm{spr}\,T$ (see Theorem 3 on p. 211 of Section VIII.2 of K. Yosida [227]).

Subsequently, when $\zeta \in \mathbb{C}$, in the expression $\zeta I - T$, I stands for the identity map of $E_{\mathbb{C}}$, the canonical complexification of E,

$$E_{\mathbb{C}} := E + iE,$$

where i is the complex imaginary unit, and T denotes the canonical extension of T to $E_{\mathbb{C}}$, i.e.,

$$T(x + iy) := Tx + iTy, \qquad x, y \in E.$$

For a given $T \in \mathcal{L}(E)$, we denote by $\sigma(T)$ the *spectrum* of T. It consists of the set of values $\lambda \in \mathbb{C}$ for which $\lambda I - T$ is not an isomorphism of $E_{\mathbb{C}}$. Also, we denote by $\varrho(T)$ the *resolvent set* of T, which is defined by

$$\varrho(T) := \mathbb{C} \setminus \sigma(T).$$

It is well known that

$$\operatorname{spr} T = \sup_{\lambda \in \sigma(T)} |\lambda| \tag{6.9}$$

(see Theorem 4 on p. 212 of Section VIII.2 of K. Yosida [227]).

Throughout the rest of this chapter, for any real Banach space E and $T \in \mathcal{L}(E)$, we denote by T^* the *adjoint operator* of T, which is defined as the unique operator $T^* \in \mathcal{L}(E')$ such that

$$T^*x'(x) = x'(Tx) \qquad \text{for every} \quad (x, x') \in E \times E'.$$

An operator $T \in \mathcal{L}(E)$ is said to be *compact* if $\overline{T(B)}$ is a compact subset of E for every bounded subset $B \subset E$. The class of compact operators of $\mathcal{L}(E)$ will be denoted throughout by $\mathcal{K}(E)$. By a celebrated result of J. Schauder,

$$T \in \mathcal{K}(E) \quad \text{if and only if} \quad T^* \in \mathcal{K}(E') \tag{6.10}$$

(see Theorem of Schauder on p. 282 of Section X.4 of K. Yosida [227]). Moreover, in such case, by the Riesz–Schauder Theory, we have that, for every $\zeta \in \mathbb{C} \setminus \{0\}$,

$$R[\zeta I - T] \quad \text{is closed}, \qquad \dim N[\zeta I - T] < \infty,$$

and

$$\begin{cases} R[\zeta I - T] = N[\zeta I - T^*]^\perp, \\[2mm] \dim N[\zeta I - T] = \dim N[\zeta I - T^*], \end{cases} \tag{6.11}$$

where, for every $M \subset E'$, we have denoted

$$M^\perp := \{\, x \in E \ : \ x'(x) = 0 \ \ \forall\, x' \in M \,\}$$

(see Section X.5 of K. Yosida [227]). The following result collects some of the most fundamental spectral properties of the compact operators.

Theorem 6.2 (of spectral structure). *Suppose $T \in \mathcal{K}(E)$. Then, the following properties hold:*

(a) $0 \in \sigma(T)$ *if* $\dim E = \infty$, $\sigma(T) \setminus \{0\}$ *is a discrete set, and there exists* $\lambda \in \sigma(T)$ *such that* $|\lambda| = \operatorname{spr} T$. *Therefore, by* (6.9),

$$\operatorname{spr} T = \max_{\lambda \in \sigma(T)} |\lambda|. \tag{6.12}$$

(b) *For every $\lambda \in \sigma(T) \setminus \{0\}$, λ is an eigenvalue of T and there exists a minimal integer number $\nu(\lambda) \geq 1$ such that*

$$N[(\lambda I - T)^{\nu(\lambda)+n}] = N[(\lambda I - T)^{\nu(\lambda)}]$$

*for all $n \in \mathbb{N}$. The integer $\nu(\lambda) \geq 1$ is the **algebraic ascent** of λ, the linear space $N[(\lambda T - T)^{\nu(\lambda)}]$ is the **ascent generalized eigenspace** associated to λ, and the **algebraic multiplicity** of λ is defined through*

$$\mathfrak{m}(\lambda; T) := \dim N[(\lambda I - T)^{\nu(\lambda)}].$$

*An eigenvalue $\lambda \in \sigma(T) \setminus \{0\}$ is said to be **algebraically simple** if*

$$\nu(\lambda) = \mathfrak{m}(\lambda; T) = 1.$$

(c) *The resolvent operator*

$$\varrho(T) \longrightarrow \mathcal{L}(E)$$
$$\zeta \mapsto \mathcal{R}(\zeta; T)$$

*is holomorphic. Moreover, for each $\lambda \in \sigma(T) \setminus \{0\}$, λ is a pole of order $\nu(\lambda)$ of $\mathcal{R}(\zeta; T)$. Therefore, it possesses a (unique) **Laurent development** at λ,*

$$\mathcal{R}(\zeta; T) = \sum_{n=-\nu(\lambda)}^{\infty} (\zeta - \lambda)^n \mathfrak{T}_n, \qquad \zeta \sim \lambda, \quad \zeta \neq \lambda, \tag{6.13}$$

where, for every $n \geq -\nu(\lambda)$ and sufficiently small $\epsilon > 0$,

$$\mathfrak{T}_n := \frac{1}{2\pi i} \int_{|\zeta - \lambda| = \epsilon} (\zeta - \lambda)^{-(n+1)} \mathcal{R}(\zeta; T) \, d\zeta \in \mathcal{L}(E). \tag{6.14}$$

Furthermore, $\mathfrak{T}_{-\nu(\lambda)} \neq 0$, $\mathfrak{T}_n$ is a finite rank operator for every $n \in \{-\nu(\lambda), ..., -1\}$, and $\mathfrak{T}_{-1}$ is a projection of E onto $N[(\lambda I - T)^{\nu(\lambda)}]$. Consequently,

$$\operatorname{tr} \mathfrak{T}_{-1} := \dim R[\mathfrak{T}_{-1}] = \dim N[(\lambda I - T)^{\nu(\lambda)}] = \mathfrak{m}(\lambda; T).$$

The proof of Theorem 6.2(a) can be found in Theorem VI.8 of H. Brézis [29]. The finite-dimensional counterpart of Theorem 6.2(b) is folklore, but, surprisingly, it might not be trivial to give a reference containing a self-contained proof of it in the general infinite-dimensional setting. Part (c) is a more specialized material, though, naturally, it is folklore for experts. A self-contained proof of Theorem 6.2(b)(c) may be built up from the results of Chapter I of T. Kato [112], through a preliminary reduction of dimension, and the results of Section VIII.8 of K. Yosida [227]. The reader should refer to Chapters 6 and 7 of [141] and Chapter 7 of [142] for the finite-dimensional theory, and to the more recent monograph of J. López-Gómez and C. Mora-Corral [153] for the general infinite-dimensional case, where, besides a complete self-contained proof of Theorem 6.2, the reader will find some very recent advances in the theory of algebraic multiplicities of eigenvalues of linear operators, as the axiomatization theorem. Classical textbooks covering most of these materials, easily accessible to beginners, are I. C. Göhberg, P. Lancaster and L. Rodman [84], and I. C. Göhberg, S. Goldberg and M. A. Kaashoek [82].

By (6.10) and (6.11), it becomes apparent from Theorem 6.2 that

$$\sigma(T) = \sigma(T^*)$$

and that

$$\mathfrak{m}(\lambda; T) = \mathfrak{m}(\lambda; T^*), \qquad \lambda \in \sigma(T) \setminus \{0\}, \tag{6.15}$$

for all $T \in \mathcal{K}(E)$. Moreover, the following fundamental identity, going back to R. S. Phillips [177], holds

$$\mathcal{R}(\zeta; T)^* = \mathcal{R}(\zeta; T^*), \qquad \zeta \in \varrho(T) = \varrho(T^*), \tag{6.16}$$

(see Theorem 2 of Section VIII.6 of K. Yosida [227]).

6.3   The Krein–Rutman theorem

Let (X, P_X) and (Y, P_Y) be two ordered vector spaces and $T : X \to Y$ a linear operator. Then,

- T is said to be *positive* if

$$T(P_X) \subset P_Y.$$

- T is said to be *strictly positive* if

$$T(P_X \setminus \{0\}) \subset P_Y \setminus \{0\}.$$

- T is said to be *strongly positive* if Y possesses a norm $\| \cdot \|_Y$ for which $(Y, \| \cdot \|_Y, P_Y)$ is an ordered Banach space with

$$\overset{\circ}{P}_Y \neq \emptyset \quad \text{and} \quad T(P_X \setminus \{0\}) \subset \overset{\circ}{P}_Y.$$

Naturally, T is said to be *strongly negative* if $-T$ is strongly positive.

The main result of this chapter can be stated as follows.

Theorem 6.3 (generalized of Krein–Rutman). *Let $(E, \| \cdot \|, P)$ be an ordered Banach space with $\overset{\circ}{P} \neq \emptyset$, and $T \in \mathcal{K}(E)$ a compact strongly positive operator, i.e.,*

$$T(P \setminus \{0\}) \subset \overset{\circ}{P}. \tag{6.17}$$

Then, the following assertions are true:

(a) $\operatorname{spr} T > 0$ *is an algebraically simple eigenvalue of T with*

$$N[\operatorname{spr} T\, I - T] = \operatorname{span}[x_0]$$

for some $x_0 \in \overset{\circ}{P}$.

(b) $\operatorname{spr} T$ *is the unique real eigenvalue of T to an eigenvector in $P \setminus \{0\}$.*

(c) $\operatorname{spr} T$ *is the unique eigenvalue of T in the spectral circle*

$$|\zeta| = \operatorname{spr} T.$$

In other words,

$$|\lambda| < \operatorname{spr} T \quad \text{for all} \quad \lambda \in \sigma(T) \setminus \{\operatorname{spr} T\}.$$

(d) *For every real number $\lambda > \operatorname{spr} T$, the resolvent operator*

$$\mathcal{R}(\lambda; T) := (\lambda I - T)^{-1} \in \mathcal{L}(E)$$

is strongly positive, i.e.,

$$\mathcal{R}(\lambda; T)(P \setminus \{0\}) \subset \overset{\circ}{P}.$$

(e) *There exist $\epsilon > 0$ and $x > 0$ such that*

$$\mathcal{R}(\lambda; T)x \ll 0 \quad \text{for all} \quad \lambda \in (\operatorname{spr} T - \epsilon, \operatorname{spr} T).$$

(f) *Suppose, in addition, that P is a normal cone. Then, the following assertions are true:*

  (a) *There exists $x_0' \in P^* \setminus \{0\}$ such that*

$$x_0'(x_0) > 0 \quad \text{and} \quad N[\operatorname{spr} T\, I - T^*] = \operatorname{span}[x_0'].$$

  (b) *For every $x \in P \setminus \{0\}$,*

$$x_0'(x) > 0$$

  and the equation

$$\operatorname{spr} T u - T u = x$$

  cannot admit a solution $u \in E$.

Remark 6.1.

(a) By the general properties of adjoint operators revisited in Section 6.2, it is apparent that $T^* \in \mathcal{K}(E')$ satisfies

$$\sigma(T) = \sigma(T^*) \quad \text{and} \quad \operatorname{spr} T = \operatorname{spr} T^*.$$

Thus, by (6.11) and Theorem 6.3(a)(c), $\operatorname{spr} T$ must be a simple eigenvalue of T^*, and it is the unique eigenvalue of T^* in the spectral circle $|\zeta| = \operatorname{spr} T$.

(b) According to Lemma 6.2, P is generating and, hence, it is total, because $\operatorname{int} P \neq \emptyset$. Thus, by Lemma 6.3, $(E', \|\cdot\|_{E'}, P^*)$ is an OBS. By definition, for every $x \in P$ and $x' \in P^*$, the following estimate holds

$$T^* x'(x) = x'(Tx) \geq 0$$

and hence $T^* x' \in P^*$. Therefore, T^* is positive with respect to the cone P^*. But T^* might not be strongly positive. Actually, it is far from obvious whether P^* has empty interior, though we are assuming that P^* is generating, because P is normal in Part (f).

6.4 Preliminaries of the proof of Theorem 6.3

This section collects some important properties needed in the proof of Theorem 6.3. Therefore, throughout it, we will assume that $(E, \|\cdot\|, P)$ is an OBS with $\operatorname{int} P \neq \emptyset$, and that $T \in \mathcal{K}(E)$ is strongly positive. Also, we will consider the sets

$$B := \{\, x \in E \; : \; \|x\| < 1 \,\}, \qquad \partial B := \{\, x \in E \; : \; \|x\| = 1 \,\},$$

$$K := \overline{T(\partial B \cap P)},$$

and, for every $\gamma > 0$,

$$M_\gamma := \{\, x \in P \setminus \{0\} \; : \; Tx \geq \gamma x \,\}$$
$$= (T - \gamma I)^{-1}(P) \cap (P \setminus \{0\}),$$

and

$$\Sigma_\gamma := \{\, x \in K \cap M_\gamma \; : \; \|x\| \geq \gamma \,\}$$
$$= K \cap (T - \gamma I)^{-1}(P) \cap P \cap \{\, x \in E \; : \; \|x\| \geq \gamma \,\}.$$

The following result collects some useful properties of these sets which are going to be used in the proof of Theorem 6.3.

Proposition 6.1. *K is a compact subset of P, and Σ_γ is a compact subset of K such that*

$$\Sigma_\gamma \subset M_\gamma$$

for all $\gamma > 0$. Moreover, the following properties are satisfied:

(a) *If $\tilde\gamma > \gamma > 0$, then*

$$M_{\tilde\gamma} \subset M_\gamma \quad \text{and} \quad \Sigma_{\tilde\gamma} \subset \Sigma_\gamma.$$

(b) *The cone P can be expressed through*

$$P \setminus \{0\} = \bigcup_{\gamma > 0} M_\gamma.$$

(c) *If $Tx > \gamma x$ for some $\gamma > 0$ and $x > 0$, then, $Tx \in M_{\tilde\gamma}$ for some $\tilde\gamma > \gamma$ and, in particular, $Tx \in M_\gamma$, by Part (a).*

(d) *Let $\gamma > 0$ and any sequence $\{\gamma_n\}_{n \geq 1}$ such that*

$$\lim_{n \to \infty} \gamma_n = \gamma, \qquad 0 < \gamma_n < \gamma_{n+1} < \gamma, \quad \text{and} \quad \Sigma_{\gamma_n} \neq \emptyset,$$

for all $n \geq 1$. Then,

$$\Sigma_\gamma = \bigcap_{n \geq 1} \Sigma_{\gamma_n} \neq \emptyset.$$

(e) *$\Sigma_\gamma = M_\gamma = \emptyset$ if $\gamma > \|T\|_{\mathcal{L}(E)}(> 0)$.*

(f) *For every $\gamma \in \left(0, \|T\|_{\mathcal{L}(E)}\right]$, $\Sigma_\gamma \neq \emptyset$ if and only if $M_\gamma \neq \emptyset$.*

Proof. By (6.17), $T(P) \subset P$ and hence,

$$T(\partial B \cap P) \subset T(P) \subset P,$$

which implies

$$K = \overline{T(\partial B \cap P)} \subset \bar{P} = P.$$

Moreover, since $\partial B \cap P$ is bounded and T is compact, $T(\partial B \cap P)$ is relatively compact. Therefore, K is a compact subset of P.

Now, fix a $\gamma > 0$. Then, since K, P, $(T - \gamma I)^{-1}(P)$ and $\{x \in E : \|x\| \geq \gamma\}$ are closed sets, it is apparent that Σ_γ is a closed subset of K. Thus, as K is compact, Σ_γ is a compact subset of K. By definition, $\Sigma_\gamma \subset M_\gamma$ for all $\gamma > 0$.

Next, we will prove each of the properties listed in the statement.

Proof of Part (a): Suppose $\tilde\gamma > \gamma > 0$ and $M_{\tilde\gamma} \neq \emptyset$. Let $x \in M_{\tilde\gamma}$. Then, by definition,

$$x > 0 \quad \text{and} \quad Tx \geq \tilde\gamma x.$$

Moreover, since $\tilde{\gamma} - \gamma > 0$ and $x > 0$, we find from $\mathbb{R}_+ P \subset P$ that $(\tilde{\gamma} - \gamma)x > 0$ and hence,

$$x > 0 \quad \text{and} \quad Tx \geq \tilde{\gamma}\,x > \gamma\,x.$$

Therefore, $x \in M_\gamma$ and, consequently, $M_{\tilde{\gamma}} \subset M_\gamma$. Obviously, this also implies that

$$\Sigma_{\tilde{\gamma}} = \{\, x \in K \cap M_{\tilde{\gamma}} \ : \ \|x\| \geq \tilde{\gamma}\,\}$$
$$\subset \{\, x \in K \cap M_\gamma \ : \ \|x\| \geq \gamma\,\} = \Sigma_\gamma,$$

which concludes the proof of Part (a).

Proof of Part (b): By definition, $M_\gamma \subset P \setminus \{0\}$ for all $\gamma > 0$ and hence,

$$\bigcup_{\gamma > 0} M_\gamma \subset P \setminus \{0\}.$$

To prove the converse, let $x > 0$. Then, by (6.17), $Tx \in \overset{\circ}{P}$ and hence,

$$\lim_{\gamma \to 0}(Tx - \gamma x) = Tx \in \overset{\circ}{P}.$$

Thus,

$$Tx - \gamma x \in P$$

for sufficiently small $\gamma > 0$, and, therefore, $x \in M_\gamma$. This ends the proof of Part (b).

Proof of Part (c): Let $\gamma > 0$ and $x > 0$ such that $Tx > \gamma x$. Then, according to (6.17), we have that

$$Tx \in \overset{\circ}{P}, \qquad T^2 x - \gamma Tx \in \overset{\circ}{P},$$

and hence, $Tx > 0$ and

$$\lim_{\epsilon \to 0}(T^2 x - \gamma Tx - \epsilon Tx) = T^2 x - \gamma Tx \in \overset{\circ}{P}.$$

Thus, for sufficiently small $\epsilon > 0$, we find that

$$Tx > 0 \quad \text{and} \quad T^2 x \geq (\gamma + \epsilon)Tx.$$

Equivalently, $Tx \in M_{\gamma + \epsilon}$. This concludes the proof of Part (c).

Proof of Part (d): Let $\gamma > 0$ and a sequence $\{\gamma_n\}_{n \geq 1}$ such that

$$\lim_{n \to \infty} \gamma_n = \gamma, \qquad 0 < \gamma_n < \gamma_{n+1} < \gamma, \quad \text{and} \quad \Sigma_{\gamma_n} \neq \emptyset,$$

for all $n \geq 1$. According to Part (a), we have that

$$\Sigma_\gamma \subset \Sigma_{\gamma_{n+1}} \subset \Sigma_{\gamma_n}$$

for all $n \geq 1$, and hence $\{\Sigma_{\gamma_n}\}_{n\geq 1}$ provides us with a non-increasing sequence of non-empty compact subsets of K. Therefore, by the Cantor principle,

$$\Sigma_\gamma \subset \bigcap_{n\geq 1} \Sigma_{\gamma_n} \neq \emptyset.$$

To show the converse inclusion, let

$$x \in \bigcap_{n\geq 1} \Sigma_{\gamma_n}.$$

Then, $x \in \Sigma_{\gamma_n}$ for all $n \geq 1$, and hence,

$$x \in K, \qquad \|x\| \geq \gamma_n, \qquad Tx \geq \gamma_n x.$$

Thus, since K is a compact subset of P, letting $n \to \infty$, we find that

$$\|x\| \geq \gamma, \qquad Tx \geq \gamma x.$$

Therefore, $x \in \Sigma_\gamma$, which completes the proof of Part (d).

Proof of Parts (e) and (f): Fix $\gamma > 0$. By definition, $\Sigma_\gamma = \emptyset$ if $M_\gamma = \emptyset$. So, suppose

$$M_\gamma \neq \emptyset$$

and pick $x \in M_\gamma$. Then,

$$x > 0 \quad \text{and} \quad Tx \geq \gamma x > 0.$$

In particular, $T \neq 0$ and $\|T\|_{\mathcal{L}(E)} > 0$.

Subsequently, for every $\zeta \in \mathbb{C}$, we consider the series

$$S(\zeta) = \sum_{n=0}^{\infty} \zeta^n T^n x.$$

It is absolutely convergent in E provided

$$|\zeta| < R(x) := \left(\limsup_{n\to\infty} \frac{\|T^{n+1}x\|}{\|T^n x\|} \right)^{-1} \leq \infty.$$

Thus, for every $\zeta \in [0, R(x))$, we have that

$$S(\zeta) = \sum_{n=0}^{\infty} \zeta^n T^n x = x + \sum_{n=1}^{\infty} \zeta^n T^n x$$

$$= x + \zeta T \sum_{n=1}^{\infty} \zeta^{n-1} T^{n-1} x = x + \zeta T \sum_{n=0}^{\infty} \zeta^n T^n x$$

$$\geq x + \zeta \gamma \sum_{n=0}^{\infty} \zeta^n T^n x = x + \zeta \gamma S(\zeta)$$

and, consequently,

$$(1 - \zeta\gamma)S(\zeta) \geq x > 0.$$

Moreover,

$$0 < x \leq S(\zeta)$$

and hence,

$$1 - \zeta\gamma > 0$$

for all $\zeta \in [0, R(x))$. Therefore,

$$0 < \gamma \leq R^{-1}(x) := \limsup_{n \to \infty} \frac{\|T^{n+1}x\|}{\|T^n x\|} \leq \|T\|_{\mathcal{L}(E)}. \qquad (6.18)$$

When P is a normal cone, owing to Theorem 6.1, the norm $\|\cdot\|$ can be chosen to be monotone and, in such case, (6.18) can be inferred directly from

$$\gamma\, T^{n-1}x \leq T^n x \in \overset{\circ}{P}, \qquad n \geq 2,$$

by taking norms on both sides of these inequalities, which gives rise to

$$\gamma \leq \frac{\|T^{n+1}x\|}{\|T^n x\|} \leq \|T\|_{\mathcal{L}(E)}, \qquad n \geq 1.$$

But in the general case when P is not normal, the previous argument cannot be shortened according to these patterns.

According to (6.18), it is apparent that $M_\gamma = \emptyset$, and hence, $\Sigma_\gamma = \emptyset$ if $\gamma > \|T\|_{\mathcal{L}(E)}$, which concludes the proof of Part (e).

Subsequently, we consider the sequence of vectors

$$y_n := T\left(\frac{T^n x}{\|T^n x\|}\right) = \frac{T^{n+1}x}{\|T^n x\|} \in E, \qquad n \geq 1.$$

By construction,

$$\frac{T^n x}{\|T^n x\|} > 0 \qquad \text{and} \qquad \left\|\frac{T^n x}{\|T^n x\|}\right\| = 1, \qquad n \geq 1.$$

Thus,

$$\frac{T^n x}{\|T^n x\|} \in \partial B \cap P, \qquad n \geq 1,$$

and hence,

$$y_n \in T(\partial B \cap P) \subset \overline{T(\partial B \cap P)} = K, \qquad n \geq 1.$$

Actually, according to (6.17), we have that

$$y_n \in \overset{\circ}{P} \cap K, \qquad n \geq 1. \tag{6.19}$$

Note that (6.18) can be rewritten as

$$0 < \gamma \leq R^{-1}(x) := \limsup_{n \to \infty} \|y_n\| \leq \|T\|_{\mathcal{L}(E)}.$$

As K is compact, we find from (6.19) that there exist $y \in K$ and a subsequence $\{y_{n_k}\}_{k \geq 1}$ of $\{y_n\}_{n \geq 1}$ such that

$$y = \lim_{k \to \infty} y_{n_k} \qquad \text{and} \qquad R^{-1}(x) = \lim_{k \to \infty} \|y_{n_k}\|.$$

Necessarily,

$$0 < \gamma \leq R^{-1}(x) = \|y\|$$

and hence, $y \in (P \setminus \{0\}) \cap K$. Also, for every $k \geq 1$,

$$T y_{n_k} = T \left(\frac{T^{n_k+1} x}{\|T^{n_k} x\|} \right) \geq \gamma \frac{T^{n_k+1} x}{\|T^{n_k} x\|} = \gamma y_{n_k},$$

because $Tx \geq \gamma x$. Therefore, letting $k \to \infty$ yields

$$Ty \geq \gamma y.$$

Consequently, $y \in \Sigma \gamma$ and hence $\Sigma_\gamma \neq \emptyset$. The converse is always true, because $\Sigma_\gamma \subset M_\gamma$ for all $\gamma > 0$ and hence $M_\gamma \neq \emptyset$ if $\Sigma_\gamma \neq \emptyset$. This concludes the proof. $\qquad\square$

6.5 Proof of Theorem 6.3

This section consists of the proof of Theorem 6.3. Therefore, it is throughout assumed that $(E, \|\cdot\|, P)$ is an OBS with $\text{int } P \neq \emptyset$ and that $T \in \mathcal{K}(E)$ is strongly positive. All notations introduced in the previous section will be maintained here. The proof of the theorem will follow after a series of lemmas.

Lemma 6.4. *There exist $x_0 \in \overset{\circ}{P}$ and $\rho > 0$ such that*

$$T x_0 = \rho x_0. \tag{6.20}$$

Proof. Subsequently, we consider the set

$$G := \{ \gamma > 0 \,:\, \Sigma_\gamma \neq \emptyset \}.$$

By Proposition 6.1(e),

$$\|T\|_{\mathcal{L}(E)} > 0 \quad \text{and} \quad G \subset (0, \|T\|_{\mathcal{L}(E)}] .$$

Moreover, due to Proposition 6.1(b), there exists $\gamma > 0$ such that $M_\gamma \neq \emptyset$. By Proposition 6.1(f), we have that $\Sigma_\gamma \neq \emptyset$ and, therefore, according to Proposition 6.1(a), we find that

$$(0, \gamma] \subset G.$$

In particular, $G \neq \emptyset$. Actually, combining Parts (a) and (d) of Proposition 6.1, it becomes apparent that there is $\rho \in [\gamma, \|T\|_{\mathcal{L}(E)}]$ such that

$$G := \{ \gamma > 0 \ : \ \Sigma_\gamma \neq \emptyset \} = (0, \rho]. \tag{6.21}$$

As $\Sigma_\rho \neq \emptyset$, there exists $x_0 > 0$ such that

$$T x_0 \geq \rho x_0.$$

Suppose $T x_0 \neq \rho x_0$. Then,

$$T x_0 > \rho x_0$$

and, owing to Proposition 6.1(c), there exists $\epsilon > 0$ such that

$$T x_0 \in M_{\rho+\epsilon}.$$

In particular, this implies $M_{\rho+\epsilon} \neq \emptyset$ and hence, by Parts (e) and (f) of Proposition 6.1, $\Sigma_{\rho+\epsilon} \neq \emptyset$, which entails $\rho + \epsilon \in G$ and contradicts (6.21). Consequently, (6.20) holds. Moreover, since $x_0 > 0$, it follows from (6.17) that

$$T x_0 = \rho x_0 \gg 0$$

and, therefore, according to Lemma 6.1, we also obtain that

$$x_0 = \rho^{-1} \rho x_0 \gg 0,$$

because $\rho^{-1} > 0$. This concludes the proof. $\qquad \Box$

Throughout the rest of this section, we fix a real number $\rho > 0$ and a vector $x_0 \gg 0$ satisfying (6.20).

Lemma 6.5. *The algebraic multiplicity of ρ as an eigenvalue of the operator T equals one. Precisely,*

$$N[T - \rho I_E] = \operatorname{span}[x_0] \tag{6.22}$$

and

$$N[(T - \rho I_E)^2] = N[T - \rho I_E]. \tag{6.23}$$

Proof. The proof of (6.22) proceeds by contradiction. Suppose there is another vector

$$x \in N[T - \rho I_E] \setminus \operatorname{span}[x_0].$$

Then, since

$$\lim_{\mu \uparrow \infty} \left(x_0 - \mu^{-1}x\right) = \lim_{\mu \uparrow \infty} \left(x_0 + \mu^{-1}x\right) = x_0 \in \overset{\circ}{P},$$

we obtain that, for sufficiently large $\mu > 0$,

$$x_0 - \mu^{-1}x \in P \quad \text{and} \quad x_0 + \mu^{-1}x \in P$$

and hence, multiplying by μ yields

$$\mu x_0 - x \in P \quad \text{and} \quad \mu x_0 + x \in P.$$

Therefore, for sufficiently large $\mu > 0$, we have that

$$-\mu x_0 \leq x \leq \mu x_0. \tag{6.24}$$

Obviously, (6.24) fails at $\mu = 0$, since $x \neq 0$. Therefore, the minimum value of μ satisfying (6.24) must be positive. Let us denote it by μ_0. Then, $\mu_0 > 0$,

$$-\mu_0 x_0 \leq x \leq \mu_0 x_0,$$

and, since x cannot be a multiple of x_0, we actually have that

$$-\mu_0 x_0 < x < \mu_0 x_0.$$

Consequently, it follows from (6.17) that

$$-\mu_0 \rho x_0 = -\mu_0 T x_0 \ll T x = \rho x \ll \mu_0 T x_0 = \mu_0 \rho x_0.$$

Therefore, according to Lemma 6.1,

$$-\mu_0 x_0 \ll x \ll \mu_0 x_0.$$

As these estimates are strict, they imply that

$$-(\mu_0 - \epsilon)x_0 \leq x \leq (\mu_0 - \epsilon)x_0$$

for sufficiently small $\epsilon > 0$. Consequently, they contradict the minimality of μ_0 and show (6.22).

The proof of (6.23) also proceeds by contradiction. Suppose there exists $x \in E$ such that

$$x \in N[(T - \rho I_E)^2] \setminus N[T - \rho I_E].$$

Then, by (6.22),

$$0 \neq (T - \rho I_E)x \in N\,[T - \rho I_E] = \text{span}\,[x_0]$$

and hence, there is $\xi \in \mathbb{R} \setminus \{0\}$ such that

$$Tx - \rho x = \xi x_0. \tag{6.25}$$

Actually, by choosing $-x$, instead of x, if necessary, we can assume, without loss of generality, that (6.25) holds with

$$\xi > 0.$$

Arguing as in the proof of (6.22) it is apparent that

$$-\mu x_0 \leq x \leq \mu x_0$$

for sufficiently large $\mu > 0$. Thus, the following minimum is well defined

$$\mu_0 := \min\{\,\mu \in \mathbb{R} \,:\, x \leq \mu x_0\,\} \in \mathbb{R}.$$

In order to show this, we will argue by contradiction. Suppose there exists a sequence $\{\mu_n\}_{n \geq 1}$ such that

$$\lim_{n \to \infty} \mu_n = \infty \quad \text{and} \quad x \leq -\mu_n x_0$$

for all $n \geq 1$. Then,

$$\mu_n^{-1} x \leq -x_0$$

for all $n \geq 1$ and hence, letting $n \to \infty$ shows that $-x_0 \geq 0$. On the other hand, we already know that $x_0 \geq 0$. Thus, $x_0 = 0$, which is a contradiction. By construction, we have that

$$x \leq \mu_0 x_0$$

and, since

$$x \notin N[T - \rho I_E] = \text{span}\,[x_0],$$

necessarily

$$x < \mu_0 x_0$$

and hence, it follows from (6.17) and (6.25) that

$$Tx = \rho x + \xi x_0 \ll \mu_0 T x_0 = \mu_0 \rho x_0.$$

Consequently,

$$x \leq \left(\mu_0 - \frac{\xi}{\rho}\right) x_0,$$

with

$$\mu_0 - \frac{\xi}{\rho} < \mu_0,$$

which contradicts the minimality of μ_0 and ends the proof. $\qquad\square$

Subsequently, we denote by $E_{\mathbb{C}}$ the canonical complexification of E,

$$E_{\mathbb{C}} := E + iE,$$

and $T_{\mathbb{C}} \in \mathcal{L}(E_{\mathbb{C}})$ stands for the operator

$$T_{\mathbb{C}}(x + iy) = Tx + iTy, \qquad x, y \in E.$$

By definition,

$$T_{\mathbb{C}} = T \quad \text{in} \quad E$$

and

$$\mathrm{Re}\,(T_{\mathbb{C}}z) = T(\mathrm{Re}\,z) \qquad \text{for all} \quad z \in E_{\mathbb{C}}.$$

Moreover, the following complex counterpart of Lemma 6.5 holds.

Lemma 6.6. *One has that*

$$N\left[(T_{\mathbb{C}} - \rho I_{E_{\mathbb{C}}})^2 \right] = N\,[T_{\mathbb{C}} - \rho I_{E_{\mathbb{C}}}] = \mathrm{span}\,[x_0].$$

Proof. Let $z = x + iy \in E_{\mathbb{C}}$ be such that

$$T_{\mathbb{C}}z = \rho z.$$

Then,

$$Tx + iTy = \rho(x + iy)$$

and hence,

$$x, y \in N\,[T - \rho I_E].$$

Thus, by (6.22), there are $\lambda, \mu \in \mathbb{R}$ such that

$$x = \lambda x_0 \quad \text{and} \quad y = \mu x_0.$$

Therefore,

$$z = x + iy = (\lambda + i\mu)x_0 \in \mathrm{span}\,[x_0].$$

Consequently,

$$N\,[T_{\mathbb{C}} - \rho I_{E_{\mathbb{C}}}] = \mathrm{span}\,[x_0]. \tag{6.26}$$

Now, let $x, y \in E$ be such that

$$z := x + iy \in N\left[(T_{\mathbb{C}} - \rho I_{E_{\mathbb{C}}})^2 \right].$$

Then,

$$(T_{\mathbb{C}} - \rho I_{E_{\mathbb{C}}})\, z \in N\,[T_{\mathbb{C}} - \rho I_{E_{\mathbb{C}}}]$$

and hence, thanks to (6.26), there are $\lambda, \mu \in \mathbb{R}$ such that

$$Tx + iTy - \rho(x + iy) = (\lambda + i\mu)x_0.$$

Equivalently,

$$(T - \rho I_E)\, x = \lambda x_0, \qquad (T - \rho I_E)\, y = \mu x_0,$$

and, so, according to Lemma 6.5, we find that

$$x, y \in N[(T - \rho I_E)^2] = \text{span}\,[x_0].$$

Consequently, there exist $\alpha, \beta \in \mathbb{R}$ such that

$$x = \alpha x_0, \qquad y = \beta x_0,$$

and, therefore,

$$z = x + iy = (\alpha + i\beta)x_0 \in N\,[T_\mathbb{C} - \rho I_{E_\mathbb{C}}] = \text{span}\,[x_0].$$

This concludes the proof. □

The next lemma establishes an important property of the eigenvalues $\lambda \neq \rho$ of $T_\mathbb{C}$.

Lemma 6.7. *Suppose*

$$\lambda \in \mathbb{C} \setminus \{\rho\}, \qquad z \in N[T_\mathbb{C} - \lambda I_{E_\mathbb{C}}].$$

Then,

$$\text{Re}\left(e^{i\theta}z\right) \neq \mu x_0 \qquad \text{for all} \quad (\theta, \mu) \in \mathbb{R} \times (\mathbb{R} \setminus \{0\}). \tag{6.27}$$

Proof. It proceeds by contradiction. Suppose

$$\text{Re}\left(e^{i\theta}z\right) = \mu x_0 \tag{6.28}$$

for some $\theta \in \mathbb{R}$, $\mu \in \mathbb{R} \setminus \{0\}$, and $z \in E_\mathbb{C} \setminus \{0\}$ such that

$$T_\mathbb{C}z = \lambda z. \tag{6.29}$$

Note that (6.28) cannot be satisfied if $z = 0$, because $\mu x_0 \neq 0$, and, therefore, (6.27) is obvious if $z = 0$.

According to (6.20), (6.28) and (6.29), it is apparent that

$$\begin{aligned}
\mu\rho x_0 = \mu T x_0 &= T\left(\text{Re}\left(e^{i\theta}z\right)\right) \\
&= \text{Re}\, T_\mathbb{C}\left(e^{i\theta}z\right) = \text{Re}\left(e^{i\theta}T_\mathbb{C}z\right) = \text{Re}\left(e^{i\theta}\lambda z\right).
\end{aligned}$$

Thus, since $\rho > 0$, we find that

$$\text{Re}\left(e^{i\theta}\rho z\right) = \text{Re}\left(e^{i\theta}\lambda z\right),$$

which can be expressed in the form

$$\text{Re}\left(\left(\frac{\lambda}{\rho}-1\right)e^{i\theta}z\right)=0,$$

or, equivalently,

$$\left(\frac{\lambda}{\rho}-1\right)e^{i\theta}z \in iE. \tag{6.30}$$

As $\lambda \neq \rho$, there is $\tau \in \mathbb{R}$ for which

$$0 \neq \frac{\lambda}{\rho}-1 = \left|\frac{\lambda}{\rho}-1\right|e^{i\tau}$$

and, so, (6.30) becomes

$$e^{i\left(\theta+\tau-\frac{\pi}{2}\right)}z \in E.$$

Consequently, there exists $t \in \mathbb{R}$ such that

$$x := e^{it}z \in E.$$

Now, going back to (6.28), yields

$$\mu x_0 = \text{Re}\left(e^{i\theta}z\right) = \text{Re}\left(e^{i(\theta-t)}e^{it}z\right)$$
$$= \text{Re}\left(e^{i(\theta-t)}x\right) = \text{Re}\left(e^{i(\theta-t)}\right)x = \cos\left(\theta-t\right)x$$

and hence,

$$\cos\left(\theta-t\right) \neq 0,$$

because $\mu \neq 0$ and $x_0 \gg 0$. Therefore,

$$z = e^{-it}x = \frac{\mu e^{-it}}{\cos\left(\theta-t\right)}x_0.$$

This is impossible, since this identity implies

$$\lambda z = T_{\mathbb{C}}z = \rho z$$

and, consequently, $\lambda = \rho$. This contradiction ends the proof. $\qquad\square$

The previous lemmas together with the next one complete the proofs of Parts (a) and (c) of Theorem 6.3.

Lemma 6.8. *One has that* $|\lambda| < \rho$ *for all* $\lambda \in \sigma(T) \setminus \{\rho\}$. *Therefore,*

$$\rho = \text{spr } T.$$

Proof.   Obviously, the estimate holds if $\lambda = 0$. So, suppose $\lambda \neq 0$ and let $z \in E_{\mathbb{C}} \setminus \{0\}$ such that

$$T_{\mathbb{C}} z = \lambda z.$$

Fix $\theta \in \mathbb{R}$. Then, since

$$\lim_{\mu \uparrow \infty} \left(x_0 - \mu^{-1} \operatorname{Re} \left(e^{i\theta} z \right) \right) = x_0 \gg 0,$$

there exists $\mu_\theta > 0$ such that

$$\operatorname{Re} \left(e^{i\theta} z \right) \ll \mu_\theta x_0 \ll \mu x_0 \qquad \text{for all} \quad \mu > \mu_\theta.$$

Moreover, as the map

$$\theta \mapsto \operatorname{Re} \left(e^{i\theta} z \right)$$

is continuous, there exists $\epsilon_\theta > 0$ such that

$$\operatorname{Re} \left(e^{it} z \right) \ll \mu_\theta x_0 \qquad \text{for all} \quad t \in I_\theta := [\theta - \epsilon_\theta, \theta + \epsilon_\theta].$$

Consequently,

$$\operatorname{Re} \left(e^{it} z \right) \ll \mu x_0 \tag{6.31}$$

for all $t \in I_\theta$ and $\mu \geq \mu_\theta$. As $[0, 2\pi]$ is compact, there exist an integer $p \geq 1$ and p points

$$\theta_j \in [0, 2\pi], \qquad j \in \{1, ..., p\},$$

such that

$$[0, 2\pi] \subset \bigcup_{j=1}^{p} I_{\theta_j}.$$

Set

$$\tilde{\mu} := \max_{1 \leq j \leq p} \mu_{\theta_j} > 0.$$

Then, by construction, the estimate (6.31) holds for all $t \in [0, 2\pi]$ and $\mu \geq \tilde{\mu}$. Therefore, by 2π-periodicity, we obtain that

$$\operatorname{Re} \left(e^{i\theta} z \right) \leq \mu x_0 \qquad \forall \ \theta \in \mathbb{R}, \ \mu \geq \tilde{\mu}. \tag{6.32}$$

Now, we will show that (6.32) fails if we take $\mu = 0$. The proof will proceed by contradiction. Suppose

$$\operatorname{Re} \left(e^{i\theta} z \right) \leq 0 \qquad \forall \ \theta \in \mathbb{R}.$$

Then, setting

$$z = x + iy, \qquad x, y \in E,$$

we obtain that

$$x \cos \theta \leq y \sin \theta, \qquad \theta \in \mathbb{R}. \tag{6.33}$$

In particular, for (6.33) at $\theta = \pm \frac{\pi}{4}$, we find that

$$x = y$$

and hence, (6.33) implies that

$$(\sin \theta - \cos \theta)x \geq 0, \qquad \theta \in \mathbb{R}. \tag{6.34}$$

For (6.34) at $\theta = \frac{\pi}{2}$ and $\theta = 0$ gives $x \geq 0$ and $-x \geq 0$. Thus,

$$x \in P \cap (-P) = \{0\}$$

and, therefore, $x = 0$, which implies $z = 0$. This is a contradiction. Consequently, (6.32) fails at $\mu = 0$, and, therefore, there exists a *minimal* $\mu_0 > 0$ such that

$$\mathrm{Re}\left(e^{i\theta}z\right) \leq \mu_0 x_0 \qquad \forall \; \theta \in \mathbb{R}.$$

According to Lemma 6.7, we must have

$$\mathrm{Re}\left(e^{i\theta}z\right) < \mu_0 x_0 \qquad \forall \; \theta \in \mathbb{R}$$

and, hence, due to (6.17) and (6.20), we find that

$$T\left(\mathrm{Re}\left(e^{i\theta}z\right)\right) \ll \mu_0 \rho x_0 \qquad \forall \; \theta \in \mathbb{R}. \tag{6.35}$$

As $\lambda \neq 0$, $|\lambda| > 0$ and there exists $\tau \in \mathbb{R}$ such that

$$\lambda = |\lambda|e^{i\tau}.$$

Thus, it follows from (6.35) that

$$\mu_0 \rho x_0 \gg T\left(\mathrm{Re}\left(e^{i\theta}z\right)\right) = \mathrm{Re}\left(e^{i\theta}T_{\mathbb{C}}z\right)$$
$$= \mathrm{Re}\left(e^{i\theta}\lambda z\right) = |\lambda|\,\mathrm{Re}\left(e^{i(\theta+\tau)}z\right)$$

for all $\theta \in \mathbb{R}$. Consequently,

$$\mathrm{Re}\left(e^{it}z\right) \ll \mu_0 \frac{\rho}{|\lambda|}x_0 \qquad \forall \; t \in \mathbb{R}.$$

By the minimality of μ_0, we must have

$$\rho \geq |\lambda|.$$

Moreover, if $\rho = |\lambda|$, then

$$\mathrm{Re}\left(e^{it}z\right) \ll \mu_0 x_0 \qquad \forall \; t \in \mathbb{R},$$

and, so, μ_0 could be shortened. Therefore, $\rho > |\lambda|$, as requested.

Finally, thanks to Theorem 6.2(a), it is obvious that

$$0 < \rho = \mathrm{spr}\, T.$$

This concludes the proof. $\qquad\qquad\qquad\qquad\qquad\qquad\qquad\qquad\qquad\qquad\qquad\square$

The next lemma shows Part (b) of Theorem 6.3.

Lemma 6.9. $\rho = \operatorname{spr} T$ *is the unique real eigenvalue of T to a positive eigenvector $x > 0$.*

Proof. Let $\lambda \in [\mathbb{R} \cap \sigma(T)] \setminus \{\rho\}$ such that

$$Tx = \lambda x$$

for some $x > 0$. Then, since T is strongly positive, we have that $Tx \gg 0$ and hence $\lambda > 0$. Thus, thanks to Lemma 6.8,

$$0 < \lambda < \rho.$$

Also, by Lemma 6.1, $x \gg 0$. Thus,

$$0 \ll x_0 \leq \mu x$$

for sufficiently large μ. Let $\mu_0 > 0$ be the minimum among all these μ's. Then,

$$0 \ll x_0 < \mu_0 x, \tag{6.36}$$

because $x_0 = \mu_0 x$ implies

$$\rho x_0 = T x_0 = \mu_0 T x = \mu_0 \lambda x = \lambda x_0,$$

and hence $\lambda = \rho$, which is a contradiction. This shows (6.36) and, consequently,

$$T x_0 = \rho x_0 \ll \mu_0 T x = \mu_0 \lambda x.$$

Therefore,

$$x_0 \ll \mu_0 \frac{\lambda}{\rho} x,$$

which contradicts the minimality of μ_0, as $\lambda < \rho$, and ends the proof. $\square$

To complete the proof of Theorem 6.3, it remains to prove Parts (d), (e) and (f). First, we will prove Part (d). According to (6.8), for every $\lambda > \operatorname{spr} T$, we have that

$$\mathcal{R}(\lambda; T) := \sum_{n=0}^{\infty} \lambda^{-(n+1)} T^n.$$

Therefore, $\mathcal{R}(\lambda; T)$ must be strongly positive, because $\lambda^{-(n+1)} > 0$ and T^n is strongly positive for all $n \geq 1$. Indeed, let $x > 0$. Then,

$$y := \lambda^{-1} x + \sum_{n=2}^{\infty} \lambda^{-(n+1)} T^n x \geq 0$$

and
$$\mathcal{R}(\lambda; T)x = \lambda^{-2}Tx + y.$$

As T is strongly positive, Lemma 6.1 implies that
$$\lambda^{-2}Tx \in \overset{\circ}{P}.$$

Thus,
$$\lambda^{-2}Tx + z \in P$$

for sufficiently small $z \in E$, and, therefore, for such z's,
$$\mathcal{R}(\lambda; T)x + z = \lambda^{-2}Tx + z + y \in P,$$

because $P + P \subset P$. This shows that
$$\mathcal{R}(\lambda; T)x \in \overset{\circ}{P}$$

and concludes the proof of Part (d).

Now, we will prove Part (e). We already know that $\operatorname{spr} T$ is a simple eigenvalue of T. Thus, by Theorem 6.2(c), there exists $\epsilon > 0$ such that
$$\mathcal{R}(\zeta; T) = \sum_{n=-1}^{\infty} (\zeta - \operatorname{spr} T)^n \mathfrak{T}_n, \qquad 0 < |\zeta - \operatorname{spr} T| \le \epsilon,$$

where
$$\mathfrak{T}_n := \frac{1}{2\pi i} \int_{|\zeta - \operatorname{spr} T| = \delta} (\zeta - \operatorname{spr} T)^{-(n+1)} \mathcal{R}(\zeta; T) \, d\zeta$$

for all $n \ge -1$ and $\delta \in (0, \epsilon]$. Moreover, $\mathfrak{T}_{-1}$ is a linear projection of E onto
$$R[\mathfrak{T}_{-1}] = N[\operatorname{spr} T I_E - T] = \operatorname{span} [x_0] \tag{6.37}$$

and, therefore,
$$\operatorname{tr} \mathfrak{T}_{-1} := \dim R[\mathfrak{T}_{-1}] = 1.$$

On the other hand, we have that
$$\mathfrak{T}_{-1} = \lim_{\substack{\lambda \in \mathbb{R} \\ \lambda \downarrow \operatorname{spr} T}} [(\lambda - \operatorname{spr} T)\, \mathcal{R}(\lambda; T)]$$

and, due to Part (d), $\mathcal{R}(\lambda; T)$ is strongly positive for all $\lambda > \operatorname{spr} T$. Thus, $\mathfrak{T}_{-1} \ge 0$ in the sense that
$$\mathfrak{T}_{-1}x \ge 0 \qquad \text{for all} \quad x \in P, \tag{6.38}$$

because $\bar{P} = P$. Moreover, there exists $x \in P$ such that
$$\mathfrak{T}_{-1}x > 0.$$

Indeed, if $\mathfrak{T}_{-1} = 0$ in P, then, since P is generating, $\mathfrak{T}_{-1} = 0$ in E, which is impossible, by (6.37). Again by (6.37), there exists $\xi > 0$ such that

$$\mathfrak{T}_{-1}x = \xi x_0 \gg 0.$$

Consequently, for sufficiently small $\lambda - \operatorname{spr} T < 0$, we find that

$$\mathcal{R}(\lambda; T)x = \frac{\xi}{\lambda - \operatorname{spr} T}\, x_0 + \sum_{n=0}^{\infty} (\lambda - \operatorname{spr} T)^n \mathfrak{T}_n x \ll 0,$$

which concludes the proof of Part (e).

Finally, we will prove Theorem 6.3(f). By the results on Section 6.2, $T^* \in \mathcal{K}(E')$, $\sigma(T) = \sigma(T^*)$, and

$$\operatorname{spr} T^* = \operatorname{spr} T$$

is a simple eigenvalue of T^*. Consequently, applying Theorem 6.2(c) to T^*, there exists $\epsilon > 0$ such that

$$\mathcal{R}(\zeta; T^*) = \sum_{n=-1}^{\infty} (\zeta - \operatorname{spr} T)^n \tilde{\mathfrak{T}}_n, \qquad 0 < |\zeta - \operatorname{spr} T| \le \epsilon,$$

where

$$\tilde{\mathfrak{T}}_n := \frac{1}{2\pi i} \int_{|\zeta - \operatorname{spr} T| = \delta} (\zeta - \operatorname{spr} T)^{-(n+1)} \mathcal{R}(\zeta; T^*)\, d\zeta, \qquad n \ge -1.$$

Thus, owing to (6.16), we have that

$$\tilde{\mathfrak{T}}_n = \frac{1}{2\pi i} \int_{|\zeta - \operatorname{spr} T| = \delta} (\zeta - \operatorname{spr} T)^{-(n+1)} \mathcal{R}(\zeta; T)^*\, d\zeta = \mathfrak{T}_n^*$$

for all $n \ge -1$ and $\delta \in (0, \epsilon]$. Moreover, $\mathfrak{T}_{-1}^*$ is a linear projection of E' onto

$$R[\mathfrak{T}_{-1}^*] = N[\operatorname{spr} T\, I_{E'} - T^*] \tag{6.39}$$

and hence, by (6.11) and Theorem 6.3(a),

$$\operatorname{tr} \mathfrak{T}_{-1}^* := \dim R[\mathfrak{T}_{-1}^*] = 1. \tag{6.40}$$

Also, by definition, for every $x' \in P^*$, we have that $x'(x) \ge 0$ for all $x \in P$ and

$$\mathfrak{T}_{-1}^* x'(x) = x'(\mathfrak{T}_{-1}x).$$

Thus, it follows from (6.38) that

$$\mathfrak{T}_{-1}^* x'(x) \ge 0 \qquad \forall\, x \in P, \ x' \in P^*,$$

and, therefore,
$$\mathfrak{T}^*_{-1}(P^*) \subset P^*.$$
As P is normal, i.e., P^* is generating, and, due to (6.40), $\mathfrak{T}^*_{-1} \neq 0$, there exists $y'_0 \in P^* \setminus \{0\}$ such that
$$x'_0 := \mathfrak{T}^*_{-1}(y'_0) \in P^* \setminus \{0\}.$$
By (6.39), we infer that
$$T^* x'_0 = \operatorname{spr} T x'_0.$$
Thus,
$$N[\operatorname{spr} T I_{E'} - T^*] = \operatorname{span}[x'_0]$$
and (6.11) implies that
$$R[\operatorname{spr} T I_E - T] = \operatorname{span}[x'_0]^\perp = \ker x'_0. \qquad (6.41)$$
Next, we will show that
$$x_0 \notin R[\operatorname{spr} T I_E - T] = \ker x'_0. \qquad (6.42)$$
The proof of (6.42) proceeds by contradiction. Suppose
$$\operatorname{spr} Tx - Tx = x_0$$
for some $x \in E$. Then,
$$(\operatorname{spr} T I_E - T)^2 x = (\operatorname{spr} T I_E - T)x_0 = 0$$
and, therefore,
$$x \in N[(\operatorname{spr} T I_E - T)^2] \setminus N[\operatorname{spr} T I_E - T]$$
which contradicts the fact that $\operatorname{spr} T$ is a simple eigenvalue of T. This shows (6.42) and hence $x'_0(x_0) \neq 0$. Necessarily, $x'_0(x_0) > 0$, because $x'_0 \in P^*$.

Finally, let $x \in P \setminus \{0\}$ be arbitrary. Then,
$$\operatorname{spr} T x'_0(x) = T^* x'_0(x) = x'_0(Tx) > 0.$$
Indeed, since $Tx \in \overset{\circ}{P}$, there exists $\epsilon > 0$ such that
$$Tx - \epsilon x_0 \in \overset{\circ}{P}$$
and hence,
$$x'_0(Tx - \epsilon x_0) \geq 0,$$
which implies
$$x'_0(Tx) \geq \epsilon x'_0(x_0) > 0$$
and, therefore, $x'_0(x) > 0$. Consequently, thanks to (6.41), we conclude that
$$x \notin R[\operatorname{spr} T I_E - T],$$
which ends the proof of Theorem 6.3.

6.6   Comments on Chapter 6

Section 6.1 introduces some basic concepts of the theory of ordered Banach spaces. It strictly contains the necessary background to prove the generalized version of the Krein–Rutman theorem studied in this chapter. As we learned most of these materials from H. Amann [8] and H. H. Schaefer [194], we have adapted the contents of these references to our purposes here, though we tidied up considerably some of the materials of H. Amann [8] (compare our proof of Lemma 6.2 with the sketch of the proof of Proposition 1.7 of H. Amann [8] through the — unnecessary — concept of *open decomposition* of E). The abstract theory of H. H. Schaefer [194], in the framework of topological vector spaces, widely surpasses the limitations imposed to this chapter.

According to Theorem V.5.5 of H. H. Schaefer [194], if E is an ordered Banach space with generating positive cone P, then, every positive linear form on E is continuous. This property, going back to V. L. Klee [114], reveals the existence of extremely sharp connections between topology and algebra within the theory of positive operators.

Although H. H. Scheafer [194] attributed to M. G. Krein [119] the proof of the fact that P^* is generating if P is normal (see Lemma V.3.2.1 of [194]), the whole characterization, P^* *is generating if and only if P is normal*, might be attributed to J. Grosberg and M. G. Krein [89].

The Krein–Rutman theorem provides us with the most natural extension of the Frobenius–Perron theorem to the general context of ordered Banach spaces. Although the original setting of the Krein–Rutman theorem was widely extended to cover abstract topological vectors spaces (see H. H. Schaefer [194] and its list of references), M. G. Krein himself and his collaborators, instead, begun a systematic program to study general spectral properties of Fredholm operators of index zero, whose highest climax was reached with the publication of the celebrated monograph of I. C. Göhberg and M. G. Krein [83] (see J. López-Gómez and C. Mora-Corral [153] for the most recent advances in this field).

The proof of Theorem 6.3(a)-(c) is based upon P. Takác [218]. Takác's proof, besides it admits a number of geometrical and analytical interpretations, as it relies upon a quite natural iterative method, it has the tremendous advantage over other available proofs that it does not make use of any sophisticated mathematical tool. As a consequence, it can be comfortably taught at undergraduate level. There are other elementary proofs in the literature, as, for instance, the dynamical one of N. D. Alikakos and G. Fusco

[6], but yet the students should be familiar with the most basic concepts of the theory of dynamical systems, like ω-limits and their properties, for reading it comfortably.

The proof of Theorem 6.3(d), though of a different nature, is also elementary. But the proof of Theorem 6.3(e) is far from elementary, as it uses some advanced mathematical tools and results from operator theory. Theorem 6.3(e) provides us with a rather abstract *anti-maximum principle* within the spirit of Theorem 4.1 of P. Takác [219]. The proofs of Theorem 6.3(e) and (f) in this book have been elaborated from the proof of the generalized version of the Krein–Rutman theorem collected on p. 265 of the Appendix of H. H. Schaefer [194], which reads as follows.

Theorem 6.4 (of M.G. Krein and M. A Rutman). *Let E be an ordered Banach space with total positive cone P, and let T be a compact positive endomorphism of E. If T has a spectral radius $\operatorname{spr} T > 0$, then $\operatorname{spr} T$ is a pole of the resolvent of maximal order on the spectral circle*

$$|\zeta| = \operatorname{spr} T,$$

with an eigenvector in P. A corresponding result holds for the adjoint T^* in E'.

Essentially, the strong positivity on T imposed in Theorem 6.3 is necessary for obtaining the simplicity of $\operatorname{spr} T$. But it is unnecessary for the existence of a positive eigenvector associated to $\operatorname{spr} T$, as it becomes apparent from Theorem 6.4.

The fact that, in the context of Theorem 6.4, $\operatorname{spr} T$ is a pole of the resolvent follows from a classical theorem by Pringsheim (see p. 262 of H. H. Schaefer [194]). The fact that T and T^* admit a positive eigenvector in P and P^*, respectively, associated to $\operatorname{spr} T$, can be obtained by easily adapting our proof of Theorem 6.3(f), as in H. H. Scheafer [194].

The proof of the fact that $x_0'(x) > 0$ for all $x \in P \setminus \{0\}$ in the final part of the proof of Theorem 6.3 was communicated by H. Amann to the author through an electronic mail.

Chapter 7

The strong maximum principle

Throughout this chapter we work under the general assumptions of Chapter 5. Thus, besides the general assumptions of Chapter 4, we suppose that $\partial\Omega$ is of class $\mathcal{C}^2$.

The main goal of the chapter is to study the scalar version of Theorem 2.1 of J. López-Gómez and M. Molina-Meyer [148], which was established for a general class of cooperative systems under Dirichlet boundary conditions. Such result characterized whether, or not, $(\mathfrak{L}, \mathfrak{B}, \Omega)$ satisfies the strong maximum principle, or, equivalently, the maximum principle, in terms of the positivity of the principal eigenvalue of the linear eigenvalue problem

$$\begin{cases} \mathfrak{L}\varphi = \tau\varphi & \text{in } \Omega, \\ \mathfrak{B}\varphi = 0 & \text{on } \partial\Omega, \end{cases} \tag{7.1}$$

as well as in terms of the existence of a positive strict supersolution h of $(\mathfrak{L}, \mathfrak{B}, \Omega)$. A function h is said to be a *strict supersolution* of $(\mathfrak{L}, \mathfrak{B}, \Omega)$ if

$$\begin{cases} \mathfrak{L}h \geq 0 & \text{in } \Omega, \\ \mathfrak{B}h \geq 0 & \text{on } \partial\Omega, \\ (\mathfrak{L}h, \mathfrak{B}h) \neq (0,0) & \text{in } \Omega \times \partial\Omega, \end{cases}$$

and $(\mathfrak{L}, \mathfrak{B}, \Omega)$ is said to satisfy the strong maximum principle if any strict supersolution h of $(\mathfrak{L}, \mathfrak{B}, \Omega)$ satisfies

$$h(x) > 0 \quad \forall\, x \in \Omega \cup \Gamma_1 \quad \text{and} \quad \frac{\partial h}{\partial\nu}(x) < 0 \quad \forall\, x \in h^{-1}(0) \cap \Gamma_0,$$

while it is said to satisfy the maximum principle when every supersolution h of $(\mathfrak{L}, \mathfrak{B}, \Omega)$ satisfies $h \geq 0$. The principal eigenvalue of (7.1) is the unique $\tau \in \mathbb{R}$ for which (7.1) admits a positive eigenfunction $\varphi > 0$. Throughout the rest of this book it will be denoted by

$$\sigma_0 := \sigma[\mathfrak{L}, \mathfrak{B}, \Omega].$$

The theorem of characterization of the strong maximum principle has shown to be a milestone for the development of the modern theory of nonlinear second order elliptic and parabolic equations via bifurcation theory and the method of sub- and supersolutions (cf. J. López-Gómez and R. M. Pardo [154], J. López-Gómez [137], M. Molina-Meyer [163–165], J. M. Fraile et al. [66], J. García-Melián et al. [76], H. Amann and J. López-Gómez [13], M. Delgado et al. [49], J. López-Gómez [138, 139, 143, 144], H. Amann [11], Y. Du [56], and the list of references therein).

As an application of the theorem of characterization of the strong maximum principle, when $\mathfrak{B} = \mathfrak{D}$ ($\Gamma_1 = \emptyset$) we will be able to re-formulate the generalized minimum principle of M. H. Protter and H. F. Weinberger (Theorem 1.7) in terms of the positivity of the principal eigenvalue σ_0, instead of on the existence of a supersolution h such that $h(x) > 0$ for all $x \in \bar{\Omega}$. This feature provides us with an extremely sharp improvement of the classical theory covered by Chapter 1, as we are actually characterizing whether the supersolution h exists. More precisely, as a result of the characterization theorem, such an h exists if and only if $\sigma_0 > 0$ and, in such case, there are infinitely many h's for which all the conclusions of Theorem 1.7 hold. Indeed, if $\sigma_0 > 0$, then, for every $f \geq 0$ and $g \geq 0$ with $\inf_{\partial\Omega} g > 0$, the unique solution of

$$\begin{cases} \mathfrak{L}h = f & \text{in } \Omega, \\ h = g & \text{on } \partial\Omega, \end{cases}$$

provides us with an admissible strict supersolution h.

This chapter is distributed as follows. Section 7.1 adapts the classical theory of Chapters 1 and 2 to deal with supersolutions $h, u \in W^{2,p}(\Omega)$, $p > N$, through the main theorem of J. M. Bony [28]. Section 7.2 derives the existence and the uniqueness of the principal eigenvalue of (7.1) from Theorem 6.3. Section 7.3 shows that any *weak eigenvalue* τ must actually be a *classical eigenvalue*, in the sense that any eigenfunction associated to τ in the weak sense must be a classical eigenfunction. Section 7.4 shows that σ_0 is *simple* and *dominant*. By dominant, it is meant that

$$\operatorname{Re} \tau \geq \sigma_0$$

for any other eigenvalue τ of (7.1). This result is imperative for ascertaining the local stability character of the solutions of large classes of parabolic quasi-linear problems through the signs of the principal eigenvalues of the associated variational problems. Section 7.5 states and proves the main theorem of this chapter. Precisely, it establishes that the following five conditions are equivalent:

- $\sigma_0 := \sigma[\mathfrak{L}, \mathfrak{B}, \Omega] > 0$.
- $(\mathfrak{L}, \mathfrak{B}, \Omega)$ admits a positive strict supersolution $h \in W^{2,p}(\Omega)$, $p > N$.
- $(\mathfrak{L}, \mathfrak{B}, \Omega)$ satisfies the strong maximum principle.
- $(\mathfrak{L}, \mathfrak{B}, \Omega)$ satisfies the maximum principle.
- The resolvent of $(\mathfrak{L}, \mathfrak{B}, \Omega)$ is strongly positive.

Finally, Section 7.6 goes back to Chapter 1 for discussing the range of validity and applicability of the classical minimum principles of E. Hopf (Theorem 1.2) and M. H. Protter and H. F. Weinberger (Theorem 1.7). Essentially, Section 7.6 polishes and updates these classical results in the light of the main theorem of this chapter.

7.1 Minimum principle of J. M. Bony

This section extends the results of Chapters 1 and 2 to cover the case when $u, h \in W^{2,p}(\Omega)$ for some $p > N$, instead of $u, h \in \mathcal{C}^2(\Omega) \cap \mathcal{C}^1(\bar{\Omega})$. Note that, according to Theorem 4.2,

$$W^{2,p}(\Omega) \subset \mathcal{C}^{1,1-\frac{N}{p}}(\bar{\Omega})$$

for all $p > N$. Also, by Theorem 4.4, any function $u \in W^{2,p}(\Omega)$, with $p > N$, is twice classically differentiable almost everywhere in Ω.

The results of this section are valid for a general uniformly elliptic operator $\mathfrak{L}$ whose coefficients satisfy

$$a_{ij} \in \mathcal{C}(\bar{\Omega}), \qquad b_j, c \in L^\infty(\Omega), \qquad \forall\, i, j \in \{1, ..., N\}, \qquad (7.2)$$

though, eventually, (7.2) could be relaxed. So, throughout this section we will impose (7.2).

A careful reading of Chapters 1 and 2 reveals that Theorems 2.1 and 2.4 are based upon Theorem 1.2, which is a direct consequence from Theorem 1.1. Consequently, thanks to the next theorem, which is the bulk of this section, the regularity requirements in the results of Chapters 1 and 2 can be substantially relaxed up to deal with supersolutions $u, h \in W^{2,p}(\Omega)$, with $p > N$.

Theorem 7.1 (Minimum principle of J. M. Bony). *Suppose $c \geq 0$ and $u \in W^{2,p}(\Omega)$, $p > N$, satisfies*

$$\inf_{K} \mathrm{ess}\, \mathfrak{L}u > 0 \qquad \textit{for every compact subset} \quad K \subset \Omega.$$

Then, u cannot attain a local minimum $m \leq 0$ in Ω.

The proof of Theorem 7.1 follows after a series of preliminary technical results. Essentially, the next lemma establishes that any sufficiently regular map sends negligible sets into negligible sets. As usual, for any given measurable set M, $|M|$ stands for the measure of M, as discussed by H. Lebesgue [124].

Lemma 7.1. *Let*

$$f : \Omega \to \mathbb{R}^N, \qquad f = (f_1, ..., f_N),$$

with $f_i \in W^{1,p}(\Omega)$, $p > N$, for all $1 \le i \le N$. Then,

$$M \subset \Omega \quad \text{and} \quad |M| = 0 \quad \Longrightarrow \quad |f(M)| = 0.$$

Proof. Subsequently, for every $x_0 \in \mathbb{R}^N$ and $\gamma > 0$, we denote by $C_\gamma(x_0)$ the closed γ-cube centered at x_0

$$C_\gamma(x_0) := x_0 + \left[-\frac{\gamma}{2}, \frac{\gamma}{2} \right]^N.$$

By Theorem 4.2, there exists a constant $K > 0$ such that

$$\max_{x,y \in C_1(0)} |g(x) - g(y)| \le K \left(\int_{C_1(0)} |\nabla g|^p \right)^{\frac{1}{p}}$$

for every $g \in W^{1,p}(C_1(0))$.

Let $x_0 \in \Omega$ and $\gamma > 0$ such that $C_\gamma(x_0) \subset \Omega$. Then, the functions

$$g_i(x) := f_i(\gamma x + x_0), \quad x \in C_1(0), \qquad 1 \le i \le N,$$

satisfy $g_i \in W^{1,p}(C_1(0))$ for all $1 \le i \le N$. Thus,

$$\max_{x,y \in C_\gamma(x_0)} |f_i(x) - f_i(y)| = \max_{x,y \in C_1(0)} |f_i(\gamma x + x_0) - f_i(\gamma y + x_0)|$$

$$= \max_{x,y \in C_1(0)} |g_i(x) - g_i(y)|$$

$$\le K \left(\int_{C_1(0)} |\nabla g_i|^p \right)^{\frac{1}{p}}$$

$$= K\gamma \left(\int_{C_1(0)} |\nabla f_i(\gamma \cdot + x_0)|^p \right)^{\frac{1}{p}}$$

$$= \gamma^{1 - \frac{N}{p}} K \left(\int_{C_\gamma(x_0)} |\nabla f_i|^p \right)^{\frac{1}{p}}$$

for all $1 \leq i \leq N$. Moreover, since

$$|\nabla f_i| = \left(\sum_{j=1}^{N} \left| \frac{\partial f_i}{\partial x_j} \right|^2 \right)^{\frac{1}{2}} \leq \sum_{j=1}^{N} \left| \frac{\partial f_i}{\partial x_j} \right| \leq \sum_{i,j=1}^{N} \left| \frac{\partial f_i}{\partial x_j} \right|,$$

we find that

$$\max_{x,y \in C_\gamma(x_0)} |f_i(x) - f_i(y)| \leq \gamma^{1-\frac{N}{p}} K \left[\int_{C_\gamma(x_0)} \left(\sum_{i,j=1}^{N} \left| \frac{\partial f_i}{\partial x_j} \right| \right)^p \right]^{\frac{1}{p}}$$

for all $1 \leq i \leq N$. Therefore,

$$|f(C_\gamma(x_0))| \leq \gamma^{N\left(1-\frac{N}{p}\right)} K^N \left[\int_{C_\gamma(x_0)} \left(\sum_{i,j=1}^{N} \left| \frac{\partial f_i}{\partial x_j} \right| \right)^p \right]^{\frac{N}{p}}. \tag{7.3}$$

As $|M| = 0$, for every $\epsilon > 0$ there exists a sequence of disjoint cubes, say

$$C_n := C_{\gamma_n}(x_n), \qquad n \geq 1,$$

such that

$$M \subset \bigcup_{n \geq 1} C_n \subset \Omega \quad \text{and} \quad \sum_{n \geq 1} |C_n| = \sum_{n \geq 1} \gamma_n^N \leq \epsilon.$$

Hence, from (7.3) and the Hölder inequality it is apparent that

$$|f(M)| \leq \sum_{n \geq 1} |f(C_n)| \leq K^N \sum_{n \geq 1} \gamma_n^{N\left(1-\frac{N}{p}\right)} \left[\int_{C_n} \left(\sum_{i,j=1}^{N} \left| \frac{\partial f_i}{\partial x_j} \right| \right)^p \right]^{\frac{N}{p}}$$

$$\leq K^N \left(\sum_{n \geq 1} \gamma_n^N \right)^{1-\frac{N}{p}} \left[\int_\Omega \left(\sum_{i,j=1}^{N} \left| \frac{\partial f_i}{\partial x_j} \right| \right)^p \right]^{\frac{N}{p}}$$

$$\leq \epsilon^{1-\frac{N}{p}} K^N \left[\int_\Omega \left(\sum_{i,j=1}^{N} \left| \frac{\partial f_i}{\partial x_j} \right| \right)^p \right]^{\frac{N}{p}}$$

for all $\epsilon > 0$. Consequently, $|f(M)| = 0$. $\square$

The next result establishes the positivity of the quadratic form $D^2 u$ at any local strict minimum when $u \in W^{2,p}(\Omega)$ for some $p > N$.

Lemma 7.2. *Let* $u \in W^{2,p}(\Omega)$, $p > N$, *and* $x_0 \in \Omega$ *such that* u *attains a local strict minimum* $m \in \mathbb{R}$ *at* x_0, *i.e., there exists* $\delta > 0$ *such that*

$$u(x) > u(x_0) = m \qquad \forall \, x \in \bar{B}_\delta(x_0) \setminus \{x_0\}. \tag{7.4}$$

Then, for every $\epsilon > 0$, there exists a measurable subset M of $B_\epsilon(x_0)$ with $|M| > 0$ such that the quadratic form

$$D^2 u(x) := \left(\frac{\partial^2 u(x)}{\partial x_i \partial x_j} \right)_{1 \leq i,j \leq N}$$

is positive definite for almost every $x \in M$.

Proof. Let S denote the $\mathcal{C}^1$ surface of $\mathbb{R}^N \times \mathbb{R}$ determined by the graph of $y = u(x)$

$$S := \{ (x, u(x)) \ : \ x \in \Omega \}.$$

Fix $\epsilon > 0$ such that $\bar{B}_\epsilon(x_0) \subset \Omega$ and let M denote the set of points $x \in B_\epsilon(x_0)$ for which S lies above the tangent hyperplane $\mathbf{T}_x$ of S at x in a neighborhood of x. Obviously, M is a closed set and hence it is measurable and $x_0 \in M$, because u attains a local strict minimum at x_0. Also, $\nabla u(x_0) = 0$ and, so, $\mathbf{T}_{x_0}$ is given through $y = m$. Now, we will prove that the next property holds:

(P) There exists $\eta > 0$ such that for every $h \in B_\eta := B_\eta(0) \subset \mathbb{R}^N$, there is some $p \in \mathbb{R}^N$ for which the hyperplane

$$y = \langle h, x \rangle + p$$

is tangent to S at some point of M.

Indeed, thanks to (7.4), for every

$$\tilde{\delta} \in (0, \min\{\epsilon, \delta\}),$$

there exists $\eta := \eta(\tilde{\delta})$ such that

$$\langle h, x - x_0 \rangle + m < u(x), \qquad \tilde{\delta} < |x - x_0| < \delta,$$

for all $h \in B_\eta$. By construction, for every $h \in B_\eta$, there must exist a parallel hyperplane to

$$y = \langle h, x - x_0 \rangle + m \tag{7.5}$$

tangent to S at some point $x \in \bar{B}_{\tilde{\delta}} \cap M$. Indeed, it is the unique hyperplane parallel to (7.5) supporting S. This shows Property (P).

Let $f : \Omega \to \mathbb{R}^N$ denote the gradient map

$$f(x) := \nabla u(x), \qquad x \in \Omega.$$

According to Property (P), for every $h \in B_\eta$, there exists $x \in M$ such that

$$f(x) = \nabla u(x) = h.$$

Thus, $B_\eta \subset f(M)$ and, consequently,

$$|f(M)| > 0.$$

On the other hand, $u \in W^{2,p}(\Omega)$ implies that

$$f_i = \frac{\partial u}{\partial x_i} \in W^{1,p}(\Omega), \qquad 1 \le i \le N,$$

and, therefore, owing to Lemma 7.1, we find that

$$|M| > 0.$$

Furthermore, as $u \in W^{2,p}(\Omega)$ with $p > N$, u is twice classically differentiable almost everywhere in Ω. Consequently, the bilinear form $D^2 u(x)$ must be positive definite for almost every $x \in M$. This ends the proof. $\square$

Finally, Theorem 7.1 is a corollary from the next result.

Proposition 7.1. *Suppose $c \ge 0$, and $u \in W^{2,p}(\Omega)$, $p > N$, possesses a local minimum $m \le 0$ at some $x_0 \in \Omega$. Then,*

$$\limsup_{x \to x_0} \operatorname{ess} \mathfrak{L}u(x) \le 0. \tag{7.6}$$

Proof. Suppose u has a local strict minimum at x_0. By Lemma 7.2,

$$\liminf_{x \to x_0} \operatorname{ess} \Delta u(x) \ge 0. \tag{7.7}$$

Thus, by performing the linear change of coordinates (1.4) of the proof of Proposition 1.1, it is apparent that (7.7) entails

$$\limsup_{x \to x_0} \operatorname{ess} \left(-\sum_{i,j=1}^{N} a_{ij}(x) \frac{\partial^2 u}{\partial x_i \partial x_j}(x) \right) \le 0.$$

Note that, according to (7.2), $a_{ij} \in \mathcal{C}(\bar{\Omega})$ and, therefore,

$$\limsup_{x \to x_0} \operatorname{ess} \mathfrak{L}u(x) \le \limsup_{x \to x_0} \operatorname{ess} \left(-\sum_{i,j=1}^{N} a_{ij}(x) \frac{\partial^2 u}{\partial x_i \partial x_j}(x) \right)$$
$$+ \limsup_{x \to x_0} \operatorname{ess} (c(x)u(x)) \le 0,$$

because $\nabla u(x_0) = 0$, $c \ge 0$ and $m \le 0$. Consequently, (7.6) holds.

In the general case when x_0 is not a strict local minimum of u, one can apply the previous result to the auxiliary function

$$v(x) := u(x) + |x - x_0|^4, \qquad x \in \Omega,$$

to get

$$\limsup_{x \to x_0} \operatorname{ess} \mathfrak{L}v(x) \le 0,$$

which implies (7.6) and concludes the proof. $\square$

By inter-exchanging the roles of Theorem 7.1 and Theorem 1.1, the proof of Theorem 1.2 can be adapted, *mutatis mutandis*, to obtain the following generalized version of the minimum principle of E. Hopf.

Theorem 7.2. *Suppose $c \geq 0$, and $u \in W^{2,p}(\Omega)$, $p > N$, satisfies*

$$\mathfrak{L}u \geq 0 \quad in \quad \Omega, \qquad and \qquad m := \inf_{\Omega} u \in (-\infty, 0].$$

Then, either $u = m$ in Ω, or $u(x) > m$ for all $x \in \Omega$. In other words, u cannot attain m in Ω, unless $u = m$ in Ω. Moreover,

$$\inf_{\bar{\Omega}} u = \inf_{\partial \Omega} u = m$$

when Ω is bounded.

Similarly, Theorem 1.3 admits the next counterpart. Note that $u \in \mathcal{C}^1(\bar{\Omega})$ if $u \in W^{2,p}(\Omega)$ with $p > N$.

Theorem 7.3. *Suppose $c \geq 0$ and $u \in W^{2,p}(\Omega)$, $p > N$, is a non-constant function satisfying*

$$\inf \mathrm{ess}\, \mathfrak{L}u \geq 0 \quad in \quad \Omega, \qquad and \quad m := \inf_{\Omega} u \in (-\infty, 0].$$

Assume, in addition, that there exists $x_0 \in \partial\Omega$ such that $u(x_0) = m$, with Ω satisfying an interior sphere property at x_0. Then,

$$\frac{\partial u}{\partial \nu}(x_0) < 0.$$

Based on Theorems 7.2 and 7.3, the proofs of Theorems 2.1 and 2.4 can be easily adapted to get the next counterparts in $W^{2,p}(\Omega)$, $p > N$.

Theorem 7.4. *Suppose $(\mathfrak{L}, \mathfrak{B}, \Omega)$ admits a supersolution $h \in W^{2,p}(\Omega)$, $p > N$, such that*

$$h(x) > 0 \qquad for \ all \ x \in \bar{\Omega}.$$

Then, any supersolution $u \in W^{2,p}(\Omega)$ of $(\mathfrak{L}, \mathfrak{B}, \Omega)$ must satisfy some of the following alternatives:

A1. $u = 0$ in Ω.
A2. $u(x) > 0$ for every $x \in \Omega \cup \Gamma_1$, and

$$\frac{\partial u}{\partial \nu}(x) < 0 \qquad for \ all \quad x \in u^{-1}(0) \cap \Gamma_0.$$

A3. *There exists a constant $m < 0$ such that*

$$u = mh \quad in \ \bar{\Omega}.$$

In such case, much like in Theorem 2.1, $u(x) < 0$ for all $x \in \bar{\Omega}$, $\Gamma_0 = \emptyset$, and $\tau = 0$ must be an eigenvalue to a positive eigenfunction (h itself) of (7.1).

Theorem 7.5. *Suppose Ω is of class $\mathcal{C}^2$ and $(\mathfrak{L}, \mathfrak{B}, \Omega)$ has a positive supersolution $h \in W^{2,p}(\Omega)$, $p > N$. Then, any supersolution $u \in W^{2,p}(\Omega)$ of $(\mathfrak{L}, \mathfrak{B}, \Omega)$ must satisfy some of the Alternatives A1, A2, or A3, of Theorem 7.4.*

Note that, thanks to Remark 2.1, Conditions i) and ii) of Theorem 2.4 hold when Ω is of class $\mathcal{C}^2$.

7.2 The existence of the principal eigenvalue

The main goal of this section is to show the existence and the uniqueness of $\tau \in \mathbb{R}$ for which (7.1) admits a positive eigenfunction φ. This fundamental property will be derived by combining Theorem 6.3 with the next invertibility and positivity result.

Throughout the rest of this book, for every $\nu \in (0,1)$, we denote by $\mathcal{C}_{\mathfrak{B}}^{1,\nu}(\bar{\Omega})$ the Banach subspace of the Hölder space $\mathcal{C}^{1,\nu}(\bar{\Omega})$ (see Section 4.1.3) consisting of all functions $u \in \mathcal{C}^{1,\nu}(\bar{\Omega})$ such that $\mathfrak{B}u = 0$ on $\partial\Omega$. Similarly,

$$\mathcal{C}_{\mathfrak{B}}^{1}(\bar{\Omega}) := \{\, u \in \mathcal{C}^1(\bar{\Omega}) \ : \ \mathfrak{B}u = 0 \ \text{on} \ \partial\Omega \,\}.$$

Theorem 7.6. *There exists $\omega_0 \in \mathbb{R}$ such that for every $\omega \geq \omega_0$ and $f \in \mathcal{C}(\bar{\Omega})$ the boundary value problem*

$$\begin{cases} (\mathfrak{L} + \omega)u = f & in \ \Omega, \\ \mathfrak{B}u = 0 & on \ \partial\Omega, \end{cases} \tag{7.8}$$

possesses a unique weak solution $u \in W_{\Gamma_0}^{1,2}(\Omega)$. Moreover,

$$u \in W^{2,\infty^-}(\Omega) \subset \mathcal{C}^{1,1^-}(\bar{\Omega}),$$

u is twice classically differentiable almost everywhere in Ω, and u is a classical solution of (7.8) (in the sense of Definition 4.1).

Also, ω_0 can be chosen so that $f > 0$ ($f \geq 0$ with $f \neq 0$) implies

$$u(x) > 0 \quad \forall\, x \in \Omega \cup \Gamma_1 \quad and \quad \frac{\partial u}{\partial \nu}(x) < 0 \quad \forall\, x \in \Gamma_0. \tag{7.9}$$

Furthermore, for every $\nu \in (0,1)$, the resolvent operator

$$\mathcal{C}(\bar{\Omega}) \xrightarrow{(\mathfrak{L}+\omega)^{-1}} \mathcal{C}_{\mathfrak{B}}^{1,\nu}(\bar{\Omega})$$

$$f \quad \mapsto \quad (\mathfrak{L}+\omega)^{-1}f := u$$

is linear and continuous. Consequently, the operator

$$\mathcal{C}(\bar{\Omega}) \xrightarrow{(\mathfrak{L}+\omega)^{-1}} \mathcal{C}_{\mathfrak{B}}^{1}(\bar{\Omega})$$

(7.10)

$$f \quad \mapsto \quad (\mathfrak{L}+\omega)^{-1}f := u$$

is linear, continuous and compact.

Proof. The existence and the uniqueness of the weak solutions for sufficiently large ω follows from Theorem 4.11. The regularity of the weak solution is a consequence from Theorem 5.11. The positivity properties established by (7.9) follow straightaway from Lemma 2.1, Proposition 2.1 and Theorem 7.4. Indeed, by Lemma 2.1 and Proposition 2.1, there exists $\omega_0 \in \mathbb{R}$ such that, for every $\omega > \omega_0$, $(\mathfrak{L}+\omega, \mathfrak{B}, \Omega)$ possesses a strict supersolution $h \in \mathcal{C}^2(\bar{\Omega})$ with $h(x) > 0$ for all $x \in \bar{\Omega}$. Thus, for sufficiently large $\omega \geq \omega_0$, the unique weak solution of (7.8) must satisfy some of the alternatives of Theorem 7.4 if $f > 0$. Clearly, Alternative A1 cannot occur because $u \neq 0$. Similarly, if $u = mh$ with $m < 0$, then

$$f = (\mathfrak{L}+\omega)u = m(\mathfrak{L}+\omega)h \leq 0,$$

which is impossible. Therefore, Alternative A2, and hence (7.9) holds.

The fact that $(\mathfrak{L}+\omega)^{-1}$ is linear follows from the linear structure of the problem and the uniqueness of the weak solution. Its continuity follows from Theorem 4.11 and the uniform elliptic estimates of Chapter 5. Indeed, let $\{f_n\}_{n\geq 1}$ be a sequence in $\mathcal{C}(\bar{\Omega})$ such that

$$\lim_{n\to\infty} \|f_n - f\|_{\mathcal{C}(\bar{\Omega})} = 0$$

(7.11)

for some $f \in \mathcal{C}(\bar{\Omega})$. Set

$$u := (\mathfrak{L}+\omega)^{-1}f, \qquad u_n = (\mathfrak{L}+\omega)^{-1}f_n, \qquad n \geq 1;$$

u is the unique weak solution of (7.8) and, for every $n \geq 1$, u_n is the unique weak solution of

$$\begin{cases} (\mathfrak{L}+\omega)u_n = f_n & \text{in } \Omega, \\ \mathfrak{B}u_n = 0 & \text{on } \partial\Omega. \end{cases}$$

By Theorem 4.11,

$$(\mathfrak{L} + \omega)^{-1} \in \mathcal{L}\left(L^2(\Omega), W^{1,2}_{\Gamma_0}(\Omega)\right),$$

and, due to (7.11),

$$\lim_{n \to \infty} \|f_n - f\|_{L^p(\Omega)} = 0 \qquad (7.12)$$

for all $p \geq 2$. Thus,

$$\lim_{n \to \infty} \|u_n - u\|_{W^{1,2}_{\Gamma_0}(\Omega)} = 0. \qquad (7.13)$$

On the other hand, for every $n, m \geq 1$, we have that

$$\begin{cases} (\mathfrak{L} + \omega)(u_n - u_m) = f_n - f_m & \text{in } \Omega, \\ \mathfrak{B}(u_n - u_m) = 0 & \text{on } \partial\Omega, \end{cases}$$

and hence, thanks to Theorem 5.10, for every $p \geq 2$, there exists a constant $C = C(p) > 0$ such that

$$\|u_n - u_m\|_{W^{2,p}(\Omega)} \leq C\|f_n - f_m\|_{L^p(\Omega)}$$

for all $n, m \geq 1$. According to (7.12), these estimates imply that $\{u_n\}_{n \geq 1}$ is a Cauchy sequence in $W^{2,p}(\Omega)$ for all $p \in [2, \infty)$. By (7.13), we must have $u \in W^{2,p}(\Omega)$ and

$$\lim_{n \to \infty} \|u_n - u\|_{W^{2,p}(\Omega)} = 0 \qquad \text{for all } p \geq 2.$$

Therefore, due to Theorem 4.2, we find that

$$\lim_{n \to \infty} \|u_n - u\|_{\mathcal{C}^{1,\nu}_{\mathfrak{B}}(\bar{\Omega})} = 0$$

for all $\nu \in (0, 1)$, which concludes the proof of the continuity.

The compactness of (7.10) is a consequence from the fact that $\mathcal{C}^{1,\nu}_{\mathfrak{B}}(\bar{\Omega})$ is compactly embedded in $\mathcal{C}^1_{\mathfrak{B}}(\bar{\Omega})$ for all $\nu \in (0, 1)$. Indeed, fix $\nu \in (0, 1)$ and let $\{u_n\}_{n \geq 1}$ be a bounded sequence of $\mathcal{C}^{1,\nu}_{\mathfrak{B}}(\bar{\Omega})$. Then, there exists a constant $C > 0$ such that

$$\|u_n\|_{\mathcal{C}^1(\bar{\Omega})} \leq C, \qquad \left|\frac{\partial u_n}{\partial x_j}(x) - \frac{\partial u_n}{\partial x_j}(y)\right| \leq C|x - y|^\nu, \qquad (7.14)$$

for all $n \geq 1$, $x, y \in \bar{\Omega}$, and $1 \leq j \leq N$. As $\{u_n\}_{n \geq 1}$ is bounded in $\mathcal{C}^1(\bar{\Omega})$, it is bounded and equicontinuous in $\mathcal{C}(\bar{\Omega})$. Thus, by the theorem of Arzela-Ascoli, there exists $u \in \mathcal{C}(\bar{\Omega})$ such that, along some subsequence relabeled by n, we have that

$$\lim_{n \to \infty} \|u_n - u\|_{\mathcal{C}(\bar{\Omega})} = 0.$$

According to (7.14), the sequence $\{\partial u_n / \partial x_1\}_{n \geq 1}$ is also bounded and equicontinuous in $\mathcal{C}(\bar{\Omega})$. Thus, there exists $v_1 \in \mathcal{C}(\bar{\Omega})$ such that, along some subsequence relabeled by n,

$$\lim_{n \to \infty} \left\| \frac{\partial u_n}{\partial x_1} - v_1 \right\|_{\mathcal{C}(\bar{\Omega})} = 0.$$

Repeating this argument N times, it is apparent that there exist $v_j \in \mathcal{C}(\bar{\Omega})$, $2 \leq j \leq N$, such that, along some subsequence relabeled by n,

$$\lim_{n \to \infty} \left\| \frac{\partial u_n}{\partial x_j} - v_j \right\|_{\mathcal{C}(\bar{\Omega})} = 0, \qquad 1 \leq j \leq N.$$

On the other hand, for every $x \in \bar{\Omega}$ and $n \geq 1$, we have that

$$u_n(x + h) - u_n(x) = \int_0^1 \langle \nabla u_n(x + th), h \rangle \, dt$$

for sufficiently small $h \in \mathbb{R}^N$. Consequently, letting $n \to \infty$, we find that

$$u(x + h) - u(x) = \int_0^1 \langle (v_1(x + th), ..., v_N(x + th)), h \rangle \, dt$$

for all $x \in \bar{\Omega}$ and sufficiently small h. From these identities it becomes apparent that $u \in \mathcal{C}^1(\bar{\Omega})$ and that

$$\nabla u = (v_1, ..., v_N) \qquad \text{in } \bar{\Omega}.$$

Therefore,

$$\lim_{n \to \infty} \| u_n - u \|_{\mathcal{C}^1(\bar{\Omega})} = 0$$

and, in particular, $\mathfrak{B}u = 0$ on $\partial \Omega$, because $\mathfrak{B}_n u = 0$ on $\partial \Omega$ for all $n \geq 1$. This ends the proof. $\square$

Throughout the rest of this chapter, we always take $\omega \geq \omega_0$. By the theorem of Arzela-Ascoli, the canonical injection

$$J : \mathcal{C}^1_{\mathfrak{B}}(\bar{\Omega}) \hookrightarrow \mathcal{C}(\bar{\Omega})$$

is a linear compact operator and, therefore,

$$R_\omega := J(\mathfrak{L} + \omega)^{-1} \; : \; \mathcal{C}(\bar{\Omega}) \longrightarrow \mathcal{C}(\bar{\Omega})$$

is a compact endomorphism of $\mathcal{C}(\bar{\Omega})$. The space $\mathcal{C}(\bar{\Omega})$ is an ordered Banach space with the natural ordering induced by its cone of non-negative functions

$$P := \{ u \in \mathcal{C}(\bar{\Omega}) \; : \; u \geq 0 \text{ in } \bar{\Omega} \}.$$

Obviously, the interior of P is given by

$$\overset{\circ}{P} = \{\, u \in \mathcal{C}(\bar{\Omega}) \;:\; u(x) > 0 \;\; \forall\, x \in \bar{\Omega} \,\} \neq \emptyset,$$

and, according to Theorem 6.1, P is normal, because the norm $\|\cdot\|_{\mathcal{C}(\bar{\Omega})}$ is monotone. But the resolvent operator R_ω is not strongly positive in $\mathcal{C}(\bar{\Omega})$ if $\Gamma_0 \neq \emptyset$, because, in such case, $R_\omega f = 0$ on Γ_0 for all $f \in P$ and, hence,

$$R_\omega f \notin \overset{\circ}{P}.$$

The Banach space $\mathcal{C}^1_{\mathfrak{B}}(\bar{\Omega})$ is also an ordered Banach space with positive cone

$$P_1 := \{\, u \in \mathcal{C}^1_{\mathfrak{B}}(\bar{\Omega}) \;:\; u \geq 0 \;\text{ in }\; \bar{\Omega} \,\} = P \cap \mathcal{C}^1_{\mathfrak{B}}(\bar{\Omega}).$$

Moreover, the interior of P_1, $\overset{\circ}{P_1}$, consists of the set of functions $u \in P_1$ satisfying (7.9). Thus,

$$R_\omega f \in \overset{\circ}{P_1} \qquad \forall\, f \in P_1 \setminus \{0\}$$

and, therefore, according to Theorem 7.6, the resolvent operator

$$R_{\omega,1} := R_\omega\big|_{\mathcal{C}^1_{\mathfrak{B}}(\bar{\Omega})} \;:\; \mathcal{C}^1_{\mathfrak{B}}(\bar{\Omega}) \longrightarrow \mathcal{C}^1_{\mathfrak{B}}(\bar{\Omega})$$

satisfies

$$R_{\omega,1}\,(P_1 \setminus \{0\}) \subset \overset{\circ}{P_1}.$$

Consequently, in this case, Theorem 6.3 applies for inferring the existence of the principal eigenvalue of (7.1), though, owing to Theorem 6.1, the cone P_1 is not normal, because the norm $\|\cdot\|_{\mathcal{C}^1_{\mathfrak{B}}(\bar{\Omega})}$ is not monotone, and hence Theorem 6.3(f) cannot be guaranteed to hold in this ordered Banach space. The lack of monotonicity of $\|\cdot\|_{\mathcal{C}^1_{\mathfrak{B}}(\bar{\Omega})}$ is inherent to the fact that it involves the derivatives of the functions on which it acts.

To overcome this shortcoming, one can proceed as follows. First, let e be the unique weak solution of

$$\begin{cases} (\mathfrak{L} + \omega)e = 1 & \text{in } \Omega, \\ \mathfrak{B}e = 0 & \text{on } \partial\Omega. \end{cases} \tag{7.15}$$

According to Theorem 7.6, we have that

$$e(x) > 0 \quad \forall\, x \in \Omega \cup \Gamma_1 \quad \text{and} \quad \frac{\partial e}{\partial \nu}(x) < 0 \quad \forall\, x \in \Gamma_0. \tag{7.16}$$

Then, consider the Banach space

$$\mathcal{C}_e(\bar{\Omega}) := \{\, u \in \mathcal{C}(\bar{\Omega}) \;:\; \exists\, \lambda > 0 \text{ such that } -\lambda e \leq u \leq \lambda e \text{ in } \bar{\Omega} \,\} \tag{7.17}$$

equipped with the Minkowski norm

$$\|u\|_e := \inf \{ \lambda > 0 \, : \, -\lambda e \le u \le \lambda e \}, \qquad u \in \mathcal{C}_e(\bar{\Omega}). \tag{7.18}$$

It is easily seen that $\left(\mathcal{C}_e(\bar{\Omega}), \|\cdot\|_e\right)$ is a Banach space whose positive cone

$$P_e := \left\{ u \in \mathcal{C}_e(\bar{\Omega}) \, : \, u \ge 0 \text{ in } \bar{\Omega} \right\} = P \cap \mathcal{C}_e(\bar{\Omega}) \tag{7.19}$$

equips it with a structure of ordered Banach space. As $\|\cdot\|_e$ is monotone, by Theorem 6.1, P_e is a normal cone. Moreover, the canonical injection

$$J_e \, : \, \mathcal{C}_e(\bar{\Omega}) \longrightarrow \mathcal{C}(\bar{\Omega})$$

is continuous, because

$$\|u\|_{\mathcal{C}(\bar{\Omega})} \le \|e\|_{\mathcal{C}(\bar{\Omega})} \|u\|_e \qquad \text{for all} \quad u \in \mathcal{C}_e(\bar{\Omega}).$$

Thus, according to Theorem 7.6, the resolvent operator

$$R_{\omega,e} := (\mathfrak{L} + \omega)^{-1} J_e \, : \, \mathcal{C}_e(\bar{\Omega}) \longrightarrow \mathcal{C}_{\mathfrak{B}}^1(\bar{\Omega}) \tag{7.20}$$

is linear and continuous. The next result shows that

$$R_{\omega,e} f \in \mathcal{C}_e(\bar{\Omega}) \qquad \text{for all} \quad f \in \mathcal{C}_e(\bar{\Omega})$$

and infers some pivotal consequences from this feature.

Proposition 7.2. $R[R_{\omega,e}] \subset \mathcal{C}_e(\bar{\Omega}) \cap \mathcal{C}_{\mathfrak{B}}^1(\bar{\Omega})$. *Therefore, $R_{\omega,e}$ can be regarded as an endomorphism of $\mathcal{C}_e(\bar{\Omega})$. Let denote it by $\mathfrak{R}_\omega$. Then, for every $f \in \mathcal{C}_e(\bar{\Omega})$,*

$$u := \mathfrak{R}_\omega f \in \mathcal{C}_e(\bar{\Omega})$$

is the unique weak solution of (7.8) and $\mathfrak{R}_\omega$ is linear, continuous and compact.

Proof. Let $f \in \mathcal{C}_e(\bar{\Omega})$. By definition,

$$u := R_{\omega,e} f$$

is the unique weak solution of (7.8). By Theorem 7.6, u is a classical solution. Moreover, since $f \in \mathcal{C}_e(\bar{\Omega})$, we have that

$$-\|f\|_e \|e\|_{\mathcal{C}(\bar{\Omega})} \le -\|f\|_e e \le f \le \|f\|_e e \le \|f\|_e \|e\|_{\mathcal{C}(\bar{\Omega})}$$

and hence,

$$-\|f\|_e \|e\|_{\mathcal{C}(\bar{\Omega})} \le (\mathfrak{L} + \omega) u \le \|f\|_e \|e\|_{\mathcal{C}(\bar{\Omega})} \quad \text{in } \Omega. \tag{7.21}$$

Subsequently, we set

$$\lambda := \|f\|_e \|e\|_{\mathcal{C}(\bar{\Omega})}, \qquad v_1 := \lambda e - u, \qquad v_2 := \lambda e + u.$$

Then, owing to (7.15) and (7.21), we obtain that

$$\begin{cases} (\mathfrak{L}+\omega)v_j \geq 0 & \text{in } \Omega, \\ \mathfrak{B}v_j = 0 & \text{on } \partial\Omega, \end{cases} \tag{7.22}$$

for each $j = 1, 2$. Thus, according to Theorem 7.6, $v_j \geq 0$ in $\bar{\Omega}$, for $j = 1, 2$, and, therefore,

$$-\lambda e \leq u \leq \lambda e \quad \text{in } \bar{\Omega}. \tag{7.23}$$

Consequently, $u \in \mathcal{C}_e(\bar{\Omega})$ and

$$\|u\|_e \leq \lambda,$$

which concludes the proof of the first assertion of the statement.

Now, let

$$\mathfrak{R}_\omega : \mathcal{C}_e(\bar{\Omega}) \longrightarrow \mathcal{C}_e(\bar{\Omega})$$

be the linear operator such that $\mathfrak{R}_\omega f$ is the unique weak solution of (7.8) for every $f \in \mathcal{C}_e(\bar{\Omega})$. The estimate (7.23) actually shows that

$$\|\mathfrak{R}_\omega f\|_e \leq \lambda = \|e\|_{\mathcal{C}(\bar{\Omega})}\|f\|_e \quad \text{for all } f \in \mathcal{C}_e(\bar{\Omega})$$

and hence,

$$\|\mathfrak{R}_\omega\|_{\mathcal{L}(\mathcal{C}_e(\bar{\Omega}))} \leq \|e\|_{\mathcal{C}(\bar{\Omega})}.$$

Consequently, $\mathfrak{R}_\omega$ is continuous. To prove the compactness of $\mathfrak{R}_\omega$, let $\{f_n\}_{n\geq 1}$ be a bounded sequence of $\mathcal{C}_e(\bar{\Omega})$. Then, there exists a constant $\lambda > 0$ such that

$$\|f_n\|_e \leq \lambda, \qquad n \geq 1. \tag{7.24}$$

As $\mathcal{C}_e(\bar{\Omega})$ is continuously embedded in $\mathcal{C}(\bar{\Omega})$, $\{f_n\}_{n\geq 1}$ is also bounded in $\mathcal{C}(\bar{\Omega})$. Subsequently, for every $n \geq 1$, we denote by $u_n \in \mathcal{C}_e(\bar{\Omega}) \cap \mathcal{C}^1_{\mathfrak{B}}(\bar{\Omega})$ the unique weak solution of

$$\begin{cases} (\mathfrak{L}+\omega)u_n = f_n & \text{in } \Omega, \\ \mathfrak{B}u_n = 0 & \text{on } \partial\Omega. \end{cases} \tag{7.25}$$

Thanks to Theorem 7.6, $\{u_n\}_{n\geq 1}$ is relatively compact in $\mathcal{C}^1_{\mathfrak{B}}(\bar{\Omega})$. Thus, there exists $u \in \mathcal{C}^1_{\mathfrak{B}}(\bar{\Omega})$ such that, along some subsequence relabeled by n,

$$\lim_{n\to\infty} \|u_n - u\|_{\mathcal{C}^1_{\mathfrak{B}}(\bar{\Omega})} = 0. \tag{7.26}$$

On the other hand, according to (7.23), it follows from (7.25) that

$$-\|f_n\|_e\|e\|_{\mathcal{C}(\bar{\Omega})}e \leq u_n \leq \|f_n\|_e\|e\|_{\mathcal{C}(\bar{\Omega})}e \quad \text{for all } n \geq 1.$$

Thus, owing to (7.24), it is apparent that

$$-\lambda \|e\|_{\mathcal{C}(\bar{\Omega})} e \leq u_n \leq \lambda \|e\|_{\mathcal{C}(\bar{\Omega})} e, \qquad n \geq 1.$$

Hence, letting $n \to \infty$, we find that

$$-\lambda \|e\|_{\mathcal{C}(\bar{\Omega})} e \leq u \leq \lambda \|e\|_{\mathcal{C}(\bar{\Omega})} e$$

and, therefore, $u \in \mathcal{C}_e(\bar{\Omega})$. Consequently, thanks to (7.26), to complete the proof of the proposition, it suffices to show that $\mathcal{C}_{\mathfrak{B}}^1(\bar{\Omega})$ is continuously embedded in $\mathcal{C}_e(\bar{\Omega})$, because this implies

$$\lim_{n \to \infty} \|u_n - u\|_{\mathcal{C}_e(\bar{\Omega})} = 0.$$

To prove this property, we first show that $\mathcal{C}_{\mathfrak{B}}^1(\bar{\Omega}) \subset \mathcal{C}_e(\bar{\Omega})$. Let $u \in \mathcal{C}_{\mathfrak{B}}^1(\bar{\Omega})$. Then, $u = 0$ on Γ_0. Moreover, by (7.15), $e = 0$ on Γ_0 and, thanks to (7.16), $\partial e(x)/\partial \nu < 0$ for all $x \in \Gamma_0$. Thus, there exists $\lambda_0 > 0$ such that

$$\lambda \frac{\partial e}{\partial \nu}(x) < \frac{\partial u}{\partial \nu}(x) < -\lambda \frac{\partial e}{\partial \nu}(x)$$

for all $x \in \Gamma_0$ and $\lambda \geq \lambda_0$. Hence, there exists $\epsilon > 0$ such that

$$-\lambda_0 e(x) < u(x) < \lambda_0 e(x) \qquad \forall\, x \in U_\epsilon, \tag{7.27}$$

where

$$U_\epsilon := \Omega \cap [\Gamma_0 + B_\epsilon(0)] = \{\, x \in \Omega \,:\, \operatorname{dist}(x, \Gamma_0) < \epsilon \,\}.$$

Moreover, according to (7.16), e must be separated away from zero in

$$\bar{\Omega} \setminus U_\epsilon = \{\, x \in \bar{\Omega} \,:\, \operatorname{dist}(x, \Gamma_0) \geq \epsilon \,\}$$

and, therefore, there exists $\lambda_1 \geq \lambda_0$ such that

$$-\lambda_1 e(x) < u(x) < \lambda_1 e(x) \qquad \forall\, x \in \bar{\Omega} \setminus U_\epsilon. \tag{7.28}$$

Obviously, (7.27) and (7.28) imply that

$$-\lambda_1 e \leq u < \lambda_1 e \qquad \text{in} \quad \bar{\Omega},$$

and, consequently, $u \in \mathcal{C}_e(\bar{\Omega})$. Now, let J denote the canonical injection

$$J \,:\, \mathcal{C}_{\mathfrak{B}}^1(\bar{\Omega}) \longrightarrow \mathcal{C}_e(\bar{\Omega}).$$

We claim that J is a linear closed operator. Indeed, let $\{u_n\}_{n \geq 1}$ be a sequence of $\mathcal{C}_{\mathfrak{B}}^1(\bar{\Omega})$, and $u \in \mathcal{C}_{\mathfrak{B}}^1(\bar{\Omega})$, $v \in \mathcal{C}_e(\bar{\Omega})$, such that

$$\lim_{n \to \infty} \|u_n - u\|_{\mathcal{C}_{\mathfrak{B}}^1(\bar{\Omega})} = 0, \qquad \lim_{n \to \infty} \|u_n - v\|_e = 0.$$

The first limit implies that $\{u_n\}_{n \geq 1}$ is point-wise convergent to u in $\bar{\Omega}$. Moreover, by definition, we have that

$$-\|u_n - v\|_e e \leq u_n - v \leq \|u_n - v\|_e e \qquad \text{in} \quad \bar{\Omega}$$

for all $n \geq 1$. In particular, $\{u_n\}_{n \geq 1}$ must be point-wise convergent to v in $\bar{\Omega}$. Consequently, $v = u$ and, therefore, J is a closed linear operator. By the closed graph theorem (see, e.g., Theorem II.7 of H. Brézis [29]), J is a linear continuous operator. This concludes the proof. $\square$

The next result shows that the operator $\mathfrak{R}_\omega$ introduced in Proposition 7.2 is strongly positive.

Proposition 7.3. *For every $\omega \geq \omega_0$, the operator*

$$\mathfrak{R}_\omega \; : \; \mathcal{C}_e(\bar{\Omega}) \longrightarrow \mathcal{C}_e(\bar{\Omega})$$

is strongly positive. In other words,

$$\mathfrak{R}_\omega \, (P_e \setminus \{0\}) \subset \mathrm{int} P_e := \overset{\circ}{P}_e.$$

Proof. Let $\omega \geq \omega_0$ and $f \in P_e \setminus \{0\}$. Then, $f \in \mathcal{C}_e(\bar{\Omega})$, $f \geq 0$, and $f \neq 0$ in Ω. Moreover, by Theorem 7.6, the solution of (7.8)

$$u := \mathfrak{R}_\omega f \in \mathcal{C}_{\mathfrak{B}}^1(\bar{\Omega}) \subset \mathcal{C}_e(\bar{\Omega})$$

satisfies (7.9). To complete the proof, it remains to show that

$$u \in \overset{\circ}{P}_e.$$

Indeed, from (7.9) and (7.16), there must exist $\epsilon > 0$ such that

$$u > \epsilon e \quad \text{in} \quad \bar{\Omega}.$$

Now, let $h \in \mathcal{C}_e(\bar{\Omega})$ be such that

$$\|h\|_e \leq \epsilon.$$

Then, by definition,

$$-\epsilon e \leq \|h\|_e e \leq h \leq \|h\|_e e \leq \epsilon e,$$

and hence,

$$u - \epsilon e \leq u + h \leq u + \epsilon e \quad \text{in} \quad \bar{\Omega}.$$

In particular,

$$u + h \geq u - \epsilon e > 0$$

and, therefore, $u + h \in P_e$ for all $h \in \mathcal{C}_e(\bar{\Omega})$ with $\|h\|_e \leq \epsilon$. Consequently, $u \in \mathrm{int} P_e$ and the proof is complete. $\qquad\square$

As P_e is a normal cone in $\mathcal{C}_e(\bar{\Omega})$, the next result is a direct consequence from Propositions 7.2, 7.3, Theorem 6.3, and Remark 6.1.

Corollary 7.1. *For every $\omega \geq \omega_0$ the following properties are satisfied:*

(a) $\mathrm{spr}\,\mathfrak{R}_\omega > 0$ *is an algebraically simple eigenvalue of $\mathfrak{R}_\omega$ and there exists $\varphi_0 \in \mathrm{int} P_e$ such that*

$$N[\mathrm{spr}\,\mathfrak{R}_\omega\, I - \mathfrak{R}_\omega] = \mathrm{span}\,[\varphi_0],$$

where I stands for the identity map of $\mathcal{C}_e(\bar{\Omega})$.

(b) spr $\mathfrak{R}_\omega$ *is the unique real eigenvalue of* $\mathfrak{R}_\omega$ *to an eigenvector in* $P_e \backslash \{0\}$.

(c) $|\lambda| <$ spr $\mathfrak{R}_\omega$ *for all* $\lambda \in \sigma(\mathfrak{R}_\omega) \backslash \{$spr $\mathfrak{R}_\omega\}$.

(d) *For every* $\lambda \in \mathbb{R}$ *with* $\lambda >$ spr $\mathfrak{R}_\omega$, *the resolvent operator*

$$(\lambda I - \mathfrak{R}_\omega)^{-1} \in \mathcal{L}\left(\mathcal{C}_e(\bar{\Omega})\right)$$

is strongly positive. Moreover, there exist $\epsilon > 0$ *and* $u > 0$ *such that*

$$(\lambda I - \mathfrak{R}_\omega)^{-1} u \ll 0 \qquad \text{for all} \quad \lambda \in (\text{spr } \mathfrak{R}_\omega - \epsilon, \text{spr } \mathfrak{R}_\omega).$$

(e) $\sigma(\mathfrak{R}_\omega) = \sigma(\mathfrak{R}_\omega^*)$ *and* spr $\mathfrak{R}_\omega =$ spr $\mathfrak{R}_\omega^*$ *is a simple eigenvalue of* $\mathfrak{R}_\omega^*$. *Moreover, it is the unique eigenvalue of* $\mathfrak{R}_\omega^*$ *in the spectral circle*

$$|\zeta| = \text{spr } \mathfrak{R}_\omega.$$

(f) *There exist* $\varphi_0^* \in P_e^* \backslash \{0\}$ *such that*

$$N[\text{spr } \mathfrak{R}_\omega \, I - \mathfrak{R}_\omega^*] = \text{span}\,[\varphi_0^*],$$

where I *stands for the identity map of* $C_e'(\bar{\Omega})$. *Moreover, for every* $x \in P_e \backslash \{0\}$, $\varphi_0^*(x) > 0$ *and*

$$\text{spr } \mathfrak{R}_\omega u - \mathfrak{R}_\omega u = x$$

cannot admit a solution $u \in \mathcal{C}_e(\bar{\Omega})$.

By Corollary 7.1(a),

$$\mathfrak{R}_\omega \varphi_0 = \text{spr } \mathfrak{R}_\omega \, \varphi_0 \qquad \text{in} \quad \mathcal{C}_e(\bar{\Omega})$$

and hence, by the definition of $\mathfrak{R}_\omega$, the eigenfunction $\varphi_0 > 0$ provides us with a weak solution of

$$\begin{cases} (\mathfrak{L} + \omega)\varphi = \frac{1}{\text{spr } \mathfrak{R}_\omega} \, \varphi & \text{in } \Omega, \\ \mathfrak{B}\varphi = 0 & \text{on } \partial\Omega. \end{cases} \tag{7.29}$$

According to Theorem 7.6, $\varphi_0 \in \mathcal{C}^{1,1^-}(\bar{\Omega})$ is twice classically differentiable almost everywhere in Ω and, actually, it is a classical solution of (7.8). Moreover, since $\varphi_0 > 0$, we find that

$$(\mathfrak{L} + \omega)\varphi_0 = \frac{1}{\text{spr } \mathfrak{R}_\omega} \, \varphi_0 > 0$$

and, therefore,

$$\varphi_0(x) > 0 \quad \forall\, x \in \Omega \cup \Gamma_1 \quad \text{and} \quad \frac{\partial \varphi_0}{\partial \nu}(x) < 0 \quad \forall\, x \in \Gamma_0. \tag{7.30}$$

As (7.29) can be equivalently expressed as

$$\begin{cases} \mathfrak{L}\varphi_0 = \left(\frac{1}{\text{spr } \mathfrak{R}_\omega} - \omega\right) \varphi_0 & \text{in } \Omega, \\ \mathfrak{B}\varphi_0 = 0 & \text{on } \partial\Omega, \end{cases}$$

it becomes apparent that there exists a value of τ for which (7.1) admits a positive eigenfunction. Namely,

$$\tau = \frac{1}{\operatorname{spr} \mathfrak{R}_\omega} - \omega.$$

Most precisely, the following fundamental result holds. As a byproduct, the previous value of τ must be independent of $\omega \geq \omega_0$.

Theorem 7.7 (Existence of the principal eigenvalue). *There is a unique value of τ for which the linear boundary value problem (7.1) admits a weak solution $\varphi \in P_e \setminus \{0\}$. Such a value of τ will be throughout denoted by*

$$\sigma_0 := \sigma[\mathfrak{L}, \mathfrak{B}, \Omega]$$

*and called the **principal eigenvalue of** (7.1), or, equivalently, the principal eigenvalue of $(\mathfrak{L}, \mathfrak{B}, \Omega)$. Any associated weak solution $\varphi_0 \in P_e \setminus \{0\}$ will be referred to as a **principal eigenfunction** of (7.1), or, equivalently, of $(\mathfrak{L}, \mathfrak{B}, \Omega)$.*

The principal eigenfunction φ_0 is unique, up to a positive multiplicative real number, it satisfies $\varphi_0 \in W^{2,\infty^-}(\Omega) \subset \mathcal{C}^{1,1^-}(\bar{\Omega})$ and (7.30), and it is a classical solution of (7.1), in the sense of Definition 4.1.

Proof. First, we will prove the uniqueness of the principal eigenvalue. Let $\sigma_1, \sigma_2 \in \mathbb{R}$ and $\varphi_1, \varphi_2 \in P_e \setminus \{0\}$ such that

$$\begin{cases} \mathfrak{L}\varphi_j = \sigma_j\varphi_j & \text{in } \Omega, \\ \mathfrak{B}\varphi_j = 0 & \text{on } \partial\Omega, \end{cases}$$

for $j = 1, 2$. Then, for sufficiently large $\omega \geq \omega_0$ and $j = 1, 2$, we have that

$$\begin{cases} (\mathfrak{L} + \omega)\varphi_j = (\sigma_j + \omega)\varphi_j > 0 & \text{in } \Omega, \\ \mathfrak{B}\varphi_j = 0 & \text{on } \partial\Omega, \end{cases}$$

and hence,

$$\mathfrak{R}_\omega\varphi_j := \frac{1}{\sigma_j + \omega}\,\varphi_j.$$

Consequently, by Corollary 7.1(b),

$$\operatorname{spr} \mathfrak{R}_\omega = \frac{1}{\sigma_1 + \omega} = \frac{1}{\sigma_2 + \omega}$$

and, therefore, $\sigma_1 = \sigma_2$, which shows the uniqueness of the principal eigenvalue.

The uniqueness of the principal eigenfunction follows by rather similar patterns. Indeed, let $\varphi_1, \varphi_2 \in P_e \setminus \{0\}$ such that

$$\begin{cases} \mathfrak{L}\varphi_j = \sigma[\mathfrak{L}, \mathfrak{B}, \Omega]\varphi_j & \text{in } \Omega, \\ \mathfrak{B}\varphi_j = 0 & \text{on } \partial\Omega, \end{cases}$$

for $j = 1, 2$. Then, for sufficiently large $\omega \geq \omega_0$ and $j = 1, 2$, we have that

$$\begin{cases} (\mathfrak{L} + \omega)\,\varphi_j = (\sigma[\mathfrak{L}, \mathfrak{B}, \Omega] + \omega)\,\varphi_j > 0 & \text{in } \Omega, \\ \mathfrak{B}\varphi_j = 0 & \text{on } \partial\Omega, \end{cases}$$

and hence,

$$\mathfrak{R}_\omega \varphi_j := \frac{1}{\sigma[\mathfrak{L}, \mathfrak{B}, \Omega] + \omega}\,\varphi_j.$$

By Corollary 7.1(b), we have that

$$\operatorname{spr} \mathfrak{R}_\omega = \frac{1}{\sigma[\mathfrak{L}, \mathfrak{B}, \Omega] + \omega}.$$

Therefore,

$$\mathfrak{R}_\omega \varphi_j := \operatorname{spr} \mathfrak{R}_\omega \, \varphi_j, \qquad j = 1, 2,$$

and, consequently, by Corollary 7.1(a), there must exist $\lambda > 0$ such that

$$\varphi_1 = \lambda \varphi_2,$$

which concludes the proof of the uniqueness of the principal eigenfunction.

The regularity of the principal eigenfunction and (7.30) follow straightaway from Theorem 7.6. This ends the proof. $\square$

7.3   Two equivalent weak eigenvalue problems

The next result establishes a bijection between the set of eigenvalues of problem (7.1) and $\sigma(\mathfrak{R}_\omega)\setminus\{0\}$, $\omega \geq \omega_0$. It also shows that any eigenfunction of (7.1) in the weak sense must be classical.

Proposition 7.4. *Let $\omega \geq \omega_0$ and*

$$\varphi \in W^{1,2}_{\Gamma_0}(\Omega) \cap \mathcal{C}_e(\bar\Omega) + i\left[W^{1,2}_{\Gamma_0}(\Omega) \cap \mathcal{C}_e(\bar\Omega)\right], \qquad \varphi \neq 0.$$

Then, the following assertions hold:

(a) *If $\tau \in \mathbb{C}$ is an eigenvalue of (7.1) and φ is a weak solution of (7.1), then, φ is a classical solution of (7.1), $\tau + \omega \neq 0$, and*

$$\frac{1}{\tau + \omega} \in \sigma(\mathfrak{R}_\omega) \setminus \{0\}.$$

(b) *Conversely, if $\sigma \in \sigma(\mathfrak{R}_\omega) \setminus \{0\}$ and φ is an eigenfunction of $\mathfrak{R}_\omega$ associated to σ, then, φ is a classical solution of (7.1) with*

$$\tau = \sigma^{-1} - \omega.$$

In particular, if we denote by $\Sigma(\mathfrak{L}, \mathfrak{B}, \Omega)$ the set of weak eigenvalues of (7.1), then, the map

$$\Sigma(\mathfrak{L}, \mathfrak{B}, \Omega) \longrightarrow \sigma(\mathfrak{R}_\omega) \setminus \{0\}$$

$$\tau \mapsto \frac{1}{\tau + \omega}$$

is a bijection, and all associated eigenfunctions are classical.

Proof. Suppose $\tau \in \mathbb{C}$ and φ is a weak solution of (7.1). Then, φ is a weak solution of

$$\begin{cases} (\mathfrak{L} + \omega)\varphi = (\tau + \omega)\varphi \in \mathcal{C}_e(\bar{\Omega}) & \text{in } \Omega, \\ \mathfrak{B}\varphi = 0 & \text{on } \partial\Omega, \end{cases} \tag{7.31}$$

and, thanks to Theorem 7.6, $\varphi \in \mathcal{C}^{1,1^-}(\bar{\Omega})$ and it is a classical solution of (7.1). Moreover, it is the unique weak solution of

$$\begin{cases} (\mathfrak{L} + \omega)u = f := (\tau + \omega)\varphi \in \mathcal{C}_e(\bar{\Omega}) & \text{in } \Omega, \\ \mathfrak{B}u = 0 & \text{on } \partial\Omega, \end{cases}$$

and hence $\tau + \omega \neq 0$, because $\varphi \neq 0$ and 0 is a weak solution if $\tau + \omega = 0$. Furthermore, by the definition of $\mathfrak{R}_\omega$, it becomes apparent that

$$\mathfrak{R}_\omega \varphi = \frac{1}{\tau + \omega}\,\varphi,$$

which concludes the proof of Part (a).

Conversely, assume that

$$\mathfrak{R}_\omega \varphi = \sigma\,\varphi$$

for some $\sigma \neq 0$. Then, by the definition of $\mathfrak{R}_\omega$, φ is a weak solution of

$$\begin{cases} (\mathfrak{L} + \omega)\varphi = \sigma^{-1}\varphi \in \mathcal{C}_e(\bar{\Omega}) & \text{in } \Omega, \\ \mathfrak{B}\varphi = 0 & \text{on } \partial\Omega, \end{cases}$$

and hence, thanks to Theorem 7.6, $\varphi \in \mathcal{C}^{1,1^-}(\bar{\Omega})$ and it is a classical solution of this problem. Therefore, φ is a classical solution of (7.1) for $\tau = \sigma^{-1} - \omega$. The proof is complete. □

Throughout the rest of this book, we will use the following concept.

Definition 7.1. *Given $\tau \in \Sigma(\mathfrak{L}, \mathfrak{B}, \Omega)$, it is said that φ is an eigenfunction associated to (7.1) if*

$$\varphi \in W^{1,2}_{\Gamma_0}(\Omega) \cap \mathcal{C}_e(\bar{\Omega}) + i\left[W^{1,2}_{\Gamma_0}(\Omega) \cap \mathcal{C}_e(\bar{\Omega})\right]$$

and φ is a weak solution of (7.1).

According to Theorem 7.6 and Proposition 7.4, all these weak eigenfunctions are classical solutions of (7.1) lying in $\mathcal{C}^{1,1^-}(\bar{\Omega})$.

7.4 Simplicity and dominance of $\sigma[\mathfrak{L}, \mathfrak{B}, \Omega]$

The following concept plays a central role in the theory of nonlinear elliptic problems involving second order operators.

Definition 7.2 (Simple eigenvalue). *Let*

$$\tau \in \mathbb{R} \cap \Sigma(\mathfrak{L}, \mathfrak{B}, \Omega)$$

for which (7.1) *admits a unique eigenfunction* φ, *up to a multiplicative constant. Then,* τ *is said to be a **simple eigenvalue** of* (7.1) *if*

$$\begin{cases} (\mathfrak{L} - \tau)u = \varphi & in \ \Omega, \\ \mathfrak{B}u = 0 & on \ \partial\Omega, \end{cases} \tag{7.32}$$

does not admit a weak solution $u \in W^{1,2}_{\Gamma_0}(\Omega) \cap \mathcal{C}_e(\bar{\Omega})$.

As for every $\omega \geq \omega_0$, any weak solution $u \in W^{1,2}_{\Gamma_0}(\Omega) \cap \mathcal{C}_e(\bar{\Omega})$ of (7.32) must be a weak solution of

$$\begin{cases} (\mathfrak{L} + \omega)u = (\omega + \tau)u + \varphi \in \mathcal{C}_e(\bar{\Omega}) & in \ \Omega, \\ \mathfrak{B}u = 0 & on \ \partial\Omega, \end{cases}$$

due to Theorem 7.6, these weak solutions are classical solutions of (7.32).

Now, we can state the main result of this section.

Theorem 7.8 (Dominance of σ_0). *The principal eigenvalue*

$$\sigma_0 := \sigma[\mathfrak{L}, \mathfrak{B}, \Omega]$$

is a simple eigenvalue of (7.1). *Moreover, it is **dominant**, in the sense that*

$$\operatorname{Re}\tau \geq \sigma_0 \quad for \ all \quad \tau \in \Sigma(\mathfrak{L}, \mathfrak{B}, \Omega). \tag{7.33}$$

In particular,

$$\tau > \sigma_0 \quad if \quad \tau \in \mathbb{R} \cap [\Sigma(\mathfrak{L}, \mathfrak{B}, \Omega) \setminus \{\sigma_0\}].$$

Proof. Let $\varphi \in P_e \setminus \{0\}$ be a principal eigenfunction of $(\mathfrak{L}, \mathfrak{B}, \Omega)$ and suppose

$$\begin{cases} (\mathfrak{L} - \sigma_0)u = \varphi & in \ \Omega, \\ \mathfrak{B}u = 0 & on \ \partial\Omega, \end{cases} \tag{7.34}$$

admits a weak solution $u \in W^{1,2}_{\Gamma_0}(\Omega) \cap \mathcal{C}_e(\bar{\Omega})$. Let $\omega \geq \omega_0$ (see Theorem 7.6) such that

$$\omega + \sigma_0 > 0.$$

Clearly, u must be a weak solution of

$$\begin{cases} (\mathfrak{L} + \omega)u = (\omega + \sigma_0)u + \varphi & \text{in } \Omega, \\ \mathfrak{B}u = 0 & \text{on } \partial\Omega, \end{cases}$$

and hence,

$$u = (\omega + \sigma_0)\mathfrak{R}_\omega u + \mathfrak{R}_\omega \varphi. \tag{7.35}$$

On the other hand, we find from Corollary 7.1(a)(b) that

$$\mathfrak{R}_\omega \varphi = \operatorname{spr} \mathfrak{R}_\omega \varphi = \frac{1}{\sigma_0 + \omega} \varphi$$

and, so, dividing (7.35) by $\omega + \sigma_0$ we are driven to

$$(\operatorname{spr} \mathfrak{R}_\omega I - \mathfrak{R}_\omega) u = \frac{1}{(\omega + \sigma_0)^2} \varphi.$$

Consequently,

$$\varphi \in R\left[\operatorname{spr} \mathfrak{R}_\omega I - \mathfrak{R}_\omega\right].$$

As this contradicts Corollary 7.1(f), σ_0 must be a simple eigenvalue of (7.1).

Now, we will prove (7.33). Let

$$\tau \in \Sigma(\mathfrak{L}, \mathfrak{B}, \Omega) \setminus \{\sigma_0\}.$$

Then, according to Proposition 7.4,

$$\frac{1}{\tau + \omega} \in \sigma(\mathfrak{R}_\omega) \qquad \text{for all} \quad \omega \geq \omega_0,$$

and hence, by Corollary 7.1(c), we find that

$$\left| \frac{1}{\tau + \omega} \right| < \operatorname{spr} \mathfrak{R}_\omega = \frac{1}{\sigma_0 + \omega} \qquad \text{for all} \quad \omega \geq \omega_0.$$

Equivalently,

$$(\operatorname{Re} \tau + \omega)^2 + (\operatorname{Im} \tau)^2 > (\sigma_0 + \omega)^2 \qquad \text{for all} \quad \omega \geq \omega_0.$$

Thus, rearranging terms, it is apparent that

$$(\operatorname{Re} \tau)^2 + 2\omega(\operatorname{Re} \tau - \sigma_0) + (\operatorname{Im} \tau)^2 - \sigma_0^2 > 0$$

for all $\omega \geq \omega_0$, and, therefore, dividing by ω and letting $\omega \to \infty$ we obtain that

$$\operatorname{Re} \tau \geq \sigma_0.$$

This proves (7.33). The fact that any real eigenvalue $\tau \neq \sigma_0$ must satisfy $\tau > \sigma_0$ is a byproduct of (7.33). The proof is complete. $\qquad \square$

The next result sharpens Theorem 7.8.

Theorem 7.9 (Strict dominance of σ_0). *The principal eigenvalue σ_0 is* **strictly dominant**, *in the sense that*

$$\operatorname{Re} \tau > \sigma_0 \qquad \text{for all} \quad \tau \in \Sigma(\mathfrak{L}, \mathfrak{B}, \Omega) \setminus \{\sigma_0\}. \tag{7.36}$$

7.4.1   *Proof of the strict dominance in case* $\Gamma_0 = \emptyset$

Suppose $\Gamma_0 = \emptyset$, pick an eigenvalue

$$\tau \in \Sigma(\mathfrak{L}, \mathfrak{B}, \Omega) \setminus \{\sigma_0\},$$

and let u be an eigenfunction associated to (7.1), in the sense of Definition
7.1. By Theorems 4.10 and 5.11,

$$u \in W^{2,p}(\Omega) \qquad \text{for all} \quad p > N.$$

Now, consider the auxiliary function

$$v := \frac{u}{\varphi_0} \qquad \text{in} \quad \Omega, \tag{7.37}$$

where $\varphi_0 \gg 0$ is a principal eigenfunction associated to σ_0. By (7.30),

$$\varphi_0(x) > 0 \quad \forall\, x \in \bar{\Omega},$$

because $\partial\Omega = \Gamma_1$. Hence, $v \in W^{2,\infty^-}(\Omega)$ and, in particular, $v \in \mathcal{C}^{1,1^-}(\bar{\Omega})$
and it is twice classically differentiable almost everywhere in $\bar{\Omega}$. Arguing as
in the proof of Theorem 1.7, a direct calculation from (7.37) shows that

$$\tau u = \mathfrak{L}u = \varphi_0 \mathfrak{L}_0 v \qquad \text{in} \quad \Omega, \tag{7.38}$$

where

$$
\begin{aligned}
\mathfrak{L}_0 &:= -\sum_{i,j=1}^{N} a_{ij} \frac{\partial^2}{\partial x_i \partial x_j} + \sum_{j=1}^{N}\left(b_j - \frac{2}{\varphi_0}\sum_{i=1}^{N} a_{ij}\frac{\partial \varphi_0}{\partial x_i}\right)\frac{\partial}{\partial x_j} + \frac{\mathfrak{L}\varphi_0}{\varphi_0} \\
&= -\sum_{i,j=1}^{N} a_{ij}\frac{\partial^2}{\partial x_i \partial x_j} + \sum_{j=1}^{N}\left(b_j - \frac{2}{\varphi_0}\sum_{i=1}^{N} a_{ij}\frac{\partial \varphi_0}{\partial x_i}\right)\frac{\partial}{\partial x_j} + \sigma_0.
\end{aligned}
$$

Thus, dividing by φ_0 the identity (7.38), yields to

$$\mathfrak{L}_0 v = \tau v$$

and hence,

$$(\mathfrak{L}_0 - \sigma_0)\, v = (\tau - \sigma_0)v \quad \text{and} \quad (\mathfrak{L}_0 - \sigma_0)\, \bar{v} = (\bar{\tau} - \sigma_0)\bar{v},$$

where $\bar{\tau}$ and $\bar{v}$ stand for the complex conjugates of τ and v, respectively.
Therefore, by a chain of direct calculations, we are driven to the identities

$$
\begin{aligned}
(\mathfrak{L}_0 - \sigma_0)\, |v|^2 &= (\mathfrak{L}_0 - \sigma_0)\, (v\bar{v}) \\
&= \bar{v}\,(\mathfrak{L}_0 - \sigma_0)\, v + v\,(\mathfrak{L}_0 - \sigma_0)\,\bar{v} - 2\sum_{i,j=1}^{N} a_{ij}\frac{\partial v}{\partial x_i}\frac{\partial \bar{v}}{\partial x_j} \\
&= \bar{v}\,(\tau - \sigma_0)\, v + v\,(\bar{\tau} - \sigma_0)\,\bar{v} - 2\sum_{i,j=1}^{N} a_{ij}\frac{\partial v}{\partial x_i}\frac{\partial \bar{v}}{\partial x_j}
\end{aligned}
$$

and, consequently,

$$(\mathfrak{L}_0 - \sigma_0) |v|^2 = 2 (\operatorname{Re} \tau - \sigma_0) |v|^2 - 2 \sum_{i,j=1}^{N} a_{ij} \frac{\partial v}{\partial x_i} \frac{\partial \bar{v}}{\partial x_j}. \tag{7.39}$$

On the other hand, setting

$$v = \psi + i\xi,$$

we have that

$$\begin{aligned}
\sum_{i,j=1}^{N} a_{ij} \frac{\partial v}{\partial x_i} \frac{\partial \bar{v}}{\partial x_j} &= \sum_{i,j=1}^{N} a_{ij} \left(\frac{\partial \psi}{\partial x_i} + i \frac{\partial \xi}{\partial x_i} \right) \left(\frac{\partial \psi}{\partial x_j} - i \frac{\partial \xi}{\partial x_j} \right) \\
&= \sum_{i,j=1}^{N} a_{ij} \left[\frac{\partial \psi}{\partial x_i} \frac{\partial \psi}{\partial x_j} + \frac{\partial \xi}{\partial x_i} \frac{\partial \xi}{\partial x_j} + i \left(\frac{\partial \xi}{\partial x_i} \frac{\partial \psi}{\partial x_j} - \frac{\partial \xi}{\partial x_j} \frac{\partial \psi}{\partial x_i} \right) \right] \\
&= \sum_{i,j=1}^{N} a_{ij} \left(\frac{\partial \psi}{\partial x_i} \frac{\partial \psi}{\partial x_j} + \frac{\partial \xi}{\partial x_i} \frac{\partial \xi}{\partial x_j} \right),
\end{aligned}$$

because $a_{ij} = a_{ji}$ for all $i, j \in \{1, ..., N\}$. Thus, by the strong ellipticity of $\mathfrak{L}$, we find that

$$\sum_{i,j=1}^{N} a_{ij} \frac{\partial v}{\partial x_i} \frac{\partial \bar{v}}{\partial x_j} \geq \mu \left(|\nabla \psi|^2 + |\nabla \xi|^2 \right), \tag{7.40}$$

where $\mu > 0$ is the ellipticity constant of $\mathfrak{L}$ in Ω.

To show that $\operatorname{Re} \tau > \sigma_0$ we proceed by contradiction. Suppose

$$\operatorname{Re} \tau \leq \sigma_0.$$

Then, owing to (7.33), we must have

$$\operatorname{Re} \tau = \sigma_0$$

and hence, it follows from (7.39) and (7.40) that

$$(\mathfrak{L}_0 - \sigma_0) \left(-|v|^2 \right) \geq 2\mu \left(|\nabla \psi|^2 + |\nabla \xi|^2 \right). \tag{7.41}$$

On the other hand, on $\partial \Omega$, we have that

$$0 = \mathfrak{B}u = \mathfrak{B}(\varphi_0 v) = \varphi_0 \frac{\partial v}{\partial \nu} + v \mathfrak{B}\varphi_0 = \varphi_0 \frac{\partial v}{\partial \nu}$$

and hence,

$$\frac{\partial v}{\partial \nu} = 0 \quad \text{on} \quad \partial \Omega, \tag{7.42}$$

because $\varphi_0(x) > 0$ for all $x \in \bar{\Omega}$. Consequently, setting

$$\phi := -|v|^2 \quad \text{in} \quad \bar{\Omega}, \tag{7.43}$$

we obtain from (7.41) and (7.42) that

$$(\mathfrak{L}_0 - \sigma_0)\phi \geq 0 \quad \text{in} \quad \Omega \quad \text{and} \quad \frac{\partial \phi}{\partial \nu} = 0 \quad \text{on} \quad \partial\Omega. \tag{7.44}$$

As the zero order term of $\mathfrak{L}_0 - \sigma_0$ vanishes and $\phi \leq 0$ in $\bar{\Omega}$, we find from Theorem 7.2 that

$$m := \inf_{\bar{\Omega}} \phi \leq 0$$

cannot be reached in Ω, unless

$$\phi = m \quad \text{in} \quad \bar{\Omega}. \tag{7.45}$$

Also, by Theorem 7.3, if m is attained at some $x \in \partial\Omega$, then

$$\frac{\partial \phi}{\partial \nu}(x) < 0,$$

which is impossible, by (7.44). Therefore, (7.45) holds and, hence, going back to (7.41), it becomes apparent that

$$0 = (\mathfrak{L}_0 - \sigma_0)\phi \geq \mu \left(|\nabla\psi|^2 + |\nabla\xi|^2 \right).$$

Consequently,

$$\nabla\psi = \nabla\xi = 0 \quad \text{in} \quad \bar{\Omega},$$

and, so, $v \neq 0$ is constant in Ω. Therefore,

$$\sigma_0 v = \mathfrak{L}_0 v = \tau v$$

and hence, $\sigma_0 = \tau$, which is impossible, because we have taken $\tau \neq \sigma_0$ from the beginning. This contradiction concludes the proof of (7.36).

7.4.2 *Proof of the strict dominance in case* $\Gamma_1 = \emptyset$

Suppose $\Gamma_1 = \emptyset$. Then $\mathfrak{B} = \mathfrak{D}$ is the Dirichlet boundary operator.

Subsequently, we consider the product open set $\tilde{\Omega} := \Omega \times \Omega$, with points $(x, y) \in \tilde{\Omega}$ for all $x, y \in \Omega$, the second order differential operator

$$\tilde{\mathfrak{L}} := -\sum_{i,j=1}^{N} \left(a_{ij}(x)\frac{\partial^2}{\partial x_i \partial x_j} + a_{ij}(y)\frac{\partial^2}{\partial y_i \partial y_j} \right)$$

$$+ \sum_{j=1}^{N} \left(b_j(x)\frac{\partial}{\partial x_j} + b_j(y)\frac{\partial}{\partial y_j} \right) + c(x) + c(y)$$

in $\tilde{\Omega}$, and the boundary operator

$$\tilde{\mathfrak{D}}u = u|_{\partial\tilde{\Omega}} \qquad \text{for all} \quad u \in \mathcal{C}(\overline{\Omega \times \Omega}).$$

Obviously, $(\tilde{\mathfrak{L}}, \tilde{\mathfrak{D}}, \tilde{\Omega})$ fits into the general setting of this chapter, as $(\mathfrak{L}, \mathfrak{D}, \Omega)$, and, in particular, it possesses a principal eigenvalue, denoted by

$$\tilde{\sigma}_0 := \sigma[\tilde{\mathfrak{L}}, \tilde{\mathfrak{D}}, \tilde{\Omega}].$$

Pick a principal eigenfunction $\varphi_0 \gg 0$ associated to σ_0 and consider the product function defined by

$$\tilde{\varphi}_0(x, y) := \varphi_0(x)\varphi_0(y), \qquad (x, y) \in \overline{\Omega \times \Omega}.$$

Then,

$$\tilde{\varphi}_0 = 0 \qquad \text{on} \quad \partial\tilde{\Omega},$$

and

$$\tilde{\mathfrak{L}}\tilde{\varphi}_0(x, y) = \varphi_0(y)\mathfrak{L}\varphi_0(x) + \varphi_0(x)\mathfrak{L}\varphi_0(y) = 2\sigma_0\tilde{\varphi}_0(x, y)$$

for all $x, y \in \Omega$. Therefore, by the uniqueness of the principal eigenvalue already established by Theorem 7.7, we have that

$$\tilde{\sigma}_0 = 2\sigma_0.$$

Now, pick

$$\tau \in \Sigma\,(\mathfrak{L}, \mathfrak{B}, \Omega) \setminus \{\sigma_0\}$$

and, given any eigenfunction $\varphi := \psi + i\xi \neq 0$ associated to τ, consider the auxiliary function $\tilde{\varphi}$ defined by

$$\begin{aligned}
\tilde{\varphi}(x, y) &:= \varphi(x)\bar{\varphi}(y) + \bar{\varphi}(x)\varphi(y) \\
&= 2\,\mathrm{Re}\,[\varphi(x)\bar{\varphi}(y)] \\
&= 2[\psi(x)\psi(y) + \xi(x)\xi(y)]
\end{aligned}$$

for all $x, y \in \bar{\Omega}$. Then, $\tilde{\varphi}$ is a real function such that $\tilde{\varphi} = 0$ on $\partial\tilde{\Omega}$ and

$$\begin{aligned}
\tilde{\mathfrak{L}}\tilde{\varphi}(x, y) &= \varphi(x)\mathfrak{L}\bar{\varphi}(y) + \bar{\varphi}(y)\mathfrak{L}\varphi(x) + \varphi(y)\mathfrak{L}\bar{\varphi}(x) + \bar{\varphi}(x)\mathfrak{L}\varphi(y) \\
&= \bar{\tau}\varphi(x)\bar{\varphi}(y) + \tau\bar{\varphi}(y)\varphi(x) + \bar{\tau}\varphi(y)\bar{\varphi}(x) + \tau\bar{\varphi}(x)\varphi(y) \\
&= (\tau + \bar{\tau})\varphi(x)\bar{\varphi}(y) + (\tau + \bar{\tau})\varphi(y)\bar{\varphi}(x) \\
&= (\tau + \bar{\tau})\tilde{\varphi}(x, y) = 2\,\mathrm{Re}\,\tau\tilde{\varphi}(x, y)
\end{aligned}$$

for all $x, y \in \Omega$. As, for every $x \in \Omega$,

$$\tilde{\varphi}(x, x) = 2\varphi(x)\bar{\varphi}(x) = 2|\varphi(x)|^2 \geq 0$$

and $\varphi \neq 0$, we have that $\tilde{\varphi} \neq 0$. Thus, $2\,\mathrm{Re}\,\tau$ is an eigenvalue of $(\tilde{\mathfrak{L}}, \tilde{\mathfrak{D}}, \tilde{\Omega})$ with associated eigenfunction $\tilde{\varphi} \neq 0$.

Suppose $\tilde{\varphi} > 0$ in $\tilde{\Omega}$. Then, according to Theorem 7.7,

$$2\,\mathrm{Re}\,\tau = \tilde{\sigma}_0 = 2\sigma_0$$

and, consequently,

$$\mathrm{Re}\,\tau = \sigma_0.$$

Moreover, there must exist $\lambda \in \mathbb{C}$ such that

$$\tilde{\varphi} = \lambda \tilde{\varphi}_0.$$

Without loss of generality, we can take $\lambda = 2$ by changing φ by an appropriate multiple, if necessary. Hence, by the definitions of $\tilde{\varphi}$ and $\tilde{\varphi}_0$,

$$\varphi(x)\bar{\varphi}(y) + \bar{\varphi}(x)\varphi(y) = 2\varphi_0(x)\varphi_0(y) \qquad \forall\, x, y \in \Omega,$$

and, consequently, at $x = y$ yields

$$|\varphi(x)| = \varphi_0(x) \qquad \forall\, x \in \Omega.$$

Thus, there exists $\theta \in \mathcal{C}(\Omega, [0, 2\pi))$ such that

$$\varphi(x) = e^{i\theta(x)}\varphi_0(x), \qquad \forall\, x \in \Omega.$$

Going back to the previous identity in $\tilde{\Omega}$, we obtain that

$$e^{i\theta(x)}\varphi_0(x)e^{-i\theta(y)}\varphi_0(y) + e^{-i\theta(x)}\varphi_0(x)e^{i\theta(y)}\varphi_0(y) = 2\varphi_0(x)\varphi_0(y)$$

and so,

$$\cos\left(\theta(x) - \theta(y)\right) = 1 \qquad \forall\, x, y \in \Omega.$$

Consequently, θ must be constant, say $\theta_0 \in [0, 2\pi)$, and hence,

$$\varphi = e^{i\theta_0}\varphi_0.$$

Thus,

$$\tau\varphi = \mathfrak{L}\varphi = e^{i\theta_0}\mathfrak{L}\varphi_0 = e^{i\theta_0}\sigma_0\varphi_0 = \sigma_0\varphi$$

and, therefore, $\tau = \sigma_0$, which is a contradiction. As this contradiction comes from the assumption that $\tilde{\varphi} > 0$ in $\tilde{\Omega}$, necessarily, $\tilde{\varphi}$ changes sign in $\tilde{\Omega}$. Then, by Theorem 7.7,

$$2\,\mathrm{Re}\,\tau \in \mathbb{R} \cap \left[\Sigma\left(\tilde{\mathfrak{L}}, \tilde{\mathfrak{B}}, \tilde{\Omega}\right) \setminus \{\tilde{\sigma}_0\}\right]$$

and, therefore, thanks to Theorem 7.8, we find that

$$2\,\mathrm{Re}\,\tau > \tilde{\sigma}_0 = 2\sigma_0,$$

which implies $\mathrm{Re}\,\tau > \sigma_0$ and ends the proof.

7.4.3 *Proof of the strict dominance in the general case*

According to Theorem 7.8, any eigenvalue

$$\tau \in \Sigma(\mathfrak{L}, \mathfrak{B}, \Omega) \setminus \{\sigma_0\}$$

satisfies

$$\operatorname{Re}\tau \geq \sigma_0.$$

Suppose that, for some of these eigenvalues,

$$\operatorname{Re}\tau = \sigma_0.$$

Then, $\operatorname{Im}\tau \neq 0$ and it follows from Theorem 2.4 of G. Greiner [88] that

$$\sigma_0 + ik\operatorname{Im}\tau \in \Sigma(\mathfrak{L}, \mathfrak{B}, \Omega)$$

for all $k \in \mathbb{Z}$. But this is impossible, because $\Sigma(\mathfrak{L}, \mathfrak{B}, \Omega)$ is contained in a symmetric sector around the real axis in the complex plane with a total angle less than π, because an appropriate L^p realization of $-\mathfrak{L}$ generates a holomorphic semigroup in $L^p(\Omega)$ and, therefore, $\mathfrak{L}$ is a *sectorial operator*, as discussed by D. Henry [93] (see Section II.3 of J. López-Gómez [134] for a rather general proof of this fact in the special case $\Gamma_1 = \emptyset$, and the argument on p. 41 of H. Amann [11] for the general case).

7.5 The strong maximum principle

This section establishes the main result of this chapter. To state it, we need to introduce some preliminary concepts.

Definition 7.3.

(a) *A function $h \in W^{2,p}(\Omega)$, $p > N$, is said to be a **supersolution** of $(\mathfrak{L}, \mathfrak{B}, \Omega)$ if*

$$\begin{cases} \mathfrak{L}h \geq 0 & in\ \Omega, \\ \mathfrak{B}h \geq 0 & on\ \partial\Omega. \end{cases}$$

*The function h is said to be a **strict supersolution** of $(\mathfrak{L}, \mathfrak{B}, \Omega)$ if, in addition, some of these inequalities is strict on a measurable set with positive measure.*

(b) *It is said that $(\mathfrak{L}, \mathfrak{B}, \Omega)$ satisfies the **strong maximum principle** (SMP) if any supersolution $u \in W^{2,p}(\Omega)$, $p > N$, $u \neq 0$, of $(\mathfrak{L}, \mathfrak{B}, \Omega)$ — in particular, any strict supersolution — satisfies*

$$u(x) > 0\ \forall\, x \in \Omega \cup \Gamma_1 \quad and \quad \frac{\partial u}{\partial \nu}(x) < 0\ \forall\, x \in u^{-1}(0) \cap \Gamma_0.$$

(c) *It is said that* $(\mathfrak{L}, \mathfrak{B}, \Omega)$ *satisfies the* **maximum principle** *(MP) if any supersolution* $u \in W^{2,p}(\Omega)$, $p > N$, *of* $(\mathfrak{L}, \mathfrak{B}, \Omega)$ *satisfies* $u(x) \geq 0$ *for all* $x \in \bar{\Omega}$.

The main theorem of this chapter reads as follows.

Theorem 7.10 (of characterization of the SMP). *The following assertions are equivalent:*

i) $\sigma_0 := \sigma[\mathfrak{L}, \mathfrak{B}, \Omega] > 0$.

ii) $(\mathfrak{L}, \mathfrak{B}, \Omega)$ *possesses a positive strict supersolution* $h \in W^{2,\infty^-}(\Omega)$.

iii) $(\mathfrak{L}, \mathfrak{B}, \Omega)$ *satisfies the strong maximum principle.*

iv) $(\mathfrak{L}, \mathfrak{B}, \Omega)$ *satisfies the maximum principle.*

v) *The resolvent of the linear boundary value problem*

$$\begin{cases} \mathfrak{L}u = f \in \mathcal{C}_e(\bar{\Omega}) & in \ \Omega, \\ \mathfrak{B}u = 0 & on \ \partial\Omega, \end{cases} \tag{7.46}$$

subsequently denoted by $\mathfrak{R}_0 : \mathcal{C}_e(\bar{\Omega}) \to \mathcal{C}_e(\bar{\Omega})$, *is well defined and it is strongly positive.*

Proof. Suppose $\sigma_0 > 0$, and let $h = \varphi_0$, where $\varphi_0 > 0$ is a principal eigenfunction associated to σ_0. Then, $\mathfrak{B}h = 0$ on $\partial\Omega$,

$$\mathfrak{L}h = \sigma_0 h > 0 \quad in \ \Omega,$$

and, according to Theorems 4.10 and 5.11, $h \in W^{2,p}(\Omega)$ for all $p > N$. Thus, $h \in W^{2,\infty^-}(\Omega)$ and it provides us with a positive strict supersolution of $(\mathfrak{L}, \mathfrak{B}, \Omega)$. Therefore, i) implies ii).

Suppose $(\mathfrak{L}, \mathfrak{B}, \Omega)$ has a positive strict supersolution $h \in W^{2,\infty^-}(\Omega)$ and let $u \in W^{2,p}(\Omega) \setminus \{0\}$, $p > N$, be a supersolution of $(\mathfrak{L}, \mathfrak{B}, \Omega)$. Then, the assumptions of Theorem 7.5 are fulfilled. Hence, some of the alternatives A1, A2, or A3, of Theorem 7.4 holds. As $u \neq 0$, Alternative 1 cannot occur. As h is a strict supersolution, Alternative 3 cannot occur either. Therefore, Alternative 2 occurs and, consequently, ii) implies iii). Obviously, iii) implies iv).

Suppose iv) and $\sigma_0 \leq 0$. Then,

$$\mathfrak{L}(-\varphi_0) = -\sigma_0 \varphi_0 \geq 0 \quad in \ \Omega$$

and $\mathfrak{B}(-\varphi_0) = 0$ on $\partial\Omega$. Thus, according to iv), we should have $-\varphi_0 \geq 0$, which is a contradiction. This contradiction concludes the proof of the equivalence of i), ii), iii) and iv).

Now, suppose $\sigma_0 > 0$ and pick $\omega > \max\{\omega_0, 0\}$. Then, $\mathfrak{L}u = f$ if, and only if,

$$(\mathfrak{L} + \omega)u = \omega u + f$$

and hence, u solves (7.46) if and only if

$$\left(\frac{1}{\omega} - \mathfrak{R}_\omega\right) u = \frac{1}{\omega}\mathfrak{R}_\omega f.$$

On the other hand, as $\sigma_0 > 0$,

$$\operatorname{spr}\mathfrak{R}_\omega = \frac{1}{\sigma_0 + \omega} < \frac{1}{\omega},$$

and, consequently, by Corollary 7.1(d), the resolvent operator

$$\left(\frac{1}{\omega} - \mathfrak{R}_\omega\right)^{-1} \in \mathcal{L}\left(\mathcal{C}_e(\bar{\Omega})\right)$$

is well defined and strongly positive. In particular, the resolvent of the problem (7.46) must be given through

$$\mathfrak{R}_0 := \frac{1}{\omega}\left(\frac{1}{\omega} - \mathfrak{R}_\omega\right)^{-1} \mathfrak{R}_\omega. \tag{7.47}$$

As, according to Proposition 7.3, $\mathfrak{R}_\omega$ is also strongly positive, it follows from (7.47) that $\mathfrak{R}_0$ must be strongly positive. Consequently, i) implies v).

Conversely, if $\mathfrak{R}_0$ is well defined and it is strongly positive, then, $h := \mathfrak{R}_0 f > 0$ for all $f > 0$ and, hence, any of these functions provides us with an admissible positive strict supersolution of $(\mathfrak{L}, \mathfrak{B}, \Omega)$. Therefore, v) implies ii). This ends the proof. $\qquad\square$

7.6 The classical minimum principles revisited

Throughout this section we suppose that $\mathfrak{B} = \mathfrak{D}$ is the Dirichlet boundary operator and, as usual, we denote

$$\sigma_0 := \sigma[\mathfrak{L}, \mathfrak{D}, \Omega].$$

The next consequence from Theorem 7.10 shows that the assumption that $\mathfrak{L}$ admits a superharmonic function h in Ω such that

$$h(x) > 0 \qquad \text{for all} \quad x \in \bar{\Omega}$$

in the generalized minimum principle of M. H. Protter and H. F. Weinberger (Theorem 1.7) is equivalent to the positivity of σ_0.

Corollary 7.2. *Suppose $\mathfrak{B} = \mathfrak{D}$. Then, conditions* i)–v) *of Theorem 7.10 are equivalent to the next two:*

vi) $(\mathfrak{L}, \mathfrak{D}, \Omega)$ *admits a positive supersolution* $h \in W^{2,\infty^-}(\Omega)$ *such that* $h(x) > 0$ *for all* $x \in \bar{\Omega}$.

vii) $(\mathfrak{L}, \mathfrak{D}, \Omega)$ *admits a positive strict supersolution* $h \in W^{2,\infty^-}(\Omega)$ *such that* $h = 0$ *on* $\partial\Omega$.

Proof.   Suppose $\sigma_0 > 0$. Then, by Theorem 7.10, the unique solution of

$$\begin{cases} \mathfrak{L}h = 0 & \text{in } \Omega, \\ h = 1 & \text{on } \partial\Omega, \end{cases}$$

provides us with a strict supersolution satisfying vi). Note that

$$h = 1 - \mathfrak{R}_0 \, c,$$

where c is the zero order term of $\mathfrak{L}$. Moreover, any principal eigenfunction $\varphi_0 > 0$ provides us with a positive strict supersolution satisfying vii).

Conversely, under any of the conditions vi) or vii), h provides us with a positive strict supersolution of $(\mathfrak{L}, \mathfrak{D}, \Omega)$ and hence, thanks to Theorem 7.10, $\sigma_0 > 0$. The proof is complete.   □

When $c \geq 0$, the constant function $h := 1$ provides us with a supersolution satisfying condition vi), and hence $\sigma_0 > 0$. Consequently, the next result provides us with a substantial generalization of the classical theorems of E. Hopf (Theorems 1.2, 1.3), and M. H. Protter and H. F. Weinberger (Theorem 1.7).

Theorem 7.11.   *Suppose* $\sigma_0 > 0$ *and* $u \in W^{2,p}(\Omega)$, $p > N$, *satisfies*

$$\mathfrak{L}u \geq 0 \quad \text{in } \Omega \quad \text{and} \quad \inf_{\Omega} u \geq 0. \qquad (7.48)$$

Then, either $u = 0$, *or*

$$u(x) > 0 \quad \forall \, x \in \Omega \quad \text{and} \quad \frac{\partial u}{\partial \nu}(x) < 0 \quad \forall \, x \in u^{-1}(0) \cap \partial\Omega.$$

If, instead of (7.48), *u satisfies*

$$\mathfrak{L}u \geq 0 \quad \text{in } \Omega \quad \text{and} \quad \inf_{\Omega} u < 0, \qquad (7.49)$$

then, for every $h \in W^{2,p}(\Omega)$ *such that*

$$\mathfrak{L}h \geq 0 \quad \text{in } \Omega \quad \text{and} \quad \inf_{\Omega} h > 0, \qquad (7.50)$$

the quotient function

$$v(x) := \frac{u(x)}{h(x)}, \qquad x \in \bar{\Omega}, \qquad (7.51)$$

satisfies

$$m := \inf_{\Omega} v < 0,$$

$$v(x) > m \quad \forall \ x \in \Omega \quad and \quad \frac{\partial v}{\partial \nu}(x) < 0 \ \forall \ x \in v^{-1}(m) \cap \partial\Omega, \qquad (7.52)$$

unless $v = m$ *in* $\bar{\Omega}$. *In particular, for every* $f \in L^{\infty}(\Omega)$, $f > 0$, *the unique weak solution* h *of the problem*

$$\begin{cases} \mathfrak{L}h = f & in \ \Omega, \\ h = 1 & on \ \partial\Omega, \end{cases} \qquad (7.53)$$

satisfies

$$\inf_{\Omega} \frac{u}{h} = \inf_{\partial\Omega} \frac{u}{h} = \inf_{\partial\Omega} u < 0, \qquad (7.54)$$

$$u(x) > \left(\inf_{\partial\Omega} u\right) h(x) \qquad for \ all \ \ x \in \Omega, \qquad (7.55)$$

and

$$\frac{\partial}{\partial \nu} \frac{u}{h}(x_0) < 0 \qquad for \ all \ \ x_0 \in u^{-1}\left(\inf_{\partial\Omega} u\right) \cap \partial\Omega. \qquad (7.56)$$

Proof. As $\sigma_0 > 0$, by Theorem 7.10, $(\mathfrak{L}, \mathfrak{D}, \Omega)$ satisfies the strong maximum principle. Suppose $u \neq 0$ satisfies (7.48). Then, u provides us with a non-zero non-negative supersolution of $(\mathfrak{L}, \mathfrak{B}, \Omega)$ and, therefore, thanks to Theorem 7.5, u satisfies the required properties.

Subsequently, we suppose that u and h satisfy (7.49) and (7.50), respectively. By Theorem 7.10, the solution of (7.53) provides us with one of these functions h for every $f > 0$. Note that

$$h = 1 + \mathfrak{R}_0(f - c).$$

Next, we consider the quotient function v defined by (7.51). As

$$\inf_{\Omega} u < 0 \quad and \quad \inf_{\Omega} h > 0,$$

we have that

$$m := \inf_{\Omega} v < 0.$$

Moreover, by Theorem 7.5, (7.52) holds, unless $v = m$ in $\bar{\Omega}$. In any of these circumstances, we have that

$$\inf_{\Omega} v = \inf_{\partial\Omega} v. \qquad (7.57)$$

Subsequently, we fix $f > 0$, $f \in L^{\infty}(\Omega)$, and suppose that h is the unique solution of (7.53). Then, since $h = 1$ on $\partial\Omega$, we have that $v = u$ on $\partial\Omega$ and hence (7.57) implies (7.54).

Suppose $v = m$ in $\bar{\Omega}$. Then, $u = mh$ and, hence,

$$0 \leq \mathfrak{L}u = m\mathfrak{L}u = mf < 0,$$

which is impossible. Therefore, (7.55) and (7.56) follow from (7.52). The proof is complete. $\qquad \square$

7.7   Comments on Chapter 7

Seemingly, the characterization of the strong maximum principle given by Theorem 7.10 goes back to Theorem 2.1 of J. López-Gómez and M. Molina-Meyer [148]. Although when [148] appeared, in early 1994, there were already available a number of preliminary results trying to establish the hidden connections between the sign of σ_0, the validity of the maximum principle, the validity of the strong maximum principle, and the existence of a positive supersolution (see G. Sweers [217], D. G. de Figueiredo and E. Mitidieri [62, 63], Lemma 3.2 of J. López-Gómez and R. M. Pardo [154], and J. Fleckinger, J. Hernández and F. de Thélin [64]), the theorem establishing the equivalence between the following five conditions

- $(\mathfrak{L}, \mathfrak{D}, \Omega)$ possesses a positive strict supersolution.
- The resolvent of $(\mathfrak{L}, \mathfrak{D}, \Omega)$ is well defined and it is strongly positive.
- $(\mathfrak{L}, \mathfrak{D}, \Omega)$ satisfies the strong maximum principle.
- $(\mathfrak{L}, \mathfrak{D}, \Omega)$ satisfies the maximum principle.
- The principal eigenvalue of $(\mathfrak{L}, \mathfrak{D}, \Omega)$, denoted by σ_0, is positive.

seems to go back to Theorem 2.1 of [148], not only for a single second order elliptic operator, but, more generally, for a rather general class of linear elliptic systems of cooperative type.

Almost simultaneously, but in this case for the scalar operator, without any regularity constraint on $\partial\Omega$, Theorem 1.1 of H. Berestycki, L. Nirenberg and S. R. S. Varadhan [27] established that $(\mathfrak{L}, \mathfrak{D}, \Omega)$ satisfies the maximum principle if and only if $\sigma_0 > 0$. Some precursors of this result had already been given by S. Agmon [3]. They also found that the maximum principle holds if $(\mathfrak{L}, \mathfrak{D}, \Omega)$ admits a positive strict supersolution (see Corollary 2.4 of [27]), but the authors did not infer from this result the strong maximum principle, possibly because of the lack of regularity of $\partial\Omega$. Incidentally, H. Berestycki, L. Nirenberg and S. R. S. Varadhan [27] might have not realized the importance of their Corollary 2.4, as it was left outside of Section 1 of [27], where the main results of their paper were collected.

The fact that the characterization of the strong maximum principle in terms of the existence of a strict positive supersolution had been left outside the general scope of H. Berestycki, L. Nirenberg and S. R. S. Varadhan [27], prompted J. López-Gómez to give a short proof of Theorem 7.10 for Dirichlet boundary conditions, in Theorem 2.5 of [137], as the author realized that even the simplest version of Theorem 2.1 of J. López-Gómez and M. Molina-Meyer [148] for the scalar operator was unknown for the more recognized specialists. The proof of Theorem 2.5 of [137] is based on Theorem 2 of W. Walter [224]; an old version, for Dirichlet boundary conditions, of our Theorem 2.4 here. All the materials covered by [137] had already been delivered by J. López-Gómez in his Ph.D. course on *Bifurcation Theory* at the University of Zürich during the summer semester of 1994 (see the Acknowledgements of [137]).

As it will become apparent throughout Chapters 8 and 9, from the point of view of the applications, the most crucial feature of Theorem 7.10 is the fact that the existence of a positive strict supersolution characterizes the strong maximum principle, as this is the usual strategy adopted in the applications to make sure that $\sigma_0 > 0$, or, equivalently, that the strong maximum principle holds. This provides Theorem 2.1 of J. López-Gómez and M. Molina-Meyer [148] with its greatest significance when it is weighted versus the results of H. Berestycki, L. Nirenberg and S. R. S. Varadhan [27].

Three years later, in March 1997, Theorem 2.4 of H. Amann and J. López-Gómez [13] generalized Theorem 2.5 of J. López-Gómez [137] up to cover the general boundary operators considered in this book. Actually, in [13], not only the existence of a smooth positive strict supersolution, but also of a positive strict supersolution in $W^{2,p}$, $p > N$, was shown to be necessary and sufficient for the validity of the strong maximum principle. The proof of H. Amann and J. López-Gómez relies on Theorem 6.1 of H. Amann [9]. Naturally, the proof of Theorem 6.1 of [9] is based upon the construction of a strict supersolution $h \in W^{2,p}$, $p > N$, satisfying $h(x) > 0$ for all $x \in \bar{\Omega}$, which is the classical assumption of the generalized minimum principle of M. H. Protter and H. F. Weinberger (Theorem 1.7), together with the minimum principle of J. M. Bony [28] (Theorem 7.1). But, incidentally, it turns out that the technical details of that construction do actually rely on the inverse trace theorem of H. Amann [9] (see Lemma 5.1 and Sections 5.5 and 5.7 of [9]) whose proof is, in words of H. Amann [11] (see p. 17 and Remark 36(a) of [11])

somewhat involved and complicated and perhaps not too transparent.

This *lack of transparency* prompted J. López-Gómez (see Theorem 6.1 of [144]) to give a new self-contained elementary proof of Theorem 2.4 of H. Amann and J. López-Gómez [13] through Theorem 2.4 of Chapter 2. At the end of the day, this has been the proof of Theorem 7.10 included in this chapter.

The relevance of Theorem 7.10 has been recognized and, possibly, enhanced by the recent monographs of H. Amann [11] and Y. Du [56]. Indeed,

Theorem 13 of H. Amann [11] replaced the $W^{2,p}$-supersolutions of Theorem 2.4 of H. Amann and J. López-Gómez [13] by L^p-supersolutions and extended to a very general class of linear cooperative systems the Theorem 2.1 of J. López-Gómez and M. Molina-Meyer [148], whereas Y. Du [56] dedicated Chapter 2, among 7 chapters, to Theorem 7.10. Incidentally, neither J. López-Gómez and M. Molina-Meyer [148], nor J. López-Gómez [144], were incorporated to the bibliography of Y. Du [56], however the main theorem of Chapter 2 of [56] goes back to [148], and the proof given by Y. Du in Chapter 2 of [56] is based on J. López-Gómez [144].

The reader should be aware of the extremely significant fact that the statements of all these characterization theorems have remained unchanged since J. López-Gómez and M. Molina-Meyer established Theorem 2.1 of [148], except for the brilliant, though natural, observation of H. Amann [11] that the classical strong maximum principle is equivalent to his *weak* and *very weak* formulations.

The results of Section 2 go back to J. M. Bony [28], whose proofs have been adapted in this book. Theorem 7.1 was later extended by P. L. Lions [132] to cover the case when $p = N$.

The proof of the existence and the uniqueness of the principal eigenvalue established by Theorem 7.7 is deeply indebted with the general theory developed by H. Amann in his monographs [7] and [8] and by P. H. Rabinowitz in his course of Paris [185]. The pivotal Proposition 7.2 is attributed to H. Amann [7] (see Lemma 5.3 there in). Corollary 7.1(d) provides us with a version of the anti-maximum principle of Ph. Clément and L. A. Peletier [43] (see P. Takác [219]).

Up to the best of our knowledge, the more pioneering results on principal eigenvalues for the Dirichlet problem associated to the Laplace equation go back to J. B. J. Fourier [65], though he possibly never referred to π^2 as the principal eigenvalue of

$$\begin{cases} -\varphi'' = \tau\varphi & \text{in } (0,1), \\ \varphi(0) = \varphi(1) = 0. \end{cases}$$

As H. Poincaré [180] established the analyticity of the resolvent of the Dirichlet problem for the Laplace operator and could then infer from it the fact that it has discrete spectrum, H. Poincaré surely knew that the lowest eigenvalue of the problem plays a significant role in potential theory, as well as J. W. S. Rayleigh [186]. Incidentally, H. Poincaré was teaching an advanced course on Potential Theory when D. Hilbert visited Paris for the first time after defending his Ph.D. Thesis. Later, R. Courant and D.

Hilbert [44] clearly realized the importance of the principal eigenvalues in Mathematical Physics.

The simplicity of the principal eigenvalue, as it has been delivered here, is indebted to M. G. Crandall and P. H. Rabinowitz [45]. Once chosen an appropriate realization for $(\mathfrak{L}, \mathfrak{B}, \Omega)$ on some subspace $X \subset W^{1,2}_{\Gamma_0}(\Omega)$, the fact that τ is a simple eigenvalue according to Definition 7.2 can be equivalently stated by simply saying that

$$N[\mathfrak{L} - \tau] = \mathrm{span}[\varphi], \qquad \varphi \notin R[\mathfrak{L} - \tau].$$

Thus, if, in addition, $\mathfrak{L} - \tau$ is a Fredholm operator of index zero, i.e.,

$$\mathrm{codim}\, R[\mathfrak{L} - \tau] = \dim N[\mathfrak{L} - \tau] = 1,$$

then

$$N[\mathfrak{L} - \tau] \oplus R[\mathfrak{L} - \tau] = X,$$

and, therefore, τ is indeed an algebraically simple eigenvalue of (7.1) in the classical sense.

In H. Amann [11], the fact that σ_0 is a simple eigenvalue of (7.1) was stated in Theorem 12 of [11] and proven on p. 40 of [11]. In some moment of the proof, after (64), it was claimed, and then used, that

The Krein–Rutman theorem guarantees also that there exists an eigenvector φ of the dual $T' \in \mathcal{L}(E')$ to the eigenvalue $r = \mathrm{spr}\, T$ satisfying $\langle \varphi, v \rangle > 0$ for $v \in E^+ \setminus \{0\}$.

But H. Amann [11] did not provide the readers with a reference for this result and we could not find it in the literature. Thus, apparently, the proof of the simplicity of the principal eigenvalue of H. Amann [11] contains a gap. After reading this book, H. Amann sent the author, through an electronic mail, a short proof of this fact, which is a slight modification of the one included in the final part of the proof of Theorem 6.3.

The proof of the dominance of σ_0 in Theorem 7.8 seems to be new, whereas the proof of Theorem 7.9 in case $\Gamma_0 = \emptyset$ is based on the proof of Theorem 1 of M. H. Protter and H. F. Weinberger [182], the proof of Theorem 7.9 in case $\Gamma_1 = \emptyset$ has been borrowed from H. Berestycki, L. Nirenberg and S. R. S. Varadhan [27], and the proof of the strict dominance in the general case has been taken from p. 41 of H. Amann [11]. Incidentally, the proof on p. 7 of Y. Du [56] is wrong, since

$$K\phi \geq 2(\mathrm{Re}\,\lambda - \lambda_1)\phi$$

cannot imply $K\phi \geq 0$ if $\mathrm{Re}\,\lambda < \lambda_1$.

Apparently, Theorem 1 of M. H. Protter and H. F. Weinberger [182] is the first dominance result available in the literature for general nonselfadjoint operators. Using the notations of this chapter, Theorem 1 of [182] established that, for any eigenvalue τ of (7.1),

$$\operatorname{Re} \tau \geq \inf_{\Omega} \frac{\mathfrak{L}u}{u} \tag{7.58}$$

for all $u \in \mathcal{C}^2(\Omega) \cap \mathcal{C}^1(\bar{\Omega})$ such that $\inf_{\Omega} u > 0$ and $\mathfrak{B}u \geq 0$ on $\partial\Omega$. In particular, $\Sigma(\mathfrak{L}, \mathfrak{B}, \Omega)$ must be contained in a half plane. Naturally, if we approximate a principal eigenfunction $\varphi_0 > 0$ by those functions, (7.58) will provide us with

$$\operatorname{Re} \tau \geq \sigma_0.$$

Therefore, $\operatorname{Re} z \geq \sigma_0$ is the optimal half-plane containing the set of eigenvalues of (7.1). Some precursors of these results, for selfadjoint operators, go back to J. Barta [23], R. J. Duffin [57], J. Hersch [94], W. W. Hooker [102] and M. H. Protter [181].

In its greatest generality, Theorem 7.9 goes back to Theorem 12.1 of H. Amann [9]. But, incidentally, according to the comments of H. Amann after the statement of Theorem 12 on p. 15 of [11], the proof of Theorem 12.1 of [9]

contained a gap since it had not been asserted that the spectrum is nonempty. Motivated by this, B. de Pagter [173] derived a general theorem on irreducible compact positive operators on Banach lattices implying that such an operator has a strictly positive spectral radius. That theorem can be used to fill the gap (cf. the proof of Theorem 2.2 of H. Amann and J. López-Gómez [13]).

Indeed, in the proof of Theorem 12.1 of H. Amann [9], the book of H. H. Schaefer [194] was invoked to assert that a positive compact irreducible linear operator on a Banach lattice has a strictly positive spectral radius, however this result cannot be found in H. H. Schaefer [194], but it is Theorem 3 of B. de Pagter [173].

Under Dirichlet boundary conditions, J. P. Gossez and E. Lami-Dozo [87] also established the strict dominance of σ_0.

The results of Section 7.6 go back to J. López-Gómez [146].

Chapter 8

Properties of the principal eigenvalue

This chapter applies Theorem 7.10 to obtain some useful properties, from the point of view of the applications, of the principal eigenvalue of (7.1)

$$\sigma_0 := \sigma[\mathfrak{L}, \mathfrak{B}, \Omega].$$

Among them, count the monotonicity properties of σ_0 with respect to $c(x)$, $\beta(x)$, and Ω, its continuity and concavity with respect to $c(x)$, its continuous dependence with respect to the variations of Ω along Γ_0, as well as with respect to β, and the crucial property that $(\mathfrak{L}, \mathfrak{B}, \Omega)$ satisfies the strong maximum principle for sufficiently large β and small $|\Omega|$.

Throughout the rest of this book, we will denote

$$W^{2,\infty^-}(\Omega) := \bigcap_{p>1} W^{2,p}(\Omega),$$

as in Theorem 5.8,

$$W_{\mathfrak{B}}^{2,p}(\Omega) := \left\{ u \in W^{2,p}(\Omega) \; : \; \mathfrak{B}u = 0 \right\}, \qquad p > 1,$$

and

$$W_{\mathfrak{B}}^{2,\infty^-}(\Omega) := \bigcap_{p>1} W_{\mathfrak{B}}^{2,p}(\Omega).$$

Since $\beta \in \mathcal{C}(\Gamma_1)$, by Theorem 4.6, we have that

$$\mathfrak{B} \in \mathcal{L}\left(W^{2,p}(\Omega), L^p(\partial\Omega)\right) \qquad \text{for all } p > 1.$$

Thus, $W_{\mathfrak{B}}^{2,p}(\Omega)$ is a closed subspace of $W^{2,p}(\Omega)$ for all $p > 1$.

Also, we will denote by

$$\varphi_0 := \varphi_{[\mathfrak{L}, \mathfrak{B}, \Omega]} > 0$$

the principal eigenfunction associated to σ_0, normalized so that

$$\int_\Omega \varphi_0^2 = 1.$$

Note that φ_0 satisfies (7.30).

Eventually, we shall emphasize the dependence of the boundary operator $\mathfrak{B}$ on the weight function $\beta \in \mathcal{C}(\Gamma_1)$ by setting

$$\mathfrak{B}[\beta] := \mathfrak{B}.$$

If not strictly necessary, such dependency will be dropped.

Suppose $\Gamma_1 \neq \emptyset$. Then, for every proper subdomain Ω_0 of class $\mathcal{C}^2$ of Ω with

$$\text{dist}\,(\Gamma_1, \partial\Omega_0 \cap \Omega) > 0, \tag{8.1}$$

we shall denote by $\mathfrak{B}[\beta, \Omega_0]$, or simply by $\mathfrak{B}[\Omega_0]$, the boundary operator

$$\mathfrak{B}[\Omega_0]u := \begin{cases} u & \text{on} \quad \partial\Omega_0 \cap \Omega, \\ \mathfrak{B}u & \text{on} \quad \partial\Omega_0 \cap \partial\Omega, \end{cases} \tag{8.2}$$

for all $u \in W^{2,p}(\Omega)$ with $p > 1$. If $\bar{\Omega}_0 \subset \Omega$, then $\partial\Omega_0 \subset \Omega$ and hence,

$$\mathfrak{B}[\Omega_0]u = u$$

for all $u \in W^{2,p}(\Omega)$ with $p > 1$, i.e., $\mathfrak{B}[\Omega_0]$ becomes the Dirichlet boundary operator $\mathfrak{D}$ in Ω_0.

If $\Gamma_1 = \emptyset$, i.e., $\mathfrak{B} = \mathfrak{D}$, then, we define

$$\mathfrak{B}[\Omega_0] := \mathfrak{D} \tag{8.3}$$

for all subdomain $\Omega_0 \subset \Omega$ of class $\mathcal{C}^2$. Note that (8.2) becomes (8.3) if $\Gamma_1 = \emptyset$. Finally, we also allow $\Omega_0 = \Omega$ by setting

$$\mathfrak{B}[\Omega] := \mathfrak{B},$$

and, throughout the rest of this book, we shall denote

$$\partial_\nu := \frac{\partial}{\partial\nu} \quad \text{and} \quad W_0^{1,p}(\Omega) = W_{\partial\Omega}^{1,p}(\Omega)$$

for all $p \geq 1$ (see (4.16)).

8.1 Monotonicity properties

This section applies Theorem 7.10 to obtain some fundamental monotonicity properties of the principal eigenvalue. Among them, its monotonicity with respect to the coefficients c and β, and with respect to the support domain Ω.

As an easy consequence of the uniqueness given by Theorem 7.7, for every subdomain $\Omega_0 \subset \Omega$ of class $\mathcal{C}^2$ the following identity holds

$$\sigma[\mathfrak{L} + s, \mathfrak{B}, \Omega_0] = \sigma[\mathfrak{L}, \mathfrak{B}, \Omega_0] + s \qquad \text{for all } s \in \mathbb{R}.$$

This identity will be used very often throughout the remaining of this book.

From the point of view of the theory of dynamical systems, the next result shows that, among all boundary operators $\mathfrak{B}$, the Dirichlet operator $\mathfrak{D}$ is the one with the strongest stabilizing effects.

Proposition 8.1. *Suppose* $\Gamma_1 \neq \emptyset$. *Then,*

$$\sigma[\mathfrak{L}, \mathfrak{B}[\beta], \Omega] < \sigma[\mathfrak{L}, \mathfrak{D}, \Omega] \qquad \text{for all } \beta \in \mathcal{C}(\Gamma_1).$$

Proof. Let $\varphi_{[\mathfrak{L},\mathfrak{B},\Omega]}$ and $\varphi_{[\mathfrak{L},\mathfrak{D},\Omega]}$ denote the normalized principal eigenfunctions associated with $\sigma[\mathfrak{L}, \mathfrak{B}, \Omega]$ and $\sigma[\mathfrak{L}, \mathfrak{D}, \Omega]$, respectively. By (7.30), we have that

$$\varphi_{[\mathfrak{L},\mathfrak{B},\Omega]}(x) > 0 \qquad \text{for all } x \in \Omega \cup \Gamma_1$$

and hence

$$\mathfrak{D}\varphi_{[\mathfrak{L},\mathfrak{B},\Omega]} = \varphi_{[\mathfrak{L},\mathfrak{B},\Omega]} > 0 \qquad \text{on } \partial\Omega.$$

Thus, $\varphi_{[\mathfrak{L},\mathfrak{B},\Omega]}$ provides us with a positive strict supersolution of

$$(\mathfrak{L} - \sigma[\mathfrak{L}, \mathfrak{B}, \Omega], \mathfrak{D}, \Omega)$$

and, therefore, according to Theorem 7.10, we find that

$$0 < \sigma[\mathfrak{L} - \sigma[\mathfrak{L}, \mathfrak{B}, \Omega], \mathfrak{D}, \Omega] = \sigma[\mathfrak{L}, \mathfrak{D}, \Omega] - \sigma[\mathfrak{L}, \mathfrak{B}, \Omega].$$

This completes the proof. $\qquad\square$

The following result shows the monotonicity of the principal eigenvalue with respect to Ω.

Proposition 8.2. *Let* Ω_0 *be a proper subdomain of* Ω *of class* $\mathcal{C}^2$ *satisfying* (8.1) *if* $\Gamma_1 \neq \emptyset$. *Then,*

$$\sigma[\mathfrak{L}, \mathfrak{B}, \Omega] < \sigma[\mathfrak{L}, \mathfrak{B}[\Omega_0], \Omega_0],$$

where $\mathfrak{B}[\Omega_0]$ *is the boundary operator defined by* (8.2).

Proof. Let $\varphi_{[\mathfrak{L},\mathfrak{B},\Omega]}$ be the normalized principal eigenfunction associated with $\sigma[\mathfrak{L}, \mathfrak{B}, \Omega]$. Then, using (7.30), it is easy to see that

$$\begin{cases} (\mathfrak{L} - \sigma[\mathfrak{L}, \mathfrak{B}, \Omega])\varphi_{[\mathfrak{L},\mathfrak{B},\Omega]} = 0 & \text{in } \Omega_0, \\ \varphi_{[\mathfrak{L},\mathfrak{B},\Omega]}(x) > 0 & \text{if } x \in \partial\Omega_0 \cap \Omega, \\ \varphi_{[\mathfrak{L},\mathfrak{B},\Omega]}(x) = 0 & \text{if } x \in \partial\Omega_0 \cap \Gamma_0, \\ \partial_\nu\varphi_{[\mathfrak{L},\mathfrak{B},\Omega]}(x) + \beta(x)\varphi_{[\mathfrak{L},\mathfrak{B},\Omega]}(x) = 0 & \text{if } x \in \partial\Omega_0 \cap \Gamma_1. \end{cases}$$

Moreover, $\partial\Omega_0 \cap \Omega \neq \emptyset$, as Ω_0 is a proper subdomain of Ω. Therefore, $\varphi_{[\mathfrak{L},\mathfrak{B},\Omega]}$ is a positive strict supersolution of

$$(\mathfrak{L} - \sigma[\mathfrak{L}, \mathfrak{B}, \Omega], \mathfrak{B}[\Omega_0], \Omega_0)$$

and, consequently, it follows from Theorem 7.10 that

$$0 < \sigma[\mathfrak{L} - \sigma[\mathfrak{L}, \mathfrak{B}, \Omega], \mathfrak{B}[\Omega_0], \Omega_0] = \sigma[\mathfrak{L}, \mathfrak{B}[\Omega_0], \Omega_0] - \sigma[\mathfrak{L}, \mathfrak{B}, \Omega].$$

This completes the proof. $\qquad\square$

The next result establishes the monotonicity of the principal eigenvalue with respect to the potential.

Proposition 8.3. *Let P_1, $P_2 \in L^\infty(\Omega)$ such that $P_1 < P_2$. Then,*

$$\sigma[\mathfrak{L} + P_1, \mathfrak{B}, \Omega] < \sigma[\mathfrak{L} + P_2, \mathfrak{B}, \Omega].$$

Proof. Set

$$\varphi_1 := \varphi_{[\mathfrak{L}+P_1, \mathfrak{B}, \Omega]}.$$

Then, by (7.30),

$$(\mathfrak{L} + P_2 - \sigma[\mathfrak{L} + P_1, \mathfrak{B}, \Omega])\varphi_1 = (P_2 - P_1)\varphi_1 > 0$$

in Ω, and hence φ_1 is a positive strict supersolution of

$$(\mathfrak{L} + P_2 - \sigma[\mathfrak{L} + P_1, \mathfrak{B}, \Omega], \mathfrak{B}, \Omega).$$

Therefore, thanks to Theorem 7.10, we find that

$$0 < \sigma[\mathfrak{L}+P_2-\sigma[\mathfrak{L}+P_1, \mathfrak{B}, \Omega], \mathfrak{B}, \Omega] = \sigma[\mathfrak{L}+P_2, \mathfrak{B}, \Omega] - \sigma[\mathfrak{L}+P_1, \mathfrak{B}, \Omega].$$

This ends the proof.   □

As an immediate consequence, from this result we can get the continuous dependence of the principal eigenvalue with respect to the potential.

Corollary 8.1. *Let $P_n \in L^\infty(\Omega)$, $n \geq 1$, be a sequence of potentials such that*

$$\lim_{n \to \infty} P_n = P \quad in \quad L^\infty(\Omega).$$

Then,

$$\lim_{n \to \infty} \sigma[\mathfrak{L} + P_n, \mathfrak{B}, \Omega] = \sigma[\mathfrak{L} + P, \mathfrak{B}, \Omega].$$

Proof. For every $\epsilon > 0$ there exists a natural number $n(\epsilon) \geq 1$ such that

$$P - \epsilon \leq P_n \leq P + \epsilon \quad in \quad \Omega$$

for all $n \geq n(\epsilon)$. Therefore, owing to Proposition 8.3,

$$\sigma[\mathfrak{L} + P, \mathfrak{B}, \Omega] - \epsilon \leq \sigma[\mathfrak{L} + P_n, \mathfrak{B}, \Omega] \leq \sigma[\mathfrak{L} + P, \mathfrak{B}, \Omega] + \epsilon$$

for all $n \geq n(\epsilon)$. This ends the proof.   □

The next proposition shows the monotonicity of the principal eigenvalue with respect to $\beta \in \mathcal{C}(\Gamma_1)$.

Proposition 8.4. *Suppose $\Gamma_1 \neq \emptyset$ and let β_1, $\beta_2 \in \mathcal{C}(\Gamma_1)$ with $\beta_1 < \beta_2$. Then,*

$$\sigma[\mathfrak{L}, \mathfrak{B}[\beta_1], \Omega] < \sigma[\mathfrak{L}, \mathfrak{B}[\beta_2], \Omega].$$

Proof. Set

$$\varphi_1 := \varphi_{[\mathfrak{L}, \mathfrak{B}[\beta_1], \Omega]}.$$

Then,

$$(\mathfrak{L} - \sigma[\mathfrak{L}, \mathfrak{B}[\beta_1], \Omega])\varphi_1 = 0 \quad \text{in } \Omega,$$

$\varphi_1 = 0$ on Γ_0, and, due to (7.30),

$$\partial_\nu\varphi_1 + \beta_2\varphi_1 = (\beta_2 - \beta_1)\varphi_1 > 0 \quad \text{on } \Gamma_1.$$

Thus, φ_1 is a positive strict supersolution of

$$(\mathfrak{L} - \sigma[\mathfrak{L}, \mathfrak{B}[\beta_1], \Omega], \mathfrak{B}[\beta_2], \Omega)$$

and, therefore, Theorem 7.10 implies that

$$0 < \sigma[\mathfrak{L} - \sigma[\mathfrak{L}, \mathfrak{B}[\beta_1], \Omega], \mathfrak{B}[\beta_2], \Omega] = \sigma[\mathfrak{L}, \mathfrak{B}[\beta_2], \Omega] - \sigma[\mathfrak{L}, \mathfrak{B}[\beta_1], \Omega].$$

This concludes the proof. □

Corollary 8.2. *Suppose $\Gamma_1 \neq \emptyset$ and Ω_0 is a subdomain of class C^2 of Ω satisfying (8.1). Let β_1, $\beta_2 \in \mathcal{C}(\Gamma_1)$ with $\beta_1 < \beta_2$. Then,*

$$\sigma[\mathfrak{L}, \mathfrak{B}[\beta_1, \Omega], \Omega] < \sigma[\mathfrak{L}, \mathfrak{B}[\beta_2, \Omega_0], \Omega_0]. \tag{8.4}$$

The same conclusion holds if $\beta_1 \leq \beta_2$ and $\Omega_0 \subsetneq \Omega$.

Proof. If $\Omega = \Omega_0$, then (8.4) becomes into

$$\sigma[\mathfrak{L}, \mathfrak{B}[\beta_1], \Omega] < \sigma[\mathfrak{L}, \mathfrak{B}[\beta_2], \Omega],$$

which is guaranteed by Proposition 8.4.

Suppose $\Omega_0 \subsetneq \Omega$. Then, according to Propositions 8.4 and 8.2, we have that

$$\sigma[\mathfrak{L}, \mathfrak{B}[\beta_1, \Omega], \Omega] \leq \sigma[\mathfrak{L}, \mathfrak{B}[\beta_2, \Omega], \Omega] < \sigma[\mathfrak{L}, \mathfrak{B}[\beta_2, \Omega_0], \Omega_0].$$

The proof is complete. □

8.2 Point-wise min-max characterizations

As a consequence from Theorem 7.10, the next point-wise min-max characterization of $\sigma[\mathfrak{L}, \mathfrak{B}, \Omega]$ follows.

Theorem 8.1. *Let $\mathcal{P}$ denote the set of functions $\psi \in W^{2,\infty^-}(\Omega)$ such that $\psi(x) > 0$ for all $x \in \Omega$ and $\mathfrak{B}\psi \geq 0$ on $\partial\Omega$. Then,*

$$\sigma_0 := \sigma[\mathfrak{L}, \mathfrak{B}, \Omega] = \sup_{\psi \in \mathcal{P}} \inf_\Omega \frac{\mathfrak{L}\psi}{\psi} = \max_{\psi \in \mathcal{P}} \inf_\Omega \frac{\mathfrak{L}\psi}{\psi}. \tag{8.5}$$

Proof. Fix $\lambda < \sigma_0$. Then,

$$\sigma[\mathfrak{L} - \lambda, \mathfrak{B}, \Omega] = \sigma_0 - \lambda > 0$$

and hence, according to Theorem 7.10, $(\mathfrak{L} - \lambda, \mathfrak{B}, \Omega)$ satisfies the strong maximum principle. Thus, the problem

$$\begin{cases} (\mathfrak{L} - \lambda)\psi = 1 & \text{in } \Omega, \\ \mathfrak{B}\psi = 0 & \text{on } \partial\Omega, \end{cases}$$

has a unique solution in $W_{\mathfrak{B}}^{2,\infty^-}(\Omega)$, denoted by ψ_1, and ψ_1 is strongly positive, in the sense that it satisfies (7.30). In particular, $\psi_1 \in \mathcal{P}$ and, so, $\mathcal{P} \neq \emptyset$. As $\psi_1(x) > 0$ for all $x \in \Omega \cup \Gamma_1$, we have that

$$\lambda < \frac{\mathfrak{L}\psi_1}{\psi_1} \qquad \text{in } \Omega.$$

Thus,

$$\lambda \leq \inf_{\Omega} \frac{\mathfrak{L}\psi_1}{\psi_1} \leq \sup_{\psi \in \mathcal{P}} \inf_{\Omega} \frac{\mathfrak{L}\psi}{\psi}. \tag{8.6}$$

As (8.6) holds for every $\lambda < \sigma_0$, it becomes apparent that

$$\sigma_0 \leq \sup_{\psi \in \mathcal{P}} \inf_{\Omega} \frac{\mathfrak{L}\psi}{\psi}.$$

To prove the equality we argue by contradiction. Suppose

$$\sigma_0 < \sup_{\psi \in \mathcal{P}} \inf_{\Omega} \frac{\mathfrak{L}\psi}{\psi}.$$

Then, there exist $\epsilon > 0$ and $\psi \in \mathcal{P}$ such that

$$\sigma_0 + \epsilon < \frac{\mathfrak{L}\psi(x)}{\psi(x)} \qquad \text{for all } x \in \Omega.$$

Hence,

$$\begin{cases} (\mathfrak{L} - \sigma_0 - \epsilon)\psi > 0 & \text{in } \Omega, \\ \mathfrak{B}\psi \geq 0 & \text{on } \partial\Omega, \end{cases}$$

and, consequently, ψ provides us with a positive strict supersolution of $(\mathfrak{L} - \sigma_0 - \epsilon, \mathfrak{B}, \Omega)$. Therefore, thanks to Theorem 7.10,

$$0 < \sigma[\mathfrak{L} - \sigma_0 - \epsilon, \mathfrak{B}, \Omega] = -\epsilon,$$

which is impossible. This contradiction shows that

$$\sigma_0 = \sup_{\psi \in \mathcal{P}} \inf_{\Omega} \frac{\mathfrak{L}\psi}{\psi}.$$

As the normalized principal eigenfunction φ_0 associated with σ_0 lies in $W_{\mathfrak{B}}^{2,\infty^-}(\Omega)$ and $\mathfrak{L}\varphi_0 = \sigma_0\varphi_0$, we also have that

$$\sigma_0 = \inf_{\Omega} \frac{\mathfrak{L}\varphi_0}{\varphi_0},$$

which concludes the proof of (8.5). $\square$

Similarly, the following min-max characterization holds.

Theorem 8.2. *Let* $\mathcal{P}_>$ *denote the set of functions* $\psi \in W^{2,\infty^-}(\Omega)$ *such that* $\psi(x) > 0$ *for all* $x \in \bar{\Omega}$ *and* $\mathfrak{B}\psi \geq 0$ *on* $\partial\Omega$. *Then,*

$$\sigma_0 := \sigma[\mathfrak{L}, \mathfrak{B}, \Omega] = \sup_{\psi \in \mathcal{P}_>} \inf_{\Omega} \frac{\mathfrak{L}\psi}{\psi}. \qquad (8.7)$$

Proof. Fix $\lambda < \sigma_0$. Then,

$$\sigma[\mathfrak{L} - \lambda, \mathfrak{B}, \Omega] = \sigma_0 - \lambda > 0$$

and hence, according to Theorem 7.10, $(\mathfrak{L} - \lambda, \mathfrak{B}, \Omega)$ satisfies the strong maximum principle. Subsequently, we consider the auxiliary problem

$$\begin{cases} (\mathfrak{L} - \lambda)\psi = 1 & \text{in } \Omega, \\ \mathfrak{B}\psi = 1 & \text{on } \partial\Omega. \end{cases} \qquad (8.8)$$

Let $h \in \mathcal{C}^2(\bar{\Omega})$ be such that

$$\mathfrak{B}h = 1 \quad \text{on } \partial\Omega.$$

Then, the change of variable

$$\psi = h + w$$

transforms (8.8) into

$$\begin{cases} (\mathfrak{L} - \lambda)w = 1 - (\mathfrak{L} - \lambda)h & \text{in } \Omega, \\ \mathfrak{B}w = 0 & \text{on } \partial\Omega. \end{cases}$$

Therefore, according to Theorem 7.10, the function

$$\psi_1 := h + (\mathfrak{L} - \lambda)^{-1}(1 - (\mathfrak{L} - \lambda)h)$$

provides us with the unique solution of (8.8) in $W^{2,\infty^-}(\Omega)$. Moreover, $\psi_1(x) > 0$ for all $x \in \bar{\Omega}$ and hence $\psi_1 \in \mathcal{P}_>$. Thus, we have that

$$\lambda < \frac{\mathfrak{L}\psi_1}{\psi_1} \quad \text{in } \Omega$$

and

$$\lambda \leq \inf_{\Omega} \frac{\mathfrak{L}\psi_1}{\psi_1} \leq \sup_{\psi \in \mathcal{P}_>} \inf_{\Omega} \frac{\mathfrak{L}\psi}{\psi}. \qquad (8.9)$$

As (8.9) holds for every $\lambda < \sigma_0$, it becomes apparent that

$$\sigma_0 \leq \sup_{\psi \in \mathcal{P}_>} \inf_{\Omega} \frac{\mathfrak{L}\psi}{\psi}.$$

The proof of (8.7) will be completed by contradiction. Suppose

$$\sigma_0 < \sup_{\psi \in \mathcal{P}_>} \inf_\Omega \frac{\mathfrak{L}\psi}{\psi}.$$

Then, there exist $\epsilon > 0$ and $\psi \in \mathcal{P}_>$ such that

$$\sigma_0 + \epsilon < \frac{\mathfrak{L}\psi(x)}{\psi(x)}$$

for all $x \in \Omega$. Hence,

$$\begin{cases} (\mathfrak{L} - \sigma_0 - \epsilon)\psi > 0 & \text{in } \Omega, \\ \mathfrak{B}\psi \geq 0 & \text{on } \partial\Omega, \end{cases}$$

and, consequently, ψ provides us with a positive strict supersolution of $(\mathfrak{L} - \sigma_0 - \epsilon, \mathfrak{B}, \Omega)$. Therefore, thanks to Theorem 7.10,

$$0 < \sigma[\mathfrak{L} - \sigma_0 - \epsilon, \mathfrak{B}, \Omega] = -\epsilon,$$

which is impossible. This contradiction concludes the proof. $\qquad\square$

8.3   Concavity with respect to the potential

This section establishes the concavity of the map

$$\begin{aligned} L^\infty(\Omega) &\longmapsto & \mathbb{R} \\ P &\to \sigma(P) := \sigma[\mathfrak{L} + P, \mathfrak{B}, \Omega] \end{aligned}$$

with respect to the potential P.

Theorem 8.3 (Concavity of σ_0 with respect to the potential).
For every P_1, $P_2 \in L^\infty(\Omega)$ and $t \in [0,1]$ the following inequality holds

$$\sigma(tP_1 + (1-t)P_2) \geq t\,\sigma(P_1) + (1-t)\,\sigma(P_2).$$

Proof.   Subsequently, we set $\xi = (\xi_1, ..., \xi_N)$ for all $\xi \in \mathbb{R}^N$. As $\mathfrak{L}$ is strongly uniformly elliptic in Ω, for every $x \in \bar{\Omega}$ the bilinear form

$$\langle \xi, \psi \rangle := \sum_{i,j=1}^N a_{ij}(x)\xi_i\psi_j, \qquad \xi, \psi \in \mathbb{R}^N,$$

defines a scalar product in $\mathbb{R}^N$. Hence, by the Cauchy–Schwarz inequality, we find that

$$2\langle \xi, \psi \rangle = 2\sum_{i,j=1}^N a_{ij}(x)\xi_i\psi_j \leq 2\,|\xi|\,|\psi| \leq |\xi|^2 + |\psi|^2$$

$$= \sum_{i,j=1}^N a_{ij}(x)\xi_i\xi_j + \sum_{i,j=1}^N a_{ij}(x)\psi_i\psi_j$$

for all $\xi, \psi \in \mathbb{R}^N$ and $x \in \bar{\Omega}$, where $|\cdot|$ stands for the norm associated to the scalar product $\langle \cdot, \cdot \rangle$. From this inequality, using (4.2) and Corollary 4.1, it is easy to see that the map

$$\mathfrak{N} : W^{2,\infty^-}(\Omega) \to \mathcal{C}^{0,1^-}(\bar{\Omega})$$

defined by

$$\mathfrak{N}(u) := -\sum_{i,j=1}^N a_{ij} \frac{\partial u}{\partial x_i} \frac{\partial u}{\partial x_j} = -\langle \nabla u, \nabla u \rangle, \qquad u \in W^{2,\infty^-}(\Omega),$$

is concave. Indeed, for every $u_1, u_2 \in W^{2,\infty^-}(\Omega)$ and $t \in [0,1]$, the following chain of inequalities holds

$$\begin{aligned}
\mathfrak{N}(tu_1 + (1-t)u_2) &= -\langle t\nabla u_1 + (1-t)\nabla u_2, t\nabla u_1 + (1-t)\nabla u_2 \rangle \\
&= t^2 \mathfrak{N}(u_1) + (1-t)^2 \mathfrak{N}(u_2) - 2t(1-t)\langle \nabla u_1, \nabla u_2 \rangle \\
&\geq t^2 \mathfrak{N}(u_1) + (1-t)^2 \mathfrak{N}(u_2) + t(1-t)\left(\mathfrak{N}(u_1) + \mathfrak{N}(u_2)\right) \\
&= t\mathfrak{N}(u_1) + (1-t)\mathfrak{N}(u_2).
\end{aligned}$$

Thus, the map G defined by

$$G(u) := (\mathfrak{L} - c)u + c + \mathfrak{N}(u), \qquad u \in W^{2,\infty^-}(\Omega),$$

is concave, because $\mathfrak{N}(u)$ is concave and the mapping $u \mapsto (\mathfrak{L}-c)u$ is linear and hence concave. Our interest in G comes from the fact that, for every $\psi \in \mathcal{P}_>$ (see Theorem 8.2), the following relationship holds

$$\frac{\mathfrak{L}\psi}{\psi} = G(\log \psi).$$

Subsequently, we consider $P_1, P_2 \in L^\infty(\Omega)$, $t \in [0,1]$, and $\psi_1, \psi_2 \in \mathcal{P}_>$ arbitrary. Taking into account that $\psi \in \mathcal{P}_>$ implies $\psi, 1/\psi \in L^\infty(\Omega)$ and $\nabla \psi \in L^\infty(\Omega, \mathbb{R}^N)$, it is easily seen that $\psi_1^t \psi_2^{1-t} \in \mathcal{P}_>$. Thus,

$$\begin{aligned}
[\mathfrak{L} + tP_1 + (1-t)P_2](\psi_1^t \psi_2^{1-t})/\psi_1^t \psi_2^{1-t} &= tP_1 + (1-t)P_2 + \frac{\mathfrak{L}(\psi_1^t \psi_2^{1-t})}{\psi_1^t \psi_2^{1-t}} \\
&= tP_1 + (1-t)P_2 + G(\log(\psi_1^t \psi_2^{1-t})) \\
&= tP_1 + (1-t)P_2 + G(t\log \psi_1 + (1-t)\log \psi_2) \\
&\geq t\,P_1 + (1-t)\,P_2 + t\,G(\log \psi_1) + (1-t)\,G(\log \psi_2) \\
&= t\,\frac{(\mathfrak{L} + P_1)\psi_1}{\psi_1} + (1-t)\,\frac{(\mathfrak{L} + P_2)\psi_2}{\psi_2} \\
&\geq t \inf_\Omega \frac{(\mathfrak{L} + P_1)\psi_1}{\psi_1} + (1-t) \inf_\Omega \frac{(\mathfrak{L} + P_2)\psi_2}{\psi_2}
\end{aligned}$$

and, therefore, it follows from Theorem 8.2 that

$$\sigma(tP_1 + (1-t)P_2) \geq t \inf_\Omega \frac{(\mathfrak{L} + P_1)\psi_1}{\psi_1} + (1-t) \inf_\Omega \frac{(\mathfrak{L} + P_2)\psi_2}{\psi_2}.$$

As this inequality holds for all ψ_1, $\psi_2 \in \mathcal{P}_>$, taking the supreme on its right-hand side with respect to ψ_1 and ψ_2 yields

$$\sigma(tP_1 + (1-t)P_2) \geq t\sigma(P_1) + (1-t)\sigma(P_2),$$

which concludes the proof. □

8.4 Stability of Ω along the Dirichlet components of $\partial\Omega$

In the next section we shall show that if $\Gamma_0 \neq \emptyset$ and Ω perturbs in such a way that Γ_1 is kept fixed, then $\sigma[\mathfrak{L}, \mathfrak{B}, \Omega]$ varies continuously with Ω. This is why we adopt the following concepts, where the portion of the boundary Γ_1 is kept fixed. If Γ_1 varies, then the results might not be true in general.

Definition 8.1 (Convergence of domains). *Let Ω_0 be a bounded domain of $\mathbb{R}^N$ with boundary*

$$\partial\Omega_0 = \Gamma_0^0 \cup \Gamma_1$$

such that $\Gamma_0^0 \cap \Gamma_1 = \emptyset$, and Ω_n, $n \geq 1$, a sequence of bounded domains of $\mathbb{R}^N$ with boundaries $\partial\Omega_n = \Gamma_0^n \cup \Gamma_1$ of class $\mathcal{C}^2$ such that

$$\Gamma_0^n \cap \Gamma_1 = \emptyset \qquad \forall\, n \geq 1.$$

Then:

(E) *It is said that Ω_n converges to Ω_0 as $n \to \infty$ from its exterior if*

$$\Omega_0 \subset \Omega_{n+1} \subset \Omega_n$$

for all $n \geq 1$ and

$$\bigcap_{n=1}^\infty \bar{\Omega}_n = \bar{\Omega}_0.$$

(I) *It is said that Ω_n converges to Ω_0 as $n \to \infty$ from its interior if*

$$\Omega_n \subset \Omega_{n+1} \subset \Omega_0$$

for all $n \geq 1$ and

$$\bigcup_{n=1}^\infty \Omega_n = \Omega_0.$$

(C) *It is said that* Ω_n *converges to* Ω_0 *as* $n \to \infty$ *if there are two sequences of bounded domains of class* $\mathcal{C}^2$, *say* Ω_n^I *and* Ω_n^E, $n \geq 1$, *such that* Ω_n^I *converges to* Ω_0 *from its interior,* Ω_n^E *converges to* Ω_0 *from its exterior,*

$$\Omega_n^I \subset \Omega_0 \cap \Omega_n \quad and \quad \Omega_0 \cup \Omega_n \subset \Omega_n^E \quad for\ all\ n \geq 1.$$

The main result of this section reads as follows. It introduces the concept of *stability* and establishes the stability of any smooth domain.

Theorem 8.4. *Let* Ω_0 *be a bounded domain of* $\mathbb{R}^N$ *with boundary* $\partial\Omega_0 = \Gamma_0^0 \cup \Gamma_1$ *of class* $\mathcal{C}^1$ *such that* $\Gamma_0^0 \cap \Gamma_1 = \emptyset$. *Then,* Ω_0 *is* **stable along** Γ_0^0 *in the sense that for any sequence of bounded domains* Ω_n, $n \geq 1$, *of class* $\mathcal{C}^2$ *converging to* Ω_0 *from its exterior (see Definition 8.1(E)), the following relation holds*

$$\bigcap_{n=1}^{\infty} W_{\Gamma_0^n}^{1,2}(\Omega_n) = W_{\Gamma_0^0}^{1,2}(\Omega_0), \tag{8.10}$$

where

$$\bigcap_{n=1}^{\infty} W_{\Gamma_0^n}^{1,2}(\Omega_n) := \left\{ u \in W_{\Gamma_1^0}^{1,2}(\Omega_1) \ : \ u|_{\Omega_n} \in W_{\Gamma_0^n}^{1,2}(\Omega_n), \ \forall\, n \geq 2 \right\}.$$

According to (4.16) and Theorem 4.7,

$$W_{\Gamma_0^0}^{1,2}(\Omega_n) = N[\mathcal{T}_{\Gamma_0^n}] = \overline{\mathcal{C}_{\Gamma_0^n}^{\infty}(\Omega_n)}^{W^{1,2}(\Omega_n)} \qquad \forall\, n \geq 0.$$

Thus, essentially, Theorem 8.4 shows that, for every $u \in W_{\Gamma_0^1}^{1,2}(\Omega_1)$,

$$\mathcal{T}_{\Gamma_0^n} u = 0 \quad \forall\, n \geq 1 \quad \Longrightarrow \quad \mathcal{T}_{\Gamma_0^0} u = 0.$$

So, it provides us with the *stability* of the null space $N[\mathcal{T}_{\Gamma_0^n}]$ as $n \uparrow \infty$.

The inclusion

$$W_{\Gamma_0^0}^{1,2}(\Omega_0) \subset \bigcap_{n=1}^{\infty} W_{\Gamma_0^n}^{1,2}(\Omega_n)$$

is understood in the sense that the function

$$\tilde{u} := \begin{cases} u & \text{in } \bar{\Omega}_0, \\ 0 & \text{in } \bar{\Omega}_1 \setminus \bar{\Omega}_0, \end{cases}$$

lies in $W_{\Gamma_0^n}^{1,2}(\Omega_n)$ for all $n \geq 1$ if $u \in W_{\Gamma_0^0}^{1,2}(\Omega_0)$.

The proof of Theorem 8.4 is based on the next result of technical nature.

Proposition 8.5. *Let* Ω *be a bounded domain of* $\mathbb{R}^N$ *of class* $\mathcal{C}^1$ *with boundary* $\partial\Omega = \Gamma_0 \cup \Gamma_1$, $\Gamma_0 \cap \Gamma_1 = \emptyset$, *and consider any proper subdomain* $\Omega_0 \subset \Omega$ *of class* $\mathcal{C}^1$ *with boundary* $\partial\Omega_0 = \Gamma_0^0 \cup \Gamma_1$, $\Gamma_0^0 \cap \Gamma_1 = \emptyset$. *Then,*

$$W_{\Gamma_0^0}^{1,2}(\Omega_0) = \left\{ u \in W^{1,2}(\Omega) \ : \ \text{supp } u \subset \bar{\Omega}_0 \right\}.$$

As above, the inclusion

$$W^{1,2}_{\Gamma^0_0}(\Omega_0) \subset \left\{ u \in W^{1,2}(\Omega) \; : \; \text{supp } u \subset \bar{\Omega}_0 \right\}$$

is understood in the sense that the function

$$\tilde{u} := \begin{cases} u & \text{in} \quad \bar{\Omega}_0, \\ 0 & \text{in} \quad \bar{\Omega} \setminus \bar{\Omega}_0, \end{cases}$$

lies in $W^{1,2}(\Omega)$ for all $u \in W^{1,2}_{\Gamma^0_0}(\Omega_0)$.

8.4.1 *Proof of Proposition 8.5*

Let $u \in W^{1,2}(\Omega)$ be such that

$$\text{supp } u \subset \bar{\Omega}_0.$$

Since Γ_1 is a common set of components of $\partial\Omega$ and $\partial\Omega_0$, and Ω_0 is a subdomain of Ω, setting

$$U^\epsilon := \Gamma_1 + B_\epsilon,$$

there exists $\epsilon > 0$ such that

$$U^\epsilon \cap \Omega \subset \Omega_0,$$

where B_ϵ stands for the ball of radius ϵ centered at the origin. Moreover, ϵ can be chosen sufficiently small so that

$$U^\epsilon \cap \Gamma^0_0 = \emptyset \tag{8.11}$$

because $\Gamma^0_0 \cap \Gamma_1 = \emptyset$. Let $\eta \in \mathcal{C}^\infty_0(U^\epsilon)$ such that

$$\eta(x) = 1 \qquad \text{for each} \quad x \in U^{\frac{\epsilon}{2}}.$$

Since $u \in W^{1,2}(\Omega)$, we have that $u \in W^{1,2}(\Omega_0)$. Moreover, by Theorem 4.1, there exists a sequence $\psi_n \in \mathcal{C}^\infty(\bar{\Omega})$, $n \geq 1$, such that

$$\lim_{n\to\infty} \|\psi_n - u\|_{W^{1,2}(\Omega)} = 0. \tag{8.12}$$

Subsequently, we consider the auxiliary functions

$$\xi_n := (\eta\psi_n)|_{\bar{\Omega}_0}, \qquad n \geq 1.$$

According to (8.11), ξ_n vanishes in a neighborhood of Γ^0_0, as $\eta \in \mathcal{C}^\infty_0(U^\epsilon)$. Hence,

$$\xi_n \in \mathcal{C}^\infty_{\Gamma^0_0}(\bar{\Omega}_0) \qquad \text{for all} \quad n \geq 1.$$

Moreover, by (8.12), we have that

$$\lim_{n\to\infty} \|\eta\psi_n - \eta u\|_{W^{1,2}(\Omega)} = 0$$

and, in particular,

$$\lim_{n\to\infty} \|\xi_n - \eta u\|_{W^{1,2}(\Omega_0)} = 0. \tag{8.13}$$

On the other hand, since Ω_0 is of class $\mathcal{C}^1$, it is easy to see that it satisfies the next *segment property*. For every $x \in \partial\Omega_0$ there exist a neighborhood U_x of x and a vector $v_x \in \mathbb{R}^N \setminus \{0\}$ such that

$$\bar{U}_x \cap \bar{\Omega}_0 + tv_x \subset \Omega_0 \tag{8.14}$$

for all $t \in (0,1)$. By the compactness of Γ_0^0, there exist a natural number $m \geq 1$ and m points $x_j \in \Gamma_0^0$, $1 \leq j \leq m$, such that

$$\Gamma_0^0 \subset \bigcup_{j=1}^m U_{x_j}.$$

Let U_{m+1} be an open set such that $\bar{U}_{m+1} \subset \Omega_0$ for which

$$\mathcal{C} := \{ U_{x_1}, ..., U_{x_m}, U_{m+1} \}$$

is a covering of $\bar{\Omega}_0 \setminus U^{\frac{\epsilon}{2}}$ and set

$$U_j := U_{x_j}, \qquad 1 \leq j \leq m.$$

Thanks to Proposition 2.2, there exists a partition of the unity

$$\{ \beta_1, ..., \beta_m, \beta_{m+1} \}$$

in $\Omega_0 \setminus \bar{U}^{\frac{\epsilon}{2}}$ subordinated to the covering $\mathcal{C}$. Then, $\beta_j \in \mathcal{C}_0^\infty(\mathbb{R}^N)$ and $\mathrm{supp}\,\beta_j \subset \bar{U}_j$ for all $1 \leq j \leq m+1$, and

$$\sum_{j=1}^{m+1} \beta_j = 1 \qquad \text{in} \quad \bar{\Omega}_0 \setminus U^{\frac{\epsilon}{2}}. \tag{8.15}$$

Subsequently, we set

$$\gamma_j := \beta_j(1-\eta)u, \qquad 1 \leq j \leq m+1,$$

and consider the translations

$$\gamma_j^t := \gamma_j(\cdot - tv_{x_j}), \qquad 1 \leq j \leq m, \quad 0 < t < 1.$$

By construction, we have that $\beta_{m+1} \neq 0$ if $\gamma_{m+1} \neq 0$, and hence,

$$\mathrm{supp}\,\gamma_{m+1} \subset \mathrm{supp}\,\beta_{m+1} \subset \bar{U}_{m+1} \subset \Omega_0.$$

Consequently,

$$\gamma_{m+1} \in W_{\partial\Omega_0}^{1,2}(\Omega_0)$$

and, therefore, there exists a sequence $\xi_n^{m+1} \in \mathcal{C}_0^\infty(\Omega_0)$, $n \geq 1$, such that

$$\lim_{n\to\infty} \|\xi_n^{m+1} - \gamma_{m+1}\|_{W^{1,2}(\Omega_0)} = 0. \tag{8.16}$$

Moreover, as we are assuming that supp $u \subset \bar{\Omega}_0$, it is apparent that

$$\text{supp } \gamma_j \subset \text{supp } \beta_j \cap \text{supp } u \subset \bar{U}_j \cap \bar{\Omega}_0$$

for all $1 \leq j \leq m$. Thus, thanks to (8.14),

$$\text{supp } \gamma_j^t \subset \bar{U}_j \cap \bar{\Omega}_0 + t v_{x_j} \subset \Omega_0, \qquad 1 \leq j \leq m, \quad 0 < t < 1,$$

and hence,

$$\gamma_j^t \in W_{\partial\Omega_0}^{1,2}(\Omega_0), \qquad 1 \leq j \leq m, \quad 0 < t < 1. \tag{8.17}$$

By the continuity of the translation operator, for each natural number $n \geq 1$ there exists $t_n \in (0,1)$ such that

$$\|\gamma_j^{t_n} - \gamma_j\|_{W^{1,2}(\Omega_0)} \leq \frac{1}{n}, \qquad 1 \leq j \leq m. \tag{8.18}$$

Moreover, thanks to (8.17), for each $n \geq 1$ and $1 \leq j \leq m$, there exists $\xi_n^j \in \mathcal{C}_0^\infty(\Omega_0)$ such that

$$\|\gamma_j^{t_n} - \xi_n^j\|_{W^{1,2}(\Omega_0)} \leq \frac{1}{n}. \tag{8.19}$$

Therefore, according to (8.18) and (8.19),

$$\lim_{n\to\infty} \|\xi_n^j - \gamma_j\|_{W^{1,2}(\Omega_0)} = 0, \qquad 1 \leq j \leq m. \tag{8.20}$$

Now, consider the sequence

$$\varphi_n := \xi_n + \sum_{j=1}^{m+1} \xi_n^j, \qquad n \geq 1.$$

By construction,

$$\varphi_n \in \mathcal{C}_{\Gamma_0^0}^\infty(\bar{\Omega}_0) \qquad \forall\, n \geq 1.$$

Moreover, by (8.13), (8.16) and (8.20), we have that

$$\lim_{n\to\infty} \varphi_n = \eta u + \sum_{j=1}^{m+1} \beta_j(1-\eta)u = \eta u + (1-\eta)u \sum_{j=1}^{m+1} \beta_j \quad \text{in } W^{1,2}(\Omega_0).$$

Consequently, according to (8.15), we have that

$$\lim_{n\to\infty} \varphi_n = u \quad \text{in } W^{1,2}(\Omega_0 \setminus \bar{U}^{\frac{\epsilon}{2}})$$

and, since $\eta = 1$ in $\bar{U}^{\frac{\epsilon}{2}}$, it becomes apparent that, actually,

$$\lim_{n \to \infty} \varphi_n = u \quad \text{in } W^{1,2}(\Omega_0).$$

Therefore, $u \in W_{\Gamma_0^0}^{1,2}(\Omega_0)$, which shows the validity of the inclusion

$$\{ u \in W^{1,2}(\Omega) \ : \ \operatorname{supp} u \subset \bar{\Omega}_0 \} \subset W_{\Gamma_0^0}^{1,2}(\Omega_0).$$

To prove the converse inclusion let $u \in W_{\Gamma_0^0}^{1,2}(\Omega_0)$. By Theorem 4.7, there exists a sequence

$$\varphi_n \in \mathcal{C}_{\Gamma_0^0}^{\infty}(\bar{\Omega}_0), \quad n \geq 1, \tag{8.21}$$

such that

$$\lim_{n \to \infty} \|\varphi_n - u\|_{W^{1,2}(\Omega_0)} = 0. \tag{8.22}$$

In particular,

$$\lim_{n \to \infty} \varphi_n = u \quad \text{almost everywhere in } \bar{\Omega}_0. \tag{8.23}$$

Next, we consider the auxiliary sequence

$$\psi_n := \begin{cases} \varphi_n & \text{in } \bar{\Omega}_0, \\ 0 & \text{in } \bar{\Omega} \setminus \bar{\Omega}_0, \end{cases} \quad n \geq 1.$$

Thanks to (8.21), we have that

$$\psi_n \in \mathcal{C}_{\partial\Omega \setminus \Gamma_1}^{\infty}(\bar{\Omega}), \quad n \geq 1.$$

Moreover, owing to (8.22), it is apparent that ψ_n, $n \geq 1$, is a Cauchy sequence in $W^{1,2}(\Omega)$. Thus, there exists $\psi \in W^{1,2}(\Omega)$ such that

$$\lim_{n \to \infty} \|\psi_n - \psi\|_{W^{1,2}(\Omega)} = 0.$$

In particular,

$$\lim_{n \to \infty} \psi_n = \psi \quad \text{almost everywhere in } \bar{\Omega} \tag{8.24}$$

and hence $\psi = 0$ in $\bar{\Omega} \setminus \bar{\Omega}_0$, since $\psi_n = 0$ in $\bar{\Omega} \setminus \bar{\Omega}_0$ for all $n \geq 1$. Consequently,

$$\operatorname{supp} \psi \subset \bar{\Omega}_0. \tag{8.25}$$

Also, from (8.23) and (8.24), it is apparent that

$$\psi = u \quad \text{in } \bar{\Omega}_0,$$

because $\varphi_n = \psi_n$ in $\bar{\Omega}_0$ for all $n \geq 1$, by definition. Therefore, the extension function

$$\psi = \begin{cases} u & \text{in } \bar{\Omega}_0, \\ 0 & \text{in } \bar{\Omega} \setminus \bar{\Omega}_0, \end{cases}$$

satisfies $\psi \in W^{1,2}(\Omega)$ and (8.25). This concludes the proof.

8.4.2   *Proof of Theorem 8.4*

Let Ω_n, $n \geq 1$, be a sequence of bounded domains of class $\mathcal{C}^2$ converging to Ω_0 from its exterior in the sense of Definition 8.1 (E). We have to prove (8.10). If $\Omega_1 = \Omega_0$, then $\Omega_n = \Omega_0$ for all $n \geq 1$ and, therefore, (8.10) holds true. So, for the rest of the proof we shall assume that Ω_0 is a proper subdomain of Ω_1. Let

$$u \in \bigcap_{n=1}^{\infty} W^{1,2}_{\Gamma_0^n}(\Omega_n).$$

Then,

$$u|_{\Omega_n} \in W^{1,2}_{\Gamma_0^n}(\Omega_n) \qquad \text{for all} \quad n \geq 1,$$

by definition. Thus, according to Proposition 8.5,

$$\operatorname{supp} u \subset \bigcap_{n=1}^{\infty} \bar{\Omega}_n = \bar{\Omega}_0$$

and, therefore, thanks again to Proposition 8.5, it follows that

$$u \in W^{1,2}_{\Gamma_0^0}(\Omega_0).$$

Conversely, let $u \in W^{1,2}_{\Gamma_0^0}(\Omega_0)$ and consider the function

$$\tilde{u} := \begin{cases} u & \text{in} \quad \bar{\Omega}_0, \\ 0 & \text{in} \quad \bar{\Omega}_1 \setminus \bar{\Omega}_0. \end{cases}$$

Owing to Proposition 8.5, it is apparent that

$$\tilde{u} \in W^{1,2}_{\Gamma_0^n}(\Omega_n)$$

for all $n \geq 1$. This completes the proof.

8.5   Continuous dependence with respect to Ω

This section establishes the continuous dependence of $\sigma[\mathfrak{L}, \mathfrak{B}, \Omega]$ with respect to a regular class of perturbations of the domain Ω around its Dirichlet boundary Γ_0.

The continuous dependence of the principal eigenvalue with respect to the domain is based on the next result, which provides us with the *continuous dependence from the exterior*.

Theorem 8.5. *Suppose*

$$a_{ij} \in \mathcal{C}^1(\bar{\Omega}), \quad b_i \in \mathcal{C}(\bar{\Omega}), \quad 1 \leq i, j \leq N, \tag{8.26}$$

and Ω_0 is a proper subdomain of Ω with boundary of class $\mathcal{C}^2$ such that

$$\partial\Omega_0 = \Gamma_0^0 \cup \Gamma_1, \qquad \Gamma_0^0 \cap \Gamma_1 = \emptyset.$$

Let Ω_n, $n \geq 1$, be a sequence of bounded domains of $\mathbb{R}^N$ of class $\mathcal{C}^2$ converging to Ω_0 from the exterior, in the sense of Definition 8.1 (E), and such that $\Omega_n \subset \Omega$ for all $n \geq 1$. For each $n \geq 0$, let $\mathfrak{B}_n$ denote the boundary operator defined by

$$\mathfrak{B}_n u := \begin{cases} u & \text{on} \quad \Gamma_0^n := \partial\Omega_n \setminus \Gamma_1, \\ \partial_\nu u + \beta u & \text{on} \quad \Gamma_1, \end{cases} \qquad (8.27)$$

and let $\varphi_n > 0$ be the principal eigenfunction associated to $\sigma[\mathfrak{L}, \mathfrak{B}_n, \Omega_n]$ normalized so that

$$\|\varphi_n\|_{W^{1,2}(\Omega_n)} = 1, \qquad n \geq 0. \qquad (8.28)$$

Then, $\varphi_0 \in W_{\mathfrak{B}_0}^{2,\infty^-}(\Omega_0)$,

$$\lim_{n\to\infty} \sigma[\mathfrak{L}, \mathfrak{B}_n, \Omega_n] = \sigma[\mathfrak{L}, \mathfrak{B}_0, \Omega_0]$$

and

$$\lim_{n\to\infty} \|\varphi_n|_{\Omega_0} - \varphi_0\|_{W^{1,2}(\Omega_0)} = 0.$$

Proof. Note that $\nu := A n \in \mathcal{C}^1(\Gamma_1)$, because $a_{ij} \in \mathcal{C}^1(\bar{\Omega})$, $1 \leq i, j \leq N$, and Γ_1 is of class $\mathcal{C}^2$. Also,

$$\mathfrak{B}_n[\Omega_{n+1}] = \mathfrak{B}_{n+1}, \quad \mathfrak{B}_n[\Omega_0] = \mathfrak{B}_0, \qquad n \geq 0, \qquad (8.29)$$

where we have used the notation introduced in (8.2).

The existence and the uniqueness of $\sigma[\mathfrak{L}, \mathfrak{B}_n, \Omega_n]$ and φ_n, $n \geq 0$, are guaranteed by Theorem 7.7. By construction, we have that

$$\Omega_0 \subset \Omega_{n+1} \subset \Omega_n \subset \Omega$$

for all $n \geq 0$, and hence, by (8.29) and Proposition 8.2, we find that

$$\sigma[\mathfrak{L}, \mathfrak{B}_n, \Omega_n] \leq \sigma[\mathfrak{L}, \mathfrak{B}_{n+1}, \Omega_{n+1}] \leq \sigma[\mathfrak{L}, \mathfrak{B}_0, \Omega_0], \qquad n \geq 1.$$

Thus, the limit

$$\sigma^E := \lim_{n\to\infty} \sigma[\mathfrak{L}, \mathfrak{B}_n, \Omega_n] \qquad (8.30)$$

is well defined. We have to prove that

$$\sigma^E = \sigma[\mathfrak{L}, \mathfrak{B}_0, \Omega_0].$$

Thanks to Theorem 7.7,

$$\varphi_n \in W_{\mathfrak{B}_n}^{2,\infty^-}(\Omega_n) \subset W^{2,2}(\Omega_n) \cap \mathcal{C}^{1,1^-}(\bar{\Omega}_n), \qquad n \geq 0.$$

Now, for every $n \geq 0$, let $\tilde{\varphi}_n$ denote the function

$$\tilde{\varphi}_n := \begin{cases} \varphi_n & \text{in} \quad \Omega_n, \\ 0 & \text{in} \quad \Omega \setminus \Omega_n, \end{cases} \qquad n \geq 0.$$

Since $\varphi_n \in W^{1,2}_{\Gamma^n_0}(\Omega_n)$, for each $n \geq 0$ we have that $\tilde{\varphi}_n \in W^{1,2}(\Omega)$. Moreover, by (8.28),

$$\|\tilde{\varphi}_n\|_{W^{1,2}(\Omega)} = \|\varphi_n\|_{W^{1,2}(\Omega_n)} = 1, \qquad n \geq 0. \tag{8.31}$$

Thus, by Theorem 4.5, there exist a subsequence of $\tilde{\varphi}_n$, $n \geq 1$, say $\{\tilde{\varphi}_{n_m}\}_{m \geq 1}$, and a function $\tilde{\varphi} \in L^2(\Omega)$ such that

$$\lim_{m \to \infty} \|\tilde{\varphi}_{n_m} - \tilde{\varphi}\|_{L^2(\Omega)} = 0. \tag{8.32}$$

In particular,

$$\lim_{m \to \infty} \tilde{\varphi}_{n_m}(x) = \tilde{\varphi}(x) \quad \text{almost everywhere in } \Omega. \tag{8.33}$$

We claim that

$$\text{supp } \tilde{\varphi} \subset \bar{\Omega}_0. \tag{8.34}$$

Indeed, pick

$$x \notin \bar{\Omega}_0 = \bigcap_{n=1}^{\infty} \bar{\Omega}_n.$$

Then, since $\bar{\Omega}_n$, $n \geq 1$, is a decreasing sequence of compact sets, there exists a natural number $n_0 \geq 1$ such that $x \notin \bar{\Omega}_n$ for all $n \geq n_0$. Thus,

$$\tilde{\varphi}_n(x) = 0, \qquad n \geq n_0,$$

and hence,

$$\lim_{n \to \infty} \tilde{\varphi}_n(x) = 0 \quad \text{if} \quad x \notin \bar{\Omega}_0.$$

Therefore, by the uniqueness of the limit in (8.33), we have that

$$\tilde{\varphi} = 0 \quad \text{in} \quad \Omega \setminus \bar{\Omega}_0,$$

which shows (8.34). Note that $\varphi_n(x) > 0$ for all $x \in \Omega_n \cup \Gamma_1$ and $n \geq 0$. In particular, $\varphi_n(x) > 0$ for every $x \in \Omega_0 \cup \Gamma_1$ and $n \geq 0$. Hence, it follows from (8.33) that

$$\tilde{\varphi} \geq 0 \quad \text{in} \quad \Omega_0. \tag{8.35}$$

Now, we will analyze the limiting behavior of the traces of φ_{n_m}, $m \geq 1$, on Γ_1. Since Ω_0 is of class $\mathcal{C}^2$, by Theorem 4.6, the trace operator $\mathcal{T}_{\Gamma_1}$ of $W^{1,2}(\Omega_0)$ on Γ_1 is well defined and, actually, according to Remark 5.1,

$$\mathcal{T}_{\Gamma_1} \in \mathcal{L}\left(W^{1,2}(\Omega_0), W^{\frac{1}{2},2}(\Gamma_1)\right).$$

For each $n \geq 1$, let i_n denote the canonical injection

$$i_n : W^{1,2}(\Omega_n) \to W^{1,2}(\Omega_0)$$

defined by

$$i_n u = u|_{\Omega_0} \quad \text{for all } u \in W^{1,2}(\Omega_n).$$

Then,

$$\|i_n\|_{\mathcal{L}(W^{1,2}(\Omega_n), W^{1,2}(\Omega_0))} \leq 1, \qquad n \geq 1, \tag{8.36}$$

and, setting

$$\mathcal{T}_n := \mathcal{T}_{\Gamma_1} \circ i_n, \qquad n \geq 1,$$

we find from (8.36) that

$$\|\mathcal{T}_n\|_{\mathcal{L}(W^{1,2}(\Omega_n), W^{\frac{1}{2},2}(\Gamma_1))} \leq \|\mathcal{T}_{\Gamma_1}\|_{\mathcal{L}(W^{1,2}(\Omega_0), W^{\frac{1}{2},2}(\Gamma_1))} \tag{8.37}$$

for all $n \geq 1$. Therefore, the sequence of operators $\{\mathcal{T}_n\}_{n \geq 1}$ is uniformly bounded. Moreover, since $\varphi_n \in W^{2,\infty^-}(\Omega_n) \subset \mathcal{C}^{1,1^-}(\bar{\Omega}_n)$ for all $n \geq 1$, we find that

$$\varphi_n|_{\Gamma_1} = \mathcal{T}_n \varphi_n \in W^{\frac{1}{2},2}(\Gamma_1)$$

and hence, it follows from (8.31) and (8.37) that

$$\|\varphi_n|_{\Gamma_1}\|_{W^{\frac{1}{2},2}(\Gamma_1)} = \|\mathcal{T}_n \varphi_n\|_{W^{\frac{1}{2},2}(\Gamma_1)} \leq \|\mathcal{T}_{\Gamma_1}\|_{\mathcal{L}(W^{1,2}(\Omega_0), W^{\frac{1}{2},2}(\Gamma_1))}$$

for all $n \geq 1$. Consequently, the sequence of traces $\{\varphi_n|_{\Gamma_1}\}_{n \geq 1}$ is bounded in $W^{\frac{1}{2},2}(\Gamma_1)$. Thus, as according to Remark 5.1, the embedding

$$W^{\frac{1}{2},2}(\Gamma_1) \hookrightarrow L^2(\Gamma_1)$$

is compact, because Γ_1 is compact, there exist a subsequence of φ_{n_m}, $m \geq 1$, relabeled by n_m, and a function $\varphi^* \in L^2(\Gamma_1)$ such that

$$\lim_{m \to \infty} \|\varphi_{n_m}|_{\Gamma_1} - \varphi^*\|_{L^2(\Gamma_1)} = 0. \tag{8.38}$$

Next, we will show that $\{\tilde{\varphi}_{n_m}\}_{m \geq 1}$ is a Cauchy sequence in $W^{1,2}(\Omega)$. By (8.32), this entails

$$\lim_{m \to \infty} \|\tilde{\varphi}_{n_m} - \tilde{\varphi}\|_{W^{1,2}(\Omega)} = 0. \tag{8.39}$$

Indeed, suppose that k and m are natural numbers such that $1 \leq k \leq m$. Then, $\Omega_{n_m} \subset \Omega_{n_k}$ and, since $\mathfrak{L}$ is strongly uniformly elliptic in Ω, integrating by parts and taking into account that $\varphi_n = 0$ on Γ_0^n for all $n \geq 1$ shows that

$$
\mu\|\nabla(\tilde{\varphi}_{n_k} - \tilde{\varphi}_{n_m})\|_{L^2(\Omega)}^2 \leq \sum_{i,j=1}^{N} \int_{\Omega} a_{ij} \frac{\partial(\tilde{\varphi}_{n_k} - \tilde{\varphi}_{n_m})}{\partial x_i} \frac{\partial(\tilde{\varphi}_{n_k} - \tilde{\varphi}_{n_m})}{\partial x_j}
$$

$$
= \sum_{i,j=1}^{N} \left(\int_{\Omega_{n_k}} a_{ij} \frac{\partial\varphi_{n_k}}{\partial x_i} \frac{\partial\varphi_{n_k}}{\partial x_j} + \int_{\Omega_{n_m}} a_{ij} \frac{\partial\varphi_{n_m}}{\partial x_i} \frac{\partial\varphi_{n_m}}{\partial x_j} \right.
$$
$$
\left. -2 \int_{\Omega_{n_m}} a_{ij} \frac{\partial\varphi_{n_k}}{\partial x_i} \frac{\partial\varphi_{n_m}}{\partial x_j} \right)
$$

$$
= -\sum_{i,j=1}^{N} \left[\int_{\Omega_{n_k}} \frac{\partial}{\partial x_j} \left(a_{ij} \frac{\partial\varphi_{n_k}}{\partial x_i} \right) \varphi_{n_k} + \int_{\Omega_{n_m}} \frac{\partial}{\partial x_j} \left(a_{ij} \frac{\partial\varphi_{n_m}}{\partial x_i} \right) \varphi_{n_m} \right.
$$
$$
\left. -2 \int_{\Omega_{n_m}} \frac{\partial}{\partial x_j} \left(a_{ij} \frac{\partial\varphi_{n_k}}{\partial x_i} \right) \varphi_{n_m} \right]
$$
$$
+ \sum_{i,j=1}^{N} \int_{\Gamma_1} a_{ij} \left(\frac{\partial\varphi_{n_k}}{\partial x_i} \varphi_{n_k} + \frac{\partial\varphi_{n_m}}{\partial x_i} \varphi_{n_m} - 2 \frac{\partial\varphi_{n_k}}{\partial x_i} \varphi_{n_m} \right) n_j,
$$

where $\mu > 0$ stands for the ellipticity constant of $\mathfrak{L}$ in Ω and

$$
\mathbf{n} = (n_1, \ldots, n_N)
$$

is the outward unit normal on Γ_1. From this relation, taking into account that φ_n is the principal eigenfunction associated with $\sigma[\mathfrak{L}, \mathfrak{B}_n, \Omega_n]$ for all $n \geq 0$, we find that

$$
\mu\|\nabla(\tilde{\varphi}_{n_k} - \tilde{\varphi}_{n_m})\|_{L^2(\Omega)}^2
$$
$$
\leq \int_{\Omega_{n_k}} (\sigma[\mathfrak{L}, \mathfrak{B}_{n_k}, \Omega_{n_k}]\varphi_{n_k} - \langle b, \nabla\varphi_{n_k}\rangle - c\varphi_{n_k}) \varphi_{n_k}
$$
$$
+ \int_{\Omega_{n_m}} (\sigma[\mathfrak{L}, \mathfrak{B}_{n_m}, \Omega_{n_m}]\varphi_{n_m} - \langle b, \nabla\varphi_{n_m}\rangle - c\varphi_{n_m}) \varphi_{n_m}
$$
$$
-2 \int_{\Omega_{n_m}} (\sigma[\mathfrak{L}, \mathfrak{B}_{n_k}, \Omega_{n_k}]\varphi_{n_k} - \langle b, \nabla\varphi_{n_k}\rangle - c\varphi_{n_k}) \varphi_{n_m}
$$
$$
+ \sum_{i,j=1}^{N} \int_{\Gamma_1} a_{ij} \left(\frac{\partial\varphi_{n_k}}{\partial x_i} \varphi_{n_k} + \frac{\partial\varphi_{n_m}}{\partial x_i} \varphi_{n_m} - 2 \frac{\partial\varphi_{n_k}}{\partial x_i} \varphi_{n_m} \right) n_j.
$$

Hence, rearranging terms yields

$$\mu\|\nabla(\tilde{\varphi}_{n_k} - \tilde{\varphi}_{n_m})\|^2_{L^2(\Omega)} \leq \sigma[\mathfrak{L}, \mathfrak{B}_{n_k}, \Omega_{n_k}] \int_{\Omega_{n_k}} \varphi_{n_k}(\varphi_{n_k} - \tilde{\varphi}_{n_m})$$

$$+ (\sigma[\mathfrak{L}, \mathfrak{B}_{n_m}, \Omega_{n_m}] - \sigma[\mathfrak{L}, \mathfrak{B}_{n_k}, \Omega_{n_k}]) \int_{\Omega_{n_m}} \varphi_{n_m}^2$$

$$+ \sigma[\mathfrak{L}, \mathfrak{B}_{n_k}, \Omega_{n_k}] \int_{\Omega_{n_m}} \varphi_{n_m}(\varphi_{n_m} - \varphi_{n_k})$$

$$+ \int_{\Omega_{n_k}} \langle b, \nabla\varphi_{n_k}\rangle(\tilde{\varphi}_{n_m} - \varphi_{n_k}) + \int_{\Omega_{n_m}} \langle b, \nabla(\varphi_{n_k} - \varphi_{n_m})\rangle\varphi_{n_m} \quad (8.40)$$

$$+ \int_{\Omega_{n_k}} c\varphi_{n_k}(\tilde{\varphi}_{n_m} - \varphi_{n_k}) + \int_{\Omega_{n_m}} c\varphi_{n_m}(\varphi_{n_k} - \varphi_{n_m})$$

$$+ \sum_{i,j=1}^N \int_{\Gamma_1} a_{ij} \left[(\varphi_{n_k} - \varphi_{n_m})\frac{\partial\varphi_{n_k}}{\partial x_i} + \varphi_{n_m}\frac{\partial(\varphi_{n_m} - \varphi_{n_k})}{\partial x_i}\right] n_j.$$

Now, we will estimate each of the terms on the right-hand side of (8.40). By (8.31), the following estimates hold

$$\|\tilde{\varphi}_n\|_{L^2(\Omega)} \leq 1, \quad \|\nabla\tilde{\varphi}_n\|_{L^2(\Omega)} \leq 1, \quad \text{for all } n \geq 0. \quad (8.41)$$

Moreover, by construction, $\sigma[\mathfrak{L}, \mathfrak{B}_n, \Omega_n]$, $n \geq 1$, is increasing and bounded above by $\sigma[\mathfrak{L}, \mathfrak{B}_0, \Omega_0]$. Thus, by Hölder inequality, (8.41) implies that

$$\left|\sigma_{n_k} \int_{\Omega_{n_k}} \varphi_{n_k}(\varphi_{n_k} - \tilde{\varphi}_{n_m})\right| \leq |\sigma_0| \|\tilde{\varphi}_{n_k} - \tilde{\varphi}_{n_m}\|_{L^2(\Omega)}, \quad (8.42)$$

where we denote

$$\sigma_h := \sigma[\mathfrak{L}, \mathfrak{B}_h, \Omega_h], \quad h \geq 0.$$

Similarly,

$$\left|(\sigma_{n_m} - \sigma_{n_k}) \int_{\Omega_{n_m}} \varphi_{n_m}^2\right| \leq |\sigma_{n_m} - \sigma_{n_k}|,$$

$$\left|\sigma_{n_k} \int_{\Omega_{n_m}} \varphi_{n_m}(\varphi_{n_m} - \varphi_{n_k})\right| \leq |\sigma_0| \|\tilde{\varphi}_{n_m} - \tilde{\varphi}_{n_k}\|_{L^2(\Omega)},$$

$$\left|\int_{\Omega_{n_k}} \langle b, \nabla\varphi_{n_k}\rangle(\tilde{\varphi}_{n_m} - \varphi_{n_k})\right| \leq \|b\|_\infty \|\tilde{\varphi}_{n_m} - \tilde{\varphi}_{n_k}\|_{L^2(\Omega)},$$

where

$$\|b\|_\infty := \max_{x \in \bar{\Omega}} \sqrt{\sum_{j=1}^N b_j^2(x)},$$

and

$$\left| \int_{\Omega_{n_k}} c\varphi_{n_k}(\tilde{\varphi}_{n_m} - \varphi_{n_k}) \right| \le \|c\|_{L^\infty(\Omega)} \|\tilde{\varphi}_{n_m} - \tilde{\varphi}_{n_k}\|_{L^2(\Omega)},$$

$$\left| \int_{\Omega_{n_m}} c\varphi_{n_m}(\varphi_{n_k} - \varphi_{n_m}) \right| \le \|c\|_{L^\infty(\Omega)} \|\tilde{\varphi}_{n_m} - \tilde{\varphi}_{n_k}\|_{L^2(\Omega)}.$$

In order to estimate the integrals on Γ_1, we will use the identities

$$\partial_\nu \varphi_n + \beta \varphi_n = 0 \quad \text{on} \quad \Gamma_1, \quad n \ge 0,$$

where

$$\nu_i := \sum_{j=1}^N a_{ij} n_j, \quad 1 \le i \le N.$$

According to them, we have that, on Γ_1,

$$\sum_{i,j=1}^N a_{ij} \frac{\partial \varphi_n}{\partial x_i} n_j = \sum_{i=1}^N \nu_i \frac{\partial \varphi_n}{\partial x_i} = \langle \nabla \varphi_n, \nu \rangle = \partial_\nu \varphi_n = -\beta \varphi_n$$

for all $n \ge 0$ and hence,

$$\sum_{i,j=1}^N a_{ij} \frac{\partial(\varphi_{n_m} - \varphi_{n_k})}{\partial x_i} n_j = -\beta (\varphi_{n_m} - \varphi_{n_k}).$$

Therefore,

$$\left| \sum_{i,j=1}^N \int_{\Gamma_1} a_{ij}(\varphi_{n_k} - \varphi_{n_m}) \frac{\partial \varphi_{n_k}}{\partial x_i} n_j \right| = \left| \int_{\Gamma_1} \beta \varphi_{n_k}(\varphi_{n_m} - \varphi_{n_k}) \right|$$

$$\le \|\beta\|_{L^\infty(\Gamma_1)} \|\varphi_{n_k}|_{\Gamma_1}\|_{L^2(\Gamma_1)} \|(\varphi_{n_k} - \varphi_{n_m})|_{\Gamma_1}\|_{L^2(\Gamma_1)}$$

and

$$\left| \sum_{i,j=1}^N \int_{\Gamma_1} a_{ij}\varphi_{n_m} \frac{\partial(\varphi_{n_m} - \varphi_{n_k})}{\partial x_i} n_j \right| = \left| \int_{\Gamma_1} \beta \varphi_{n_m}(\varphi_{n_m} - \varphi_{n_k}) \right| \quad (8.43)$$

$$\le \|\beta\|_{L^\infty(\Gamma_1)} \|\varphi_{n_m}|_{\Gamma_1}\|_{L^2(\Gamma_1)} \|(\varphi_{n_k} - \varphi_{n_m})|_{\Gamma_1}\|_{L^2(\Gamma_1)}.$$

It remains to estimate the integrals

$$I_{mk} := \int_{\Omega_{n_m}} \langle b, \nabla(\varphi_{n_k} - \varphi_{n_m}) \rangle \varphi_{n_m}. \qquad (8.44)$$

Since $b_i \in \mathcal{C}(\bar{\Omega})$, in order to integrate by parts in (8.44) we must first approach each of the components b_i, $1 \le i \le N$, by a sequence of smooth coefficients, say b_{in}, $n \ge 1$. This regularization can be done as follows.

Fix $\delta > 0$ and consider the δ-neighborhood of Ω

$$\Omega_\delta := \bar{\Omega} + B_\delta(0).$$

For every $1 \le i \le N$, let $\hat{b}_i$ be a continuous extension of b_i to $\mathbb{R}^N$ such that

$$\hat{b}_i \in \mathcal{C}_0(\Omega_\delta), \qquad \|\hat{b}_i\|_{L^\infty(\mathbb{R}^N)} = \|b_i\|_{L^\infty(\Omega)}. \qquad (8.45)$$

Now, consider the function

$$\rho(x) := \begin{cases} e^{\frac{1}{|x|^2-1}} & \text{if } |x| < 1, \\ 0 & \text{if } |x| \ge 1, \end{cases}$$

as well as its associated *approximation of the identity*

$$\rho_n := (\int_{\mathbb{R}^N} \rho)^{-1} n^N \rho(n \cdot), \qquad n \in \mathbb{N} \setminus \{0\}.$$

The function ρ_n satisfies

$$\rho_n \in \mathcal{C}_0^\infty(\mathbb{R}^N), \quad \text{supp}\, \rho_n \subset \bar{B}_{1/n}(0), \quad \rho_n \ge 0, \quad \|\rho_n\|_{L^1(\mathbb{R}^N)} = 1,$$

for all $n \ge 1$. Moreover, for every $1 \le i \le N$, the new sequence

$$b_{in} := \rho_n * \hat{b}_i = \int_{\mathbb{R}^N} \rho_n(\cdot - y) \hat{b}_i(y) \, dy, \qquad n \ge 1,$$

is of class $\mathcal{C}_0^\infty(\mathbb{R}^N)$ and it approximates $\hat{b}_i$ uniformly in compact subsets of $\mathbb{R}^N$ as $n \uparrow \infty$ (see, e.g., Theorem 8.1.3 of M. Guzmán and B. Rubio [91]). In particular,

$$\lim_{n \to \infty} \|b_{in}|_\Omega - b_i\|_{L^\infty(\Omega)} = 0, \qquad 1 \le i \le N, \qquad (8.46)$$

because $\hat{b}_i|_\Omega = b_i$. Moreover, thanks to (8.45), it follows from Young's inequality that

$$\|b_{in}\|_{L^\infty(\mathbb{R}^N)} \le \|\rho_n\|_{L^1(\mathbb{R}^N)} \|\hat{b}_i\|_{L^\infty(\mathbb{R}^N)} = \|b_i\|_{L^\infty(\Omega)} \qquad (8.47)$$

and

$$\|\frac{\partial b_{in}}{\partial x_j}\|_{L^\infty(\mathbb{R}^N)} \le \|\frac{\partial \rho_n}{\partial x_j}\|_{L^1(\mathbb{R}^N)} \|b_i\|_{L^\infty(\Omega)} \qquad (8.48)$$

for all $1 \leq i, j \leq N$ and $n \geq 1$, since

$$\frac{\partial b_{in}}{\partial x_j} = \frac{\partial \rho_n}{\partial x_j} * \hat{b}_i.$$

On the other hand, we find from the definition of ρ_n that

$$\|\frac{\partial \rho_n}{\partial x_j}\|_{L^1(\mathbb{R}^N)} = \left(\int_{\mathbb{R}^N} \rho\right)^{-1} n \|\frac{\partial \rho}{\partial x_j}\|_{L^1(\mathbb{R}^N)}$$

for all $1 \leq j \leq N$ and $n \geq 1$, and hence, it is apparent from (8.48) that

$$\|\frac{\partial b_{in}}{\partial x_j}\|_{L^\infty(\mathbb{R}^N)} \leq \left(\int_{\mathbb{R}^N} \rho\right)^{-1} n \|\frac{\partial \rho}{\partial x_j}\|_{L^1(\mathbb{R}^N)} \|b_i\|_{L^\infty(\Omega)} \qquad (8.49)$$

for all $1 \leq i, j \leq N$ and $n \geq 1$.

Now, going back to (8.44), it follows that

$$I_{mk} = \int_{\Omega_{n_m}} \langle b - \mathbf{b}_n, \nabla(\varphi_{n_k} - \varphi_{n_m}) \rangle \varphi_{n_m} + \int_{\Omega_{n_m}} \langle \mathbf{b}_n, \nabla(\varphi_{n_k} - \varphi_{n_m}) \rangle \varphi_{n_m}, \quad (8.50)$$

where

$$\mathbf{b}_n := (b_{1n}, b_{2n}, ..., b_{Nn}), \qquad n \geq 1.$$

We now estimate each of the terms on the right-hand side of (8.50). Subsequently, we will set

$$I_{mk,1} := \int_{\Omega_{n_m}} \langle b - \mathbf{b}_n, \nabla(\varphi_{n_k} - \varphi_{n_m}) \rangle \varphi_{n_m},$$

$$I_{mk,2} := \int_{\Omega_{n_m}} \langle \mathbf{b}_n, \nabla(\varphi_{n_k} - \varphi_{n_m}) \rangle \varphi_{n_m}.$$

By Hölder inequality,

$$|I_{mk,1}| \leq \int_{\Omega_{n_m}} (|b - \mathbf{b}_n| |\nabla(\varphi_{n_k} - \varphi_{n_m})| \varphi_{n_m})$$

$$\leq \|b - \mathbf{b}_n\|_\infty \|\nabla(\tilde{\varphi}_{n_k} - \tilde{\varphi}_{n_m})\|_{L^2(\Omega)} \|\tilde{\varphi}_{n_m}\|_{L^2(\Omega)},$$

where we have denoted

$$\|b - \mathbf{b}_n\|_\infty := \max_{x \in \bar{\Omega}} \left(\sum_{i=1}^N (b_i(x) - b_{in}(x))^2\right)^{1/2}.$$

Thus, according to (8.41), we find that

$$|I_{mk,1}| \leq 2\|b - \mathbf{b}_n\|_\infty, \qquad n \geq 1. \qquad (8.51)$$

To estimate $I_{mk,2}$ we first integrate by parts

$$I_{mk,2} = \sum_{i=1}^{N} \int_{\Omega_{n_m}} b_{in}\varphi_{n_m} \frac{\partial(\varphi_{n_k} - \varphi_{n_m})}{\partial x_i}$$

$$= -\sum_{i=1}^{N} \int_{\Omega_{n_m}} (\varphi_{n_k} - \varphi_{n_m}) \frac{\partial(b_{in}\varphi_{n_m})}{\partial x_i}$$

$$+ \int_{\Gamma_1} \varphi_{n_m}(\varphi_{n_k} - \varphi_{n_m})\langle \mathbf{b}_n, \mathbf{n}\rangle$$

and, then, we estimate

$$|I_{mk,2}| \leq \left(\sum_{i=1}^{N} \|b_{in}\|_{L^\infty(\mathbb{R}^N)}\right) \|\nabla\tilde{\varphi}_{n_m}\|_{L^2(\Omega)} \|\tilde{\varphi}_{n_k} - \tilde{\varphi}_{n_m}\|_{L^2(\Omega)}$$

$$+ \left(\sum_{i=1}^{N} \|\frac{\partial b_{in}}{\partial x_i}\|_{L^\infty(\mathbb{R}^N)}\right) \|\tilde{\varphi}_{n_m}\|_{L^2(\Omega)} \|\tilde{\varphi}_{n_k} - \tilde{\varphi}_{n_m}\|_{L^2(\Omega)}$$

$$+ \|\mathbf{b}_n\|_\infty \|\varphi_{n_m}\|_{L^2(\Gamma_1)} \|\varphi_{n_k} - \varphi_{n_m}\|_{L^2(\Gamma_1)}.$$

Consequently, taking into account (8.41), (8.47) and (8.49), it becomes apparent that

$$|I_{mk,2}| \leq \left(\sum_{i=1}^{N} \|b_i\|_{L^\infty(\Omega)}\right) \|\tilde{\varphi}_{n_k} - \tilde{\varphi}_{n_m}\|_{L^2(\Omega)}$$

$$+ \frac{n}{\int_{\mathbb{R}^N} \rho} \left(\sum_{i=1}^{N} \|\frac{\partial\rho}{\partial x_i}\|_{L^1(\mathbb{R}^N)} \|b_i\|_{L^\infty(\Omega)}\right) \|\tilde{\varphi}_{n_k} - \tilde{\varphi}_{n_m}\|_{L^2(\Omega)}$$

$$+ \|\mathbf{b}_n\|_\infty \|\varphi_{n_m}\|_{L^2(\Gamma_1)} \|\varphi_{n_k} - \varphi_{n_m}\|_{L^2(\Gamma_1)}.$$

Therefore, combining this estimate with (8.51) we obtain that

$$|I_{mk}| \leq 2\|b - \mathbf{b}_n\|_\infty + \|\mathbf{b}_n\|_\infty \|\varphi_{n_m}\|_{L^2(\Gamma_1)} \|\varphi_{n_k} - \varphi_{n_m}\|_{L^2(\Gamma_1)}$$

$$+ \sum_{i=1}^{N} \left(1 + \frac{n}{\int_{\mathbb{R}^N} \rho} \|\frac{\partial\rho}{\partial x_i}\|_{L^1(\mathbb{R}^N)}\right) \|b_i\|_{L^\infty(\Omega)} \|\tilde{\varphi}_{n_k} - \tilde{\varphi}_{n_m}\|_{L^2(\Omega)} \quad (8.52)$$

for all $n \geq 1$.

Now, fix $\epsilon > 0$. Thanks to (8.46), there is $n = n(\epsilon) \geq 1$ such that

$$2\|b - \mathbf{b}_n\|_\infty \leq \frac{\epsilon}{6}. \quad (8.53)$$

Subsequently, we fix one of those n's. According to (8.32), there exists $n_0 \geq 1$ such that

$$\sum_{i=1}^{N} \left(1 + \frac{n}{\int_{\mathbb{R}^N} \rho} \|\frac{\partial\rho}{\partial x_i}\|_{L^1(\mathbb{R}^N)}\right) \|b_i\|_{L^\infty(\Omega)} \|\tilde{\varphi}_{n_k} - \tilde{\varphi}_{n_m}\|_{L^2(\Omega)} \leq \frac{\epsilon}{6} \quad (8.54)$$

for all $n_0 \leq k \leq m$. By (8.38), $n_0 \geq 1$ can be enlarged so that

$$\|\mathbf{b}_n\|_\infty \|\varphi_{n_m}\|_{L^2(\Gamma_1)} \|\varphi_{n_k} - \varphi_{n_m}\|_{L^2(\Gamma_1)} \leq \frac{\epsilon}{6} \qquad (8.55)$$

for all $n_0 \leq k \leq m$. Thus, substituting (8.53), (8.54) and (8.55) into (8.52) we find that

$$|I_{mk}| \leq \frac{\epsilon}{2} \qquad \text{for all} \ \ n_0 \leq k \leq m. \qquad (8.56)$$

Therefore, substituting all the estimates found in between (8.42) and (8.43), as well as (8.56), in (8.40), and using (8.30), (8.32), (8.38) and (8.56), it becomes apparent that n_0 can be enlarged, if necessary, so that

$$\mu \|\nabla (\tilde{\varphi}_{n_k} - \tilde{\varphi}_{n_m})\|_{L^2(\Omega)}^2 \leq \epsilon$$

for all $m \geq k \geq n_0$, which concludes the proof of (8.39).

According to (8.39), we find from (8.31) that

$$\|\tilde{\varphi}\|_{W^{1,2}(\Omega)} = \lim_{m \to \infty} \|\tilde{\varphi}_{n_m}\|_{W^{1,2}(\Omega)} = 1. \qquad (8.57)$$

Moreover, if we denote by $\mathcal{T}_{\Gamma_1}^\Omega$ the trace operator of $W^{1,2}(\Omega)$ on Γ_1, we also obtain that

$$\lim_{m \to \infty} \varphi_{n_m}|_{\Gamma_1} = \lim_{m \to \infty} \mathcal{T}_{\Gamma_1}^\Omega (\tilde{\varphi}_{n_m}) = \mathcal{T}_{\Gamma_1}^\Omega (\tilde{\varphi}) \qquad \text{in} \ L^2(\Gamma_1)$$

and, hence, thanks to (8.38), it is apparent that

$$\mathcal{T}_{\Gamma_1}^\Omega (\tilde{\varphi}) = \varphi^*.$$

Subsequently, we set

$$\varphi := \tilde{\varphi}|_{\Omega_0}.$$

Since $\tilde{\varphi} \in W^{1,2}(\Omega)$ and, due to (8.34), supp $\tilde{\varphi} \subset \bar{\Omega}_0$, it follows from Proposition 8.5 that

$$\tilde{\varphi} \in W_{\Gamma_0^0}^{1,2}(\Omega_0)$$

and, therefore,

$$\varphi := \tilde{\varphi}|_{\Omega_0} \in W_{\Gamma_0^0}^{1,2}(\Omega_0).$$

Moreover, according to (8.57), we have that

$$\|\varphi\|_{W^{1,2}(\Omega_0)} = \|\tilde{\varphi}\|_{W^{1,2}(\Omega)} = 1,$$

and, consequently, by (8.35),

$$\varphi > 0 \qquad \text{in} \ \ \Omega_0.$$

Also, note that, thanks to (8.39),

$$\lim_{m \to \infty} \|\varphi_{n_m}|_{\Omega_0} - \varphi\|_{W^{1,2}(\Omega_0)} = 0, \qquad (8.58)$$

because $\varphi_n|_{\Omega_0} = \tilde{\varphi}_n|_{\Omega_0}$ for all $n \geq 1$.

Next, we will show that φ is a weak solution of

$$\begin{cases} \mathfrak{L}\varphi = \sigma^E \varphi & \text{in} \quad \Omega_0, \\ \mathfrak{B}_0\varphi = 0 & \text{on} \quad \partial\Omega_0, \end{cases} \qquad (8.59)$$

where σ^E is the limit (8.30). We already know that

$$\varphi_n \in W^{2,\infty^-}_{\mathfrak{B}_n}(\Omega_n)$$

provides us with a classical — and, hence, weak — solution of

$$\begin{cases} \mathfrak{L}\varphi_n = \sigma_n\varphi_n & \text{in} \quad \Omega_n, \\ \mathfrak{B}_n\varphi_n = 0 & \text{on} \quad \partial\Omega_n, \end{cases}$$

for all $n \geq 1$. Equivalently,

$$\mathfrak{a}_n(\varphi_n, \phi) = \sigma_n \int_{\Omega_n} \varphi_n\phi \qquad \text{for all} \quad \phi \in \mathcal{C}^\infty_{\Gamma^n_0}(\bar{\Omega}_n) \qquad (8.60)$$

and any $n \geq 1$, where

$$\mathfrak{a}_n(u, v) := \int_{\Omega_n} \langle A\nabla u, \nabla v \rangle + \int_{\Omega_n} (\langle b, \nabla u \rangle + cu)\, v + \int_{\Gamma_1} \beta uv\, dS$$

for every $u, v \in W^{1,2}_{\Gamma^n_0}(\Omega)$. Now, pick

$$\xi \in \mathcal{C}^\infty_{\Gamma^0_0}(\bar{\Omega}_0)$$

and set

$$\phi = \phi_m := \begin{cases} \xi & \text{in} \quad \bar{\Omega}_0, \\ 0 & \text{in} \quad \Omega_{n_m} \setminus \Omega_0, \end{cases} \qquad m \geq 1,$$

in (8.60). Then,

$$\begin{aligned} \int_{\Omega_0} \langle A\nabla\varphi_{n_m}, \nabla\xi \rangle &+ \int_{\Omega_0} (\langle b, \nabla\varphi_{n_m} \rangle + c\varphi_{n_m})\xi \\ &+ \int_{\Gamma_1} \beta\varphi_{n_m}\xi\, dS = \sigma_{n_m} \int_{\Omega_0} \varphi_{n_m}\xi \end{aligned} \qquad (8.61)$$

for all $m \geq 1$. According to (8.58), letting $m \to \infty$ in (8.61) we find from the Lebesgue dominated convergence theorem that

$$\int_{\Omega_0} \langle A\nabla\varphi, \nabla\xi \rangle + \int_{\Omega_0} (\langle b, \nabla\varphi \rangle + c\varphi)\xi + \int_{\Gamma_1} \beta\varphi\xi\, dS = \sigma^E \int_{\Omega_0} \varphi\xi$$

for all $\xi \in \mathcal{C}^\infty_{\Gamma^0_0}(\bar{\Omega}_0)$. Therefore, $\varphi > 0$ is a weak solution of (8.59). By Theorem 7.7,

$$\sigma^E := \sigma[\mathfrak{L}, \mathfrak{B}_0, \Omega_0]$$

and φ must be a principal eigenfunction associated to it. This concludes the proof of the theorem, as the same compactness argument works out along any subsequence of Ω_n, $n \geq 1$. $\qquad \square$

The following result provides us with the *continuous dependence of the principal eigenvalue from the interior* of the domain.

Theorem 8.6. *Suppose* (8.26), *and let* Ω_0 *be a proper subdomain of* Ω *with boundary of class* C^2 *such that*

$$\partial\Omega_0 = \Gamma_0^0 \cup \Gamma_1, \qquad \Gamma_0^0 \cap \Gamma_1 = \emptyset.$$

Let Ω_n, $n \geq 1$, *be a sequence of bounded domains of* $\mathbb{R}^N$ *of class* C^2 *converging from the interior to* Ω_0 *as* $n \to \infty$ *in the sense of Definition 8.1(I). For every* $n \geq 0$, *let* $\mathfrak{B}_n$ *denote the boundary operator defined by* (8.27) *and* φ_n *the unique principal eigenfunction associated to* $\sigma[\mathfrak{L}, \mathfrak{B}_n, \Omega_n]$ *for which* (8.28) *holds. Then,* $\varphi_0 \in W^{2,\infty^-}_{\mathfrak{B}_0}(\Omega_0)$,

$$\lim_{n\to\infty} \sigma[\mathfrak{L}, \mathfrak{B}_n\Omega_n] = \sigma[\mathfrak{L}, \mathfrak{B}_0, \Omega_0] \quad and \quad \lim_{n\to\infty} \|\tilde{\varphi}_n - \varphi_0\|_{W^{1,2}(\Omega_0)} = 0,$$

where

$$\tilde{\varphi}_n := \begin{cases} \varphi_n & in \ \Omega_n, \\ 0 & in \ \Omega_0 \setminus \Omega_n, \end{cases} \qquad n \geq 1.$$

Proof. As in the proof of Theorem 8.5, the existence and the uniqueness of $\sigma[\mathfrak{L}, \mathfrak{B}_n, \Omega_n]$ and φ_n, $n \geq 0$, are guaranteed by Theorem 7.7.

According to Definition 8.1(I), we have that

$$\Omega_n \subset \Omega_{n+1} \subset \Omega_0 \qquad \text{for all } n \geq 1$$

and hence, by Proposition 8.2, we find that

$$\sigma[\mathfrak{L}, \mathfrak{B}_n, \Omega_n] \geq \sigma[\mathfrak{L}, \mathfrak{B}_{n+1}, \Omega_{n+1}] \geq \sigma[\mathfrak{L}, \mathfrak{B}_0, \Omega_0], \qquad n \geq 1,$$

because

$$\mathfrak{B}_{n+1}[\Omega_n] = \mathfrak{B}_n, \qquad \mathfrak{B}_0[\Omega_n] = \mathfrak{B}_n, \qquad n \geq 1.$$

Consequently, the limit

$$\sigma^I := \lim_{n\to\infty} \sigma[\mathfrak{L}, \mathfrak{B}_n, \Omega_n]$$

is well defined. One has to prove that

$$\sigma^I = \sigma[\mathfrak{L}, \mathfrak{B}_0, \Omega_0].$$

The proof of Theorem 8.5 can be adapted to show that there exist $\tilde{\varphi} \in W^{1,2}(\Omega_0)$, $\tilde{\varphi} > 0$, and a subsequence $\tilde{\varphi}_{n_m}$, $m \geq 1$, of $\tilde{\varphi}_n$, $n \geq 1$, such that

$$\lim_{m\to\infty} \|\tilde{\varphi}_{n_m} - \tilde{\varphi}\|_{W^{1,2}(\Omega_0)} = 0. \tag{8.62}$$

Since $\varphi_n \in W^{1,2}_{\Gamma^n_0}(\Omega_n)$ for all $n \geq 1$, it becomes apparent that $\tilde{\varphi}_n \in W^{1,2}_{\Gamma^0_0}(\Omega_0)$ for all $n \geq 1$ and, consequently, we can infer from (8.62) that

$$\tilde{\varphi} \in W^{1,2}_{\Gamma^0_0}(\Omega_0).$$

Similarly, it is easily seen that $\tilde{\varphi}$ is a weak positive solution of

$$\begin{cases} \mathfrak{L}\tilde{\varphi} = \sigma^I \tilde{\varphi} & \text{in } \Omega_0, \\ \mathfrak{B}_0 \tilde{\varphi} = 0 & \text{on } \partial\Omega_0. \end{cases}$$

Therefore, owing to Theorem 7.7,

$$\tilde{\varphi} = \varphi_0, \qquad \sigma^I = \sigma[\mathfrak{L}, \mathfrak{B}_0, \Omega_0],$$

which concludes the proof, as the same argument works out along any subsequence of φ_n, $n \geq 1$. □

As an immediate consequence from Theorems 8.5 and 8.6 one can infer the continuous dependence of the principal eigenvalue with respect to the domain, which can be stated as follows.

Theorem 8.7. *Suppose (8.26), and let Ω_0 be a proper subdomain of Ω with boundary of class $\mathcal{C}^2$ such that*

$$\partial\Omega_0 = \Gamma^0_0 \cup \Gamma_1, \qquad \Gamma^0_0 \cap \Gamma_1 = \emptyset.$$

Let Ω_n, $n \geq 1$, be a sequence of bounded domains of Ω of class $\mathcal{C}^2$ converging to Ω_0 as $n \to \infty$ in the sense of Definition 8.1(C). For every $n \geq 0$, let $\mathfrak{B}_n$ denote the boundary operator defined by (8.27). Then,

$$\lim_{n \to \infty} \sigma[\mathfrak{L}, \mathfrak{B}_n, \Omega_n] = \sigma[\mathfrak{L}, \mathfrak{B}_0, \Omega_0]. \tag{8.63}$$

Proof. By Definition 8.1(C), there are two sequences of bounded domains of class $\mathcal{C}^2$, Ω^I_n and Ω^E_n, $n \geq 1$, such that Ω^I_n converges to Ω_0 from its interior, Ω^E_n coverges to Ω_0 from its exterior, as $n \to \infty$, and

$$\Omega^I_n \subset \Omega_0 \cap \Omega_n, \qquad \Omega_0 \cup \Omega_n \subset \Omega^E_n, \qquad n \geq 1.$$

In particular,

$$\Omega^I_n \subset \Omega_n \subset \Omega^E_n$$

for all $n \geq 1$. Subsequently, for every $n \geq 1$, we denote by $\mathfrak{B}^E_n$ and $\mathfrak{B}^I_n$ the boundary operators defined through (8.27) from Ω^E_n and Ω^I_n, respectively. Then, owing to Proposition 8.2, we find that

$$\sigma[\mathfrak{L}, \mathfrak{B}^I_n, \Omega^I_n] \geq \sigma[\mathfrak{L}, \mathfrak{B}_n, \Omega_n] \geq \sigma[\mathfrak{L}, \mathfrak{B}^E_n, \Omega^E_n] \tag{8.64}$$

for all $n \geq 1$, because

$$\mathfrak{B}_n[\Omega_n^I] = \mathfrak{B}_n^I \quad \text{and} \quad \mathfrak{B}_n^E[\Omega_n] = \mathfrak{B}_n.$$

Moreover, thanks to Theorems 8.5 and 8.6, it becomes apparent that

$$\lim_{n\to\infty} \sigma[\mathfrak{L}, \mathfrak{B}_n^I, \Omega_n^I] = \sigma[\mathfrak{L}, \mathfrak{B}_0, \Omega_0],$$

(8.65)

$$\lim_{n\to\infty} \sigma[\mathfrak{L}, \mathfrak{B}_n^E, \Omega_n^E] = \sigma[\mathfrak{L}, \mathfrak{B}_0, \Omega_0].$$

Finally, letting $n \to \infty$ in (8.64), (8.63) holds from (8.65).   □

8.6   Continuous dependence with respect to $\beta(x)$

This section varies the coefficient $\beta \in \mathcal{C}(\Gamma_1)$ in the boundary operator

$$\mathfrak{B}[\beta] := \mathfrak{B} = \begin{cases} \mathfrak{D} & \text{on } \Gamma_0, \\ \partial_\nu + \beta & \text{on } \Gamma_1, \end{cases}$$

to analyze the continuous dependence of

$$\sigma(\beta) := \sigma[\mathfrak{L}, \mathfrak{B}[\beta], \Omega] \tag{8.66}$$

with respect to β.

Subsequently, we denote by

$$\sigma\left(L^\infty(\Gamma_1), L^1(\Gamma_1)\right)$$

the weak $*$ topology of $L^\infty(\Gamma_1)$. Let $\beta, \beta_n \in L^\infty(\Gamma_1)$, $n \geq 1$. Then, according to Proposition III.12(i) of H. Brézis [29], it is well known that

$$\lim_{n\to\infty} \beta_n = \beta \quad \text{in } \sigma\left(L^\infty(\Gamma_1), L^1(\Gamma_1)\right)$$

if, and only if,

$$\lim_{n\to\infty} \int_{\Gamma_1} \beta_n \xi = \int_{\Gamma_1} \beta \xi \quad \text{for every } \xi \in L^1(\Gamma_1). \tag{8.67}$$

When this occurs, it is simply said that

$$\beta_n \overset{*}{\rightharpoonup} \beta \quad \text{as } n \to \infty.$$

Obviously, when

$$\lim_{n\to\infty} \|\beta_n - \beta\|_{L^\infty(\Omega)} = 0,$$

then, $\beta_n \overset{*}{\rightharpoonup} \beta$ as $n \to \infty$.

The main result of this section reads as follows.

Theorem 8.8. *Suppose* $\Gamma_1 \neq \emptyset$ *and* (8.26), *and let* $\beta, \beta_n \in \mathcal{C}(\Gamma_1)$, $n \geq 1$, *such that* $\beta_n \overset{*}{\rightharpoonup} \beta$ *as* $n \to \infty$. *For every* $n \geq 1$, *let* φ_n *denote the unique principal eigenfunction associated with* $\sigma(\beta_n)$ *normalized so that*

$$\|\varphi_n\|_{W^{1,2}(\Omega)} = 1, \qquad n \geq 1. \tag{8.68}$$

Then,

$$\lim_{n \to \infty} \sigma(\beta_n) = \sigma(\beta) \quad and \quad \lim_{n \to \infty} \|\varphi_n - \varphi\|_{W^{1,2}(\Omega)} = 0,$$

where φ *stands for the unique principal eigenfunction associated with* $\sigma(\beta)$ *normalized so that* $\|\varphi\|_{W^{1,2}(\Omega)} = 1$.

Proof. The existence and the uniqueness of $(\sigma(\beta_n), \varphi_n)$, $n \geq 1$, and $(\sigma(\beta), \varphi)$ are guaranteed by Theorem 7.7. Moreover, since $\beta_n \overset{*}{\rightharpoonup} \beta$ as $n \to \infty$, the Banach–Steinhaus theorem shows the existence of a constant $C > 0$ such that

$$\|\beta_n\|_{L^\infty(\Gamma_1)} \leq C \qquad \text{for all} \quad n \geq 1 \tag{8.69}$$

(see, e.g., Proposition III.12(iii) of H. Brézis [29]). Thus, by Proposition 8.4, we find that

$$\sigma(-C) \leq \sigma(\beta_n) \leq \sigma(C) \tag{8.70}$$

for all $n \geq 1$ and hence there exist a subsequence of $\sigma(\beta_n)$, $n \geq 1$, relabeled by n, and a $\sigma^\infty \in \mathbb{R}$ such that

$$\sigma^\infty := \lim_{n \to \infty} \sigma(\beta_n).$$

Moreover, by Theorem 7.7, we also have that

$$\varphi_n \in W^{2,\infty^-}_{\mathfrak{B}[\beta_n]}(\Omega) \subset W^{2,2}(\Omega) \cap \mathcal{C}^{1,1^-}(\bar{\Omega})$$

for all $n \geq 1$, and, so, since $W^{1,2}(\Omega)$ is compactly embedded in $L^2(\Omega)$, there exist a subsequence of φ_n, $n \geq 1$, again labeled by n, and a function $\varphi_\infty \in L^2(\Omega)$ such that

$$\lim_{n \to \infty} \|\varphi_n - \varphi_\infty\|_{L^2(\Omega)} = 0. \tag{8.71}$$

As the previous argument is valid along any subsequence and, due to Theorem 7.7, the principal eigenpair $(\sigma(\beta), \varphi)$ is unique, to complete the proof of this theorem it suffices to show that

$$\sigma^\infty = \sigma(\beta), \qquad \varphi_\infty = \varphi,$$

and that, actually,

$$\lim_{n \to \infty} \|\varphi_n - \varphi\|_{W^{1,2}(\Omega)} = 0.$$

Thanks to (8.71), we have that

$$\lim_{n\to\infty} \varphi_n = \varphi_\infty \quad \text{a.e. in } \Omega \quad \text{and} \quad \varphi_\infty \geq 0 \quad \text{in } \Omega,$$

because $\varphi_n > 0$ for all $n \geq 1$.

As far as the traces of φ_n, $n \geq 1$, on Γ_1 are concerned, the same argument of the proof of Theorem 8.5 shows that there exist a subsequence of φ_n, $n \geq 1$, relabeled by n, and a function $\varphi_* \in L^2(\Gamma_1)$ such that

$$\lim_{n\to\infty} \|\varphi_n|_{\Gamma_1} - \varphi_*\|_{L^2(\Gamma_1)} = 0. \tag{8.72}$$

In particular, there exists a constant $C_1 > 0$ such that

$$\|\varphi_n|_{\Gamma_1}\|_{L^2(\Gamma_1)} \leq C_1, \qquad n \geq 1. \tag{8.73}$$

Next, we will show that φ_n, $n \geq 1$, is a Cauchy sequence in $W^{1,2}(\Omega)$. Indeed, arguing as in the proof of Theorem 8.5, we find that, for any natural numbers $1 \leq k \leq m$ and every $n \geq 1$,

$$
\begin{aligned}
\mu\|\nabla(\varphi_k-\varphi_m)\|^2_{L^2(\Omega)} \leq{} & \sigma(\beta_k)\int_\Omega \varphi_k(\varphi_k-\varphi_m) + [\sigma(\beta_m)-\sigma(\beta_k)]\int_\Omega \varphi_m^2 \\
& + \sigma(\beta_k)\int_\Omega \varphi_m(\varphi_m-\varphi_k) + \int_\Omega \langle b, \nabla\varphi_k\rangle(\varphi_m-\varphi_k) \\
& + \int_\Omega \langle b-\mathbf{b}_n, \nabla(\varphi_k-\varphi_m)\rangle\varphi_m + \int_\Omega \langle \mathbf{b}_n, \nabla(\varphi_k-\varphi_m)\rangle\varphi_m \\
& + \int_\Omega c\varphi_k(\varphi_m-\varphi_k) + \int_\Omega c\varphi_m(\varphi_k-\varphi_m) \\
& + \sum_{i,j=1}^N \int_{\Gamma_1} a_{ij}\left[(\varphi_k-\varphi_m)\frac{\partial\varphi_k}{\partial x_i} + \varphi_m\frac{\partial(\varphi_m-\varphi_k)}{\partial x_i}\right] n_j,
\end{aligned}
\tag{8.74}
$$

where $\mathbf{b}_n \in \mathcal{C}_0^\infty(\mathbb{R}^N, \mathbb{R}^N)$, $n \geq 1$, is any sequence satisfying

$$\lim_{n\to\infty} \|\mathbf{b}_n - b\|_{L^\infty(\Omega,\mathbb{R}^N)} = 0.$$

Actually, we suppose that $\mathbf{b}_n$, $n \geq 1$, has been constructed as in the proof of Theorem 8.5. Thanks to (8.68), we have that

$$\|\varphi_n\|_{L^2(\Omega)} \leq 1 \quad \text{and} \quad \|\nabla\varphi_n\|_{L^2(\Omega)} \leq 1 \tag{8.75}$$

for all $n \geq 1$. Therefore, arguing as in the proof of Theorem 8.5, we find from (8.70) and (8.75) the following estimates

$$\left|\sigma(\beta_k)\int_\Omega \varphi_k(\varphi_k-\varphi_m)\right| \leq \sigma(C)\|\varphi_k-\varphi_m\|_{L^2(\Omega)}, \tag{8.76}$$

$$\left|[\sigma(\beta_m)-\sigma(\beta_k)]\int_\Omega \varphi_m^2\right| \leq |\sigma(\beta_m)-\sigma(\beta_k)|,$$

$$\left| \sigma(\beta_k) \int_\Omega \varphi_m(\varphi_m - \varphi_k) \right| \leq \sigma(C) \|\varphi_m - \varphi_k\|_{L^2(\Omega)},$$

$$\left| \int_\Omega \langle b, \nabla\varphi_k \rangle(\varphi_m - \varphi_k) \right| \leq \|b\|_\infty \|\varphi_m - \varphi_k\|_{L^2(\Omega)},$$

$$\left| \int_\Omega c\varphi_k(\varphi_m - \varphi_k) \right| \leq \|c\|_{L^\infty(\Omega)} \|\varphi_m - \varphi_k\|_{L^2(\Omega)},$$

$$\left| \int_\Omega c\varphi_m(\varphi_k - \varphi_m) \right| \leq \|c\|_{L^\infty(\Omega)} \|\varphi_m - \varphi_k\|_{L^2(\Omega)},$$

$$\left| \int_\Omega \langle b - \mathbf{b}_n, \nabla(\varphi_k - \varphi_m) \rangle \varphi_m \right| \leq 2\|b - \mathbf{b}_n\|_\infty,$$

and

$$\left| \int_\Omega \langle \mathbf{b}_n, \nabla(\varphi_k - \varphi_m) \rangle \varphi_m \right| \leq \|\mathbf{b}_n\|_\infty \|\varphi_m\|_{L^2(\Gamma_1)} \|\varphi_k - \varphi_m\|_{L^2(\Gamma_1)} \\ + \sum_{i=1}^N \left(1 + \frac{n}{\int_{\mathbb{R}^N} \rho} \left\| \frac{\partial\rho}{\partial x_i} \right\|_{L^1(\mathbb{R}^N)} \right) \|b_i\|_{L^\infty(\Omega)} \|\varphi_k - \varphi_m\|_{L^2(\Omega)} \tag{8.77}$$

for all $n \geq 1$.

In order to estimate the integrals over Γ_1, we will use the identities

$$\partial_\nu \varphi_n + \beta_n \varphi_n = 0 \quad \text{on} \quad \Gamma_1, \quad n \geq 1,$$

where

$$\nu_i := \sum_{j=1}^N a_{ij} n_j, \quad 1 \leq i \leq N.$$

According to them, we have that

$$\sum_{i,j=1}^N a_{ij} \frac{\partial\varphi_k}{\partial x_i} n_j = \sum_{i=1}^N \nu_i \frac{\partial\varphi_k}{\partial x_i} = \langle \nu, \nabla\varphi_k \rangle = \partial_\nu \varphi_k = -\beta_k \varphi_k,$$

$$\sum_{i,j=1}^N a_{ij} \frac{\partial(\varphi_m - \varphi_k)}{\partial x_i} n_j = -\beta_m \varphi_m + \beta_k \varphi_k.$$

Thus,

$$\sum_{i,j=1}^N \int_{\Gamma_1} a_{ij}(\varphi_k - \varphi_m) \frac{\partial\varphi_k}{\partial x_i} n_j = \int_{\Gamma_1} \beta_k \varphi_k(\varphi_m - \varphi_k), \tag{8.78}$$

$$\sum_{i,j=1}^{N} \int_{\Gamma_1} a_{ij}\varphi_m \frac{\partial(\varphi_m - \varphi_k)}{\partial x_i} n_j = \int_{\Gamma_1} (\beta_k\varphi_k - \beta_m\varphi_m)\varphi_m. \tag{8.79}$$

Moreover, owing to (8.69) and (8.73), we obtain that

$$\left| \int_{\Gamma_1} \beta_k\varphi_k(\varphi_m - \varphi_k) \right| \leq C_2 \|(\varphi_m - \varphi_k)|_{\Gamma_1}\|_{L^2(\Gamma_1)}, \tag{8.80}$$

for some constant $C_2 > 0$ independent of k and m. Similarly,

$$\left| \int_{\Gamma_1} (\beta_k\varphi_k - \beta_m\varphi_m)\varphi_m \right| \leq \left| \int_{\Gamma_1} \beta_k\varphi_m(\varphi_k - \varphi_m) \right| + \left| \int_{\Gamma_1} \varphi_m^2(\beta_k - \beta_m) \right|$$
$$\leq C_2\|\varphi_m - \varphi_k\|_{L^2(\Gamma_1)} + \left| \int_{\Gamma_1} \varphi_m^2(\beta_k - \beta_m) \right| \tag{8.81}$$

and, according to (8.69), we also have that

$$\left| \int_{\Gamma_1} \varphi_m^2(\beta_k - \beta_m) \right| \leq \left| \int_{\Gamma_1} (\varphi_m^2 - \varphi_*^2)(\beta_k - \beta_m) \right| + \left| \int_{\Gamma_1} \varphi_*^2(\beta_k - \beta_m) \right|$$
$$\leq 2C\|\varphi_m + \varphi_*\|_{L^2(\Gamma_1)}\|\varphi_m - \varphi_*\|_{L^2(\Gamma_1)} + \left| \int_{\Gamma_1} \varphi_*^2(\beta_k - \beta_m) \right|.$$

Thus, according to (8.67), (8.72) and (8.73), we find from (8.81) that, for any $\epsilon > 0$, there exists a natural number $n_0 \geq 1$ such that

$$\left| \int_{\Gamma_1} (\beta_k\varphi_k - \beta_m\varphi_m)\varphi_m \right| \leq \frac{\epsilon}{2} \tag{8.82}$$

for all $k, m \geq n_0$. Finally, owing to (8.78) and (8.79) and substituting the estimates between (8.76) and (8.77), as well as (8.80) and (8.82), in (8.74), it is easily seen that there exists $k_0 \geq n_0$ such that

$$\mu \|\nabla(\varphi_k - \varphi_m)\|_{L^2(\Omega)}^2 \leq \epsilon \qquad \text{for all } k, m \geq k_0.$$

Thanks to (8.71), this shows that

$$\lim_{n\to\infty} \|\varphi_n - \varphi_\infty\|_{W^{1,2}(\Omega)} = 0. \tag{8.83}$$

Subsequently, we denote by $\mathcal{T}_{\Gamma_1}$ the trace operator of $W^{1,2}(\Omega)$ on Γ_1. Then, there exists a constant $C_3 > 0$ such that

$$\|\varphi_n - \mathcal{T}_{\Gamma_1}\varphi_\infty\|_{L^2(\Gamma_1)} = \|\mathcal{T}_{\Gamma_1}(\varphi_n - \varphi_\infty)\|_{L^2(\Gamma_1)} \leq C_3\|\varphi_n - \varphi_\infty\|_{W^{1,2}(\Omega)}$$

for all $n \geq 1$. Hence, (8.83) implies that

$$\lim_{n\to\infty} \|\varphi_n - \mathcal{T}_{\Gamma_1}\varphi_\infty\|_{L_2(\Gamma_1)} = 0 \tag{8.84}$$

and, therefore, thanks to (8.72), we find that

$$\mathcal{T}_{\Gamma_1}\varphi_\infty = \varphi_*.$$

Similarly, since $\varphi_n|_{\Gamma_0} = 0$ for all $n \geq 1$, by the continuity of the trace operator $\mathcal{T}_{\Gamma_0}$ of $W^{1,2}(\Omega)$ on Γ_0 we find that $\mathcal{T}_{\Gamma_0}\varphi_\infty = 0$ and, consequently,

$$\varphi_\infty \in W^{1,2}_{\Gamma_0}(\Omega).$$

Moreover, thanks to (8.83), (8.68) guarantees that

$$\|\varphi_\infty\|_{W^{1,2}(\Omega)} = 1.$$

Thus, since $\varphi_n > 0$ for all $n \geq 1$, we have that

$$\varphi_\infty \geq 0, \qquad \varphi_\infty \neq 0.$$

Next, we will show that φ_∞ provides us with a weak solution of

$$\begin{cases} \mathfrak{L}\varphi_\infty = \sigma^\infty \varphi_\infty & \text{in } \Omega, \\ \mathfrak{B}(\beta)\varphi_\infty = 0 & \text{on } \partial\Omega; \end{cases} \tag{8.85}$$

it should be remembered that

$$\sigma^\infty = \lim_{n \to \infty} \sigma(\beta_n).$$

By the assumptions, φ_n is a weak solution of

$$\begin{cases} \mathfrak{L}\varphi_n = \sigma(\beta_n)\varphi_n & \text{in } \Omega, \\ \mathfrak{B}(\beta_n)\varphi_n = 0 & \text{on } \partial\Omega, \end{cases}$$

for all $n \geq 1$. Thus, for every $\xi \in W^{1,2}_{\Gamma_0}(\Omega)$, we have that

$$\int_\Omega \langle A\nabla\varphi_n, \nabla\xi\rangle + \int_\Omega (\langle b, \nabla\varphi_n\rangle + c\varphi_n)\xi + \int_{\Gamma_1} \beta_n \varphi_n \xi = \sigma(\beta_n) \int_\Omega \varphi_n \xi \tag{8.86}$$

for all $n \geq 1$. According to (8.67), we find from (8.84) that

$$\lim_{n \to \infty} \int_{\Gamma_1} (\beta_n \xi \varphi_n) = \int_{\Gamma_1} (\beta\xi\mathcal{T}_{\Gamma_1}\varphi_\infty).$$

Therefore, letting $n \to \infty$ in (8.86), the dominated convergence theorem of Lebesgue implies

$$\int_\Omega \langle A\nabla\varphi_\infty, \nabla\xi\rangle + \int_\Omega (\langle b, \nabla\varphi_\infty\rangle + c\varphi_\infty)\xi + \int_{\Gamma_1} \beta\varphi_\infty \xi = \sigma^\infty \int_\Omega \varphi_\infty \xi$$

for all $\xi \in W^{1,2}_{\Gamma_0}(\Omega)$ and, therefore, $\varphi_\infty \in W^{1,2}_{\Gamma_0}(\Omega)$ is a weak positive solution of (8.85). Theorem 7.7 concludes the proof, as it shows that $\sigma^\infty = \sigma(\beta)$ and that φ_∞ is a principal eigenfunction associated to it. $\qquad\square$

8.7   Asymptotic behavior of $\sigma(\beta)$ as $\min_{\Gamma_1}\beta\uparrow\infty$

This section analyzes the behavior of the principal eigenvalue (8.66) as $\min_{\Gamma_1}\beta\uparrow\infty$. Its main result establishes that it converges to the principal eigenvalue of the Dirichlet problem in Ω and it reads as follows.

Theorem 8.9. *Suppose $\Gamma_1\neq\emptyset$ and (8.26), and let $\beta_n\in\mathcal{C}(\Gamma_1)$, $n\geq 1$, be an arbitrary sequence such that*

$$\lim_{n\to\infty}\min_{\Gamma_1}\beta_n=\infty. \tag{8.87}$$

Then,

$$\lim_{n\to\infty}\sigma[\mathfrak{L},\mathfrak{B}[\beta_n],\Omega]=\sigma[\mathfrak{L},\mathfrak{D},\Omega]. \tag{8.88}$$

Moreover, if φ_n stands for the principal eigenfunction associated with

$$\sigma(\beta_n):=\sigma[\mathfrak{L},\mathfrak{B}[\beta_n],\Omega],\qquad n\geq 1,$$

normalized so that

$$\|\varphi_n\|_{W^{1,2}(\Omega)}=1,\qquad n\geq 1, \tag{8.89}$$

then

$$\lim_{n\to\infty}\|\varphi_n-\varphi_0\|_{W^{1,2}(\Omega)}=0, \tag{8.90}$$

where φ_0 is the corresponding principal eigenfunction of $\sigma[\mathfrak{L},\mathfrak{D},\Omega]$.

Proof.   By (8.87), we can assume, without loss of generality, that

$$\ell_n:=\min_{\Gamma_1}\beta_n>0,\qquad n\geq 1. \tag{8.91}$$

Clearly, according to Propositions 8.1 and 8.4, it follows from (8.91) that

$$\sigma(0):=\sigma[\mathfrak{L},\mathfrak{B}[0],\Omega]<\sigma[\mathfrak{L},\mathfrak{B}[\beta_n],\Omega]<\sigma[\mathfrak{L},\mathfrak{D},\Omega] \tag{8.92}$$

for all $n\geq 1$. Thus, there exist

$$\sigma^{\infty}\in[\sigma(0),\sigma[\mathfrak{L},\mathfrak{D},\Omega]]$$

and a subsequence of β_n, $n\geq 1$, relabeled by n, such that

$$\sigma^{\infty}:=\lim_{n\to\infty}\sigma[\mathfrak{L},\mathfrak{B}[\beta_n],\Omega].$$

To prove (8.88), we should show that

$$\sigma^{\infty}=\sigma[\mathfrak{L},\mathfrak{D},\Omega]. \tag{8.93}$$

By Theorem 7.7, we have that

$$\varphi_n \in W^{2,\infty^-}_{\mathfrak{B}[\beta_n]}(\Omega) \subset W^{2,2}(\Omega) \cap \mathcal{C}^{1,1^-}(\bar{\Omega}) \tag{8.94}$$

for all $n \geq 1$, and, in particular, since $W^{1,2}(\Omega)$ is compactly embedded in $L^2(\Omega)$, there exist a subsequence of φ_n, $n \geq 1$, again labeled by n, and a function $\varphi \in L^2(\Omega)$ such that

$$\lim_{n \to \infty} \|\varphi_n - \varphi\|_{L^2(\Omega)} = 0. \tag{8.95}$$

Necessarily,

$$\lim_{n \to \infty} \varphi_n = \varphi \quad \text{a.e. in} \quad \Omega \quad \text{and} \quad \varphi \geq 0.$$

Consequently, to complete the proof of the theorem it suffices to show (8.93), $\varphi = \varphi_0$ and (8.90), since the previous argument is valid along any subsequence of β_n, $n \geq 1$.

By the definition of the φ_n's, we have that

$$\mathfrak{L}\varphi_n = \sigma(\beta_n)\varphi_n \quad \text{in} \quad \Omega$$

for all $n \geq 1$. Moreover, owing to (8.89) and (8.92), it becomes apparent that

$$\|\sigma(\beta_n)\varphi_n\|_{L^2(\Omega)} \leq |\sigma(\beta_n)| \leq C_1, \qquad n \geq 1, \tag{8.96}$$

for some constant $C_1 > 0$. On the other hand, thanks to Theorem 5.10, there exists a constant $C_2 > 0$ such that

$$\|\varphi_n\|_{W^{2,2}(\Omega)} \leq C_2 \|\mathfrak{L}\varphi_n\|_{L^2(\Omega)}$$

for all $n \geq 1$. Therefore, by (8.96), we find that

$$\|\varphi_n\|_{W^{2,2}(\Omega)} \leq C_2 C_1 \quad \text{for all} \quad n \geq 1. \tag{8.97}$$

Subsequently, we denote by j_1 and j_2 the compact injections

$$j_1 : W^{\frac{3}{2},2}(\Gamma_1) \hookrightarrow L^2(\Gamma_1), \qquad j_2 : W^{\frac{1}{2},2}(\Gamma_1) \hookrightarrow L^2(\Gamma_1),$$

and by

$$\mathcal{T}_1 \in \mathcal{L}(W^{2,2}(\Omega), W^{\frac{3}{2},2}(\Gamma_1)), \qquad \mathcal{T}_2 \in \mathcal{L}(W^{1,2}(\Omega), W^{\frac{1}{2},2}(\Gamma_1)),$$

the corresponding trace operators on Γ_1. According to (8.94), we obtain that

$$\varphi_n|_{\Gamma_1} = \mathcal{T}_1 \varphi_n \in W^{\frac{3}{2},2}(\Gamma_1) \overset{j_1}{\hookrightarrow} L^2(\Gamma_1),$$

$$\nabla\varphi_n|_{\Gamma_1} = \mathcal{T}_2 \nabla\varphi_n \in W^{\frac{1}{2},2}(\Gamma_1) \overset{j_2}{\hookrightarrow} L^2(\Gamma_1),$$

for all $n \geq 1$. Hence, it follows from (8.97) that

$$
\begin{aligned}
\|\varphi_n|_{\Gamma_1}\|_{L^2(\Gamma_1)} &= \|j_1 T_1 \varphi_n\|_{L^2(\Gamma_1)} \\
&\leq \|j_1 T_1\|_{\mathcal{L}(W^{2,2}(\Omega), L^2(\Gamma_1))} \|\varphi_n\|_{W^{2,2}(\Omega)} \\
&\leq \|j_1 T_1\|_{\mathcal{L}(W^{2,2}(\Omega), L^2(\Gamma_1))} C_2 C_1
\end{aligned}
$$

for all $n \geq 1$. Similarly,

$$
\begin{aligned}
\|\nabla \varphi_n|_{\Gamma_1}\|_{L^2(\Gamma_1)} &\leq \|j_2 T_2\|_{\mathcal{L}(W^{1,2}(\Omega), L^2(\Gamma_1))} \|\nabla \varphi_n\|_{W^{1,2}(\Omega)} \\
&\leq \|j_2 T_2\|_{\mathcal{L}(W^{1,2}(\Omega), L^2(\Gamma_1))} \|\varphi_n\|_{W^{2,2}(\Omega)} \\
&\leq \|j_2 T_2\|_{\mathcal{L}(W^{1,2}(\Omega), L^2(\Gamma_1))} C_2 C_1
\end{aligned}
$$

for all $n \geq 1$. Consequently, there exists a constant $C_3 > 0$ such that

$$
\|\varphi_n|_{\Gamma_1}\|_{L^2(\Gamma_1)} \leq C_3 \quad \text{and} \quad \|\nabla \varphi_n|_{\Gamma_1}\|_{L^2(\Gamma_1)} \leq C_3 \quad \forall\, n \geq 1. \tag{8.98}
$$

On the other hand, by (8.91), we find from

$$
\partial_\nu \varphi_n = -\beta_n \varphi_n \quad \text{on} \quad \Gamma_1, \qquad n \geq 1,
$$

that

$$
(\partial_\nu \varphi_n)^2 = \beta_n^2 \varphi_n^2 \geq \ell_n^2 \varphi_n^2 \quad \text{on} \quad \Gamma_1, \qquad n \geq 1,
$$

and hence,

$$
\varphi_n^2|_{\Gamma_1} \leq \ell_n^{-2}(\partial_\nu \varphi_n)^2|_{\Gamma_1} \leq \ell_n^{-2}|\nu|^2|\nabla \varphi_n|_{\Gamma_1}|^2, \qquad n \geq 1.
$$

Thus, (8.98) implies that

$$
\|\varphi_n|_{\Gamma_1}\|_{L^2(\Gamma_1)}^2 \leq \ell_n^{-2}|\nu|^2 \|\nabla \varphi_n|_{\Gamma_1}\|_{L^2(\Gamma_1)}^2 \leq \ell_n^{-2}|\nu|^2 C_3^2, \qquad n \geq 1,
$$

and, consequently, we find from (8.87) that

$$
\lim_{n \to \infty} \|\varphi_n|_{\Gamma_1}\|_{L^2(\Gamma_1)} = 0. \tag{8.99}
$$

In particular,

$$
\lim_{n \to \infty} \varphi_n|_{\Gamma_1} = 0 \quad \text{a.e. in} \quad \Gamma_1.
$$

Also, by construction, we already know that $\varphi_n|_{\Gamma_0} = 0$ for all $n \geq 1$. Therefore, (8.99) entails that

$$
\lim_{n \to \infty} \|\varphi_n|_{\partial\Omega}\|_{L^2(\partial\Omega)} = 0. \tag{8.100}
$$

Now, we show that φ_n, $n \geq 1$, is a Cauchy sequence in $W^{1,2}(\Omega)$. Combining this fact with (8.89) and (8.95) yields

$$
\lim_{n \to \infty} \|\varphi_n - \varphi\|_{W^{1,2}(\Omega)} = 0, \qquad \|\varphi\|_{W^{1,2}(\Omega)} = 1. \tag{8.101}
$$

Indeed, arguing as in the proof of Theorems 8.5 and 8.8, for every $1 \le k \le m$ the estimate (8.74) holds, as well as the estimates (8.75)–(8.77). Moreover, by (8.26) and (8.98), there exists a constant $C_4 > 0$ such that

$$\left| \sum_{i,j=1}^{N} \int_{\Gamma_1} a_{ij}(\varphi_k - \varphi_m) \frac{\partial \varphi_k}{\partial x_i} n_j \right| \le C_4 \|(\varphi_k - \varphi_m)|_{\Gamma_1}\|_{L^2(\Gamma_1)}$$

and

$$\left| \sum_{i,j=1}^{N} \int_{\Gamma_1} a_{ij}\varphi_m \frac{\partial(\varphi_m - \varphi_k)}{\partial x_i} n_j \right| \le C_4 \|\varphi_m|_{\Gamma_1}\|_{L^2(\Gamma_1)}.$$

Therefore, owing (8.100), for every $\epsilon > 0$ there exists a natural number $n_0 = n_0(\epsilon) \ge 0$ such that

$$\left| \sum_{i,j=1}^{N} \int_{\Gamma_1} a_{ij} \left[(\varphi_k - \varphi_m) \frac{\partial \varphi_k}{\partial x_i} + \varphi_m \frac{\partial(\varphi_m - \varphi_k)}{\partial x_i} \right] n_j \right| \le \frac{\epsilon}{2} \qquad (8.102)$$

for all $k, m \ge n_0$. Finally, substituting (8.76)–(8.77) and (8.102) into (8.74), it is easily realized that there exists $k_0 \ge n_0$ such that

$$\mu \|\nabla(\varphi_k - \varphi_m)\|_{L^2(\Omega)}^2 \le \epsilon, \qquad k, m \ge k_0.$$

This completes the proof of (8.101) and, in particular, it shows that $\varphi \in W^{1,2}(\Omega)$ and that $\varphi > 0$.

We now ascertain the behavior of φ on $\partial\Omega$. Let j denote the compact injection

$$j : W^{\frac{1}{2},2}(\partial\Omega) \hookrightarrow L^2(\partial\Omega)$$

and

$$\mathcal{T} \in \mathcal{L}(W^{1,2}(\Omega), W^{\frac{1}{2},2}(\partial\Omega))$$

the trace operator on $\partial\Omega$. Since $\varphi_n - \varphi \in W^{1,2}(\Omega)$, we have that

$$\mathcal{T}(\varphi_n - \varphi) = \varphi_n - \mathcal{T}\varphi \in W^{\frac{1}{2},2}(\partial\Omega)$$

for all $n \ge 1$, and hence,

$$\|\varphi_n - \mathcal{T}\varphi\|_{L^2(\partial\Omega)} \le \|j\mathcal{T}\|_{\mathcal{L}(W^{1,2}(\Omega), L^2(\partial\Omega))} \|\varphi_n - \varphi\|_{W^{1,2}(\Omega)}.$$

Thus, by (8.101), we find that

$$\lim_{n \to \infty} \|\varphi_n - \mathcal{T}\varphi\|_{L^2(\partial\Omega)} = 0$$

and, therefore, thanks to (8.100), we obtain that $\mathcal{T}\varphi = 0$. Equivalently,

$$\varphi \in W_0^{1,2}(\Omega).$$

Finally, the same argument used in the proofs of Theorems 8.5 and 8.8 shows that φ is a weak positive solution of

$$\begin{cases} \mathfrak{L}\varphi = \sigma^{\infty}\varphi & \text{in } \Omega, \\ \varphi = 0 & \text{on } \partial\Omega. \end{cases}$$

Therefore, according to Theorem 7.7, we find that

$$(\sigma^{\infty}, \varphi) = (\sigma[\mathfrak{L}, \mathfrak{D}, \Omega], \varphi_0).$$

This completes the proof. $\square$

As an immediate consequence of Theorem 8.9, the next result holds.

Corollary 8.3. *Suppose* (8.26). *Then,*

$$\sigma[\mathfrak{L}, \mathfrak{D}, \Omega] = \sup_{\beta \in \mathcal{C}(\Gamma_1)} \sigma[\mathfrak{L}, \mathfrak{B}[\beta], \Omega].$$

Proof.   Thanks to Proposition 8.1,

$$\sup_{\beta \in \mathcal{C}(\Gamma_1)} \sigma[\mathfrak{L}, \mathfrak{B}[\beta], \Omega] \le \sigma[\mathfrak{L}, \mathfrak{D}, \Omega].$$

Moreover, according to Theorem 8.9,

$$\lim_{n \to \infty} \sigma[\mathfrak{L}, \mathfrak{B}(n), \Omega] = \sigma[\mathfrak{L}, \mathfrak{D}, \Omega].$$

This completes the proof. $\square$

8.8   Lower estimates of $\sigma[\mathfrak{L}, \mathfrak{D}, \Omega]$ in terms of $|\Omega|$

As usual, we denote by $|\cdot|$ the Lebesgue measure of $\mathbb{R}^N$. The main result of this section establishes that if $|\Omega|$ is sufficiently small and $\min_{\Gamma_1} \beta$ is sufficiently large, then

$$\sigma[\mathfrak{L}, \mathfrak{B}, \Omega] > 0.$$

Consequently, according to Theorem 7.10, $(\mathfrak{L}, \mathfrak{B}, \Omega)$ satisfies the strong maximum principle. This result is based on Theorem 8.9 and on the lower estimates of $\sigma[\mathfrak{L}, \mathfrak{D}, \Omega]$ in terms of $|\Omega|$ given by Proposition 8.6 below.

In the special case when

$$\mathfrak{L} = -\Delta$$

the celebrated inequality of C. Faber [61] and E. Krahn [117] establishes that, among all domains with a fixed Lebesgue measure, $|\Omega|$, the ball $B_R :=$

$B_R(0)$ has the smallest principal eigenvalue under homogeneous Dirichlet boundary conditions. In other words, for every $R > 0$ and Ω such that

$$|\Omega| = |B_R| = R^N |B_1| \qquad (8.103)$$

the following estimate holds

$$\sigma[-\Delta, \mathfrak{D}, \Omega] \geq \sigma[-\Delta, \mathfrak{D}, B_R] = \sigma[-\Delta, \mathfrak{D}, B_1] R^{-2}. \qquad (8.104)$$

Therefore, setting

$$\Sigma := \sigma[-\Delta, \mathfrak{D}, B_1],$$

it becomes apparent from (8.103) and (8.104) that

$$\sigma[-\Delta, \mathfrak{D}, \Omega] \geq \Sigma |B_1|^{\frac{2}{N}} |\Omega|^{-\frac{2}{N}}. \qquad (8.105)$$

Consequently,

$$\liminf_{|\Omega| \downarrow 0} \left(\sigma[-\Delta, \mathfrak{D}, \Omega] |\Omega|^{\frac{2}{N}} \right) \geq \Sigma |B_1|^{\frac{2}{N}}. \qquad (8.106)$$

The next result provides us with a substantial extension of these lower estimates to cover the case when $\mathfrak{L}$ is a general second order uniformly elliptic operator in Ω.

Proposition 8.6. *Suppose*

$$|\Omega| \leq \left(\frac{\mu \sqrt{\Sigma}}{\|b\|_\infty} \right)^N |B_1|, \qquad (8.107)$$

where $\mu > 0$ is the ellipticity constant of $\mathfrak{L}$ in Ω, and

$$\|b\|_\infty := \sup_\Omega \left(\sum_{i=1}^N b_i^2 \right)^{1/2}.$$

Then,

$$\sigma[\mathfrak{L}, \mathfrak{D}, \Omega] \geq \mu \Sigma |B_1|^{\frac{2}{N}} |\Omega|^{-\frac{2}{N}} - \|b\|_\infty \sqrt{\Sigma} |B_1|^{\frac{1}{N}} |\Omega|^{-\frac{1}{N}} + \inf_\Omega c. \qquad (8.108)$$

In particular,

$$\liminf_{|\Omega| \downarrow 0} \left(\sigma[\mathfrak{L}, \mathfrak{D}, \Omega] |\Omega|^{\frac{2}{N}} \right) \geq \mu \Sigma |B_1|^{\frac{2}{N}}, \qquad (8.109)$$

which is optimal in the light of (8.106).

Proof. Let $\varphi > 0$ denote a principal eigenfunction of $\sigma[\mathfrak{L}, \mathfrak{D}, \Omega]$. Then, multiplying by φ the differential equation

$$\mathfrak{L}\varphi = \sigma[\mathfrak{L}, \mathfrak{D}, \Omega]\varphi,$$

integrating in Ω and applying the formula of integration by parts yields

$$\sigma[\mathfrak{L}, \mathfrak{D}, \Omega] \int_\Omega \varphi^2 = \int_\Omega \langle A\nabla\varphi, \nabla\varphi\rangle + \int_\Omega \langle b, \nabla\varphi\rangle\varphi + \int_\Omega c\varphi^2$$

$$\geq \mu \int_\Omega |\nabla\varphi|^2 + \int_\Omega \langle b, \nabla\varphi\rangle\varphi + \inf_\Omega c \int_\Omega \varphi^2.$$

Moreover,

$$\left| \int_\Omega \langle b, \nabla\varphi\rangle\varphi \right| \leq \int_\Omega |b||\nabla\varphi|\varphi \leq \|b\|_\infty \|\varphi\|_{L^2(\Omega)} \|\nabla\varphi\|_{L^2(\Omega)}$$

and hence,

$$\int_\Omega \langle b, \nabla\varphi\rangle\varphi \geq -\|b\|_\infty \|\varphi\|_{L^2(\Omega)} \|\nabla\varphi\|_{L^2(\Omega)}.$$

Consequently,

$$\sigma[\mathfrak{L}, \mathfrak{D}, \Omega] \int_\Omega \varphi^2 \geq \mu \int_\Omega |\nabla\varphi|^2 - \|b\|_\infty \|\varphi\|_{L^2(\Omega)} \|\nabla\varphi\|_{L^2(\Omega)} + \inf_\Omega c \int_\Omega \varphi^2,$$

or, equivalently,

$$\sigma[\mathfrak{L}, \mathfrak{D}, \Omega] \geq \frac{\|\nabla\varphi\|_{L^2(\Omega)}}{\|\varphi\|_{L^2(\Omega)}} \left(\mu \frac{\|\nabla\varphi\|_{L^2(\Omega)}}{\|\varphi\|_{L^2(\Omega)}} - \|b\|_\infty \right) + \inf_\Omega c. \qquad (8.110)$$

On the other hand, according to the variational characterization of the principal eigenvalue $\sigma[-\Delta, \mathfrak{D}, \Omega]$ as

$$\sigma[-\Delta, \mathfrak{D}, \Omega] = \inf_{\psi \in W_0^{1,2}(\Omega)} \frac{\int_\Omega |\nabla\psi|^2}{\int_\Omega \psi^2} \qquad (8.111)$$

(see, e.g., p. 336 of L. C. Evans [60]), it follows from (8.105) that

$$\frac{\int_\Omega |\nabla\varphi|^2}{\int_\Omega \varphi^2} \geq \sigma[-\Delta, \mathfrak{D}, \Omega] \geq \Sigma|B_1|^{\frac{2}{N}} |\Omega|^{-\frac{2}{N}}. \qquad (8.112)$$

Therefore, it follows from (8.112) and (8.107) that

$$\mu \frac{\|\nabla\varphi\|_{L^2(\Omega)}}{\|\varphi\|_{L^2(\Omega)}} \geq \mu\sqrt{\Sigma}|B_1|^{\frac{1}{N}} |\Omega|^{-\frac{1}{N}} \geq \|b\|_\infty.$$

Finally, (8.108) follows by substituting (8.112) into (8.110). $\qquad \square$

Finally, from Theorem 8.9 and Proposition 8.6, the next result holds.

Corollary 8.4. *Suppose* (8.26). *Then,*

$$\liminf_{|\Omega|\downarrow 0} \lim_{\substack{\beta \in \mathcal{C}(\Gamma_1) \\ \min_{\Gamma_1} \beta \uparrow \infty}} \left(\sigma[\mathfrak{L}, \mathfrak{B}[\beta], \Omega]|\Omega|^{\frac{2}{N}} \right) \geq \mu\Sigma|B_1|^{\frac{2}{N}}.$$

In particular, $\sigma[\mathfrak{L}, \mathfrak{B}[\beta], \Omega]$ can be as large as we wish by choosing Ω with $|\Omega|$ sufficiently small and $\beta \in \mathcal{C}(\Gamma_1)$ with $\min_{\Gamma_1} \beta$ sufficiently large.

8.9   Comments on Chapter 8

Although the monotonicity properties established in Section 8.1 are folklore under Dirichlet boundary conditions, in the general setting of this chapter they might go back to S. Cano-Casanova and J. López-Gómez [39]. The proofs of the monotonicity results in case $\mathfrak{B} = \mathfrak{D}$, through Theorem 7.10, seem to go back to J. López-Gómez [137]. Most of the available proofs in the literature dealt with the selfadjoint case ($b = 0$) by means of the J. W. S. Rayleigh quotients (8.111).

Theorem 1 of M. H. Protter and H. F. Weinberger [182] might be considered not only as a precursor of Theorem 7.10 but also of the min-max characterization of the principal eigenvalue provided by Theorem 8.1. Some years later, M. D. Donsker and S. R. S Varadhan [54, 55], and M. Venturino [222] gave some closely related general variational formulas for $\sigma[\mathfrak{L}, \mathfrak{D}, \Omega]$ through (8.5). The proof of the min-max characterization (8.5) given in this chapter, based on Theorem 7.10, has been adapted from the proofs of Theorem 3.1 of J. López-Gómez [137] and Theorem 4.1 of S. Cano-Casanova and J. López-Gómez [39].

Seemingly, the most pioneering version of the concavity of the principal eigenvalue established by Theorem 8.3 goes back to T. Kato [111]. Some further extensions of Kato's result were given by A. Beltramo and P. Hess [25], for some periodic-parabolic counterparts of the original problem, by P. Hess [96], and in Proposition 2.1 of H. Berestycki, L. Nirenberg and S. R. S. Varadhan [27] in the special case when $\mathfrak{B} = \mathfrak{D}$. In its full generality, Theorem 8.3 goes back to Theorem 5.1 of S. Cano-Casanova and J. López-Gómez [39]. The proof of Theorem 8.3 given here follows the general scheme of the proof of Theorem 3.3 in J. López-Gómez [137] and it uses a device coming from the proof of Lemma 5.1 of H. Berestycki, L. Nirenberg and S. R. S. Varadhan [27]. Based on (iii) of p. 102 in P. Hess [95], one might think that the original idea of the concavity of the principal eigenvalue with respect to the potential goes back, actually, to an observation of H. Berestycki and P. L. Lions.

The concept of stability of a domain through (8.10), as discussed in the statement of Theorem 8.4, goes back to I. Babuška [19] and I. Babuška and R. Vyborny [20], where it was introduced to generalize the pioneering results of R. Courant and D. Hilbert [44] on the continuous variation with respect to the domain Ω of the eigenvalues of a self-adjoint differential operator $\mathfrak{L}$ ($b = 0$) under homogeneous Dirichlet boundary conditions ($\mathfrak{B} = \mathfrak{D}$). The concept of stability has shown to play a central rol in potential theory, as it provides with all domains for which the Dirichlet problem is well posed (cf. D. R. Adams and L. I. Hedberg [2] and the references therein). Rather surprisingly, H. Berestycki, L. Nirenberg and S. R. S. Varadhan [27] did not impose any stability condition on Ω.

Proposition 8.5 is a sharp version of Theorem 3.7 in J. Wloka [226]. It is a pivotal result to get the continuous variation of the principal eigenvalue with respect to the *exterior approximations* of Ω along Γ_0, which has been established by Theorem 8.5. More general results, within the spirit of Proposition 8.5, for Dirichlet boundary conditions, have been established by J. L. Lions and E. Magenes [130].

As R. Courant and D. Hilbert [44] and I. Babuška and R. Vyborny [20] focused their attention on the special case when $b = 0$ and $\mathfrak{B} = \mathfrak{D}$, and E. N. Dancer [46] dealt with the special case when

$$\mathfrak{L} = -\Delta + \langle b, \cdot \rangle + c$$

under Dirichlet boundary conditions, it seems that, in case $\mathfrak{B} = \mathfrak{D}$, Theorem 8.7 goes back to Theorem 4.2 of J. López-Gómez [137], though the regularity requirements on the coefficients of $\mathfrak{L}$ in this book are substantially weaker than those of [137]. In its greatest generality, Theorem 8.7 goes back to Theorem 7.4 of S. Cano-Casanova and J. López-Gómez [39].

We conjecture that, under the general assumptions of Theorem 8.7, the spectrum of $(\mathfrak{L}, \mathfrak{B}_n, \Omega_n)$ does actually approximate the spectrum of $(\mathfrak{L}, \mathfrak{B}_0, \Omega_0)$ as $n \to \infty$. R. Courant and D. Hilbert [44] observed that the continuous dependence of the spectrum with respect to the domain may fail when dealing with Neumann boundary conditions ($\Gamma_0 = \emptyset$, $\beta = 0$ and $\nu = n$). This explains why in the results of Section 8.5 the portion Γ_1 of $\partial\Omega$ remained unchanged. The celebrated example of R. Courant and D. Hilbert [44] is the following. For any $\sigma > 0$, let

$$\Omega_\sigma := \left\{ (x,y) \in \mathbb{R}^2 \ : \ |x| < \sigma/2, \ |y| < \sigma/2 \right\}$$

the square of area $\sigma^2/4$ centered at the origin. Now, for any $\epsilon > 0$ and $\tau > 0$, let

$$R_{\epsilon,\tau} := \left\{ (x,y) \in \mathbb{R}^2 \ : \ 0 < x < \epsilon, \ |y| < \tau/2 \right\}$$

and consider

$$\Omega_{\epsilon,\tau} := \Omega_1 \cup [R_{\epsilon,\tau} + (1/2, 0)] \cup [\Omega_\epsilon + (1/2 + \epsilon, 0)].$$

For $\tau = \epsilon^4$, the domain $\Omega_{\epsilon,\epsilon^4}$ can be viewed as a $\mathcal{C}^0$-perturbation of Ω_0, but not a $\mathcal{C}^1$-perturbation, as discussed by J. K. Hale [92]. For every $\epsilon > 0$, let

$$A_\epsilon : \mathcal{D}(A_\epsilon) \to L^2\left(\Omega_{\epsilon,\epsilon^4}\right)$$

the operator defined by

$$A_\epsilon u := -\Delta u \qquad \text{for all} \quad u \in \mathcal{D}(A_\epsilon),$$

where

$$\mathcal{D}(A_\epsilon) := \{u \in W^{1,2}(\Omega_{\epsilon,\epsilon^4}) \ : \ \partial_\mathbf{n} u = 0 \text{ on } \partial\Omega_{\epsilon,\epsilon^4}\}.$$

Let σ_k^ϵ, $k \geq 0$, denote the ordered eigenvalues of A_ϵ. Then, $\sigma_0^\epsilon = 0$ for all $\epsilon \geq 0$, $\sigma_2^\epsilon > 0$ for all $\epsilon > 0$, and it is shown by R. Courant and D. Hilbert [44] that

$$\lim_{\epsilon \downarrow 0} \sigma_2^\epsilon = 0.$$

Consequently, the eigenvalues exhibit a singular behavior at $\epsilon = 0$, in the sense that the second eigenvalue for $\epsilon > 0$, σ_2^ϵ, is bounded away from the second eigenvalue for $\epsilon = 0$, $\sigma_2^0 > 0$, because $\sigma_2^\epsilon \sim 0$ if $\epsilon \sim 0$. In this example, the principal eigenvalues for $\epsilon \geq 0$ equal zero.

Even if in the statement of Theorem 8.6 no regularity requirement is imposed to Γ_0^0, the limit

$$\sigma^I := \lim_{n \to \infty} \sigma[\mathfrak{L}, \mathfrak{B}_n, \Omega_n]$$

is well defined and it provides us with an eigenvalue to a weak positive solution of $(\mathfrak{L}, \mathfrak{B}_0, \Omega_0)$. This follows easily by adapting the proof of Theorem 8.6 (see Theorem 7.2 of S. Cano-Casanova and J. López-Gómez [39]). But, except when Γ_0^0 is of class $\mathcal{C}^2$, it is unknown whether or not σ^E is the unique eigenvalue of $(\mathfrak{L}, \mathfrak{B}_0, \Omega_0)$ to a weak positive eigenfunction.

Although, in order to define $W_{\Gamma_0^0}^{1,p}(\Omega_0)$, $p \geq 1$, through (4.16), this book requires Γ_0^0 to be of class $\mathcal{C}^1$, it is also possible to define

$$W_{\Gamma_0^0}^{1,p}(\Omega_0) = \overline{\mathcal{C}_{\Gamma_0^0}^\infty(\Omega_0)}^{W^{1,p}(\Omega_0)}$$

and, in such case, no a priori regularity requirement on Γ_0^0 is needed to introduce the concept of stability of Ω_0 along Γ_0^0 by (8.10). Suppose Ω_0 is stable in this sense. Then, the value

$$\sigma^E := \lim_{n \to \infty} \sigma[\mathfrak{L}, \mathfrak{B}_n, \Omega_n]$$

constructed in the proof of Theorem 8.5 also provides us with an eigenvalue to a weak positive eigenfunction of $(\mathfrak{L}, \mathfrak{B}_0, \Omega_0)$. According to Theorem 8.7, we have that

$$\sigma^I = \sigma^E \tag{8.113}$$

if Γ_0^0 is of class $\mathcal{C}^2$. Naturally, it would be of the greatest interest to characterize the class of domains Ω_0 saisfying (8.113). We conjecture that (8.113) holds if, and only if, Ω_0 is stable along Γ_0^0.

Theorem 8.8 is Theorem 8.2 of S. Cano-Casanova and J. López-Gómez [39]. The proof of Theorem 8.8 also provides us with the existence of a principal eigenvalue, not necessarily unique, for a large class of weight functions $\beta \in L^\infty(\Gamma_1)$. Indeed, if β is the point-wise limit of a (uniformly) bounded sequence $\beta_n \in \mathcal{C}(\Gamma_1)$, $n \geq 1$, then, it follows from the dominated convergence theorem of Lebesgue that $\beta_n \xrightarrow{*} \beta$ as $n \to \infty$ and that

$$\sigma^\infty := \lim_{n \to \infty} \sigma(\beta_n)$$

is well defined. Moreover, the argument of the proof of Theorem 8.8 also shows that σ^∞ is an eigenvalue of $(\mathfrak{L}, \mathfrak{B}[\beta], \Omega)$ to a positive eigenfunction.

Theorem 8.9 goes back to Theorem 9.1 of S. Cano-Casanova and J. López-Gómez [39]. When, instead of (8.87), the following holds

$$\lim_{n \to \infty} \max_{\Gamma_1} \beta_n = -\infty,$$

then, the behavior of the principal eigenvalue may change drastically as the next simple one-dimensional example shows. Suppose

$$N = 1, \quad \Omega = (0,1), \quad \Gamma_0 = \{0\}, \quad \Gamma_1 = \{1\}, \quad \mathfrak{L} = -\frac{d^2}{dx^2},$$

and $\beta \in \mathbb{R}$. Then, λ is an eigenvalue of $(\mathfrak{L}, \mathfrak{B}[\beta], \Omega)$ if, and only if, it is an eigenvalue of problem

$$\begin{cases} -u''(x) = \lambda u(x), & 0 < x < 1, \\ u(0) = 0, \quad u'(1) + \beta u(1) = 0. \end{cases} \tag{8.114}$$

Let us denote by $\sigma_k[\beta]$, $k \geq 0$, the increasing sequence of eigenvalues of problem (8.114). According to Theorem 8.9, we have that

$$\lim_{\beta \uparrow \infty} \sigma_0[\beta] = \pi^2$$

equals the principal eigenvalue of

$$\begin{cases} -u''(x) = \lambda u(x), & 0 < x < 1, \\ u(0) = 0, \quad u(1) = 0. \end{cases} \tag{8.115}$$

But

$$\lim_{\beta \downarrow -\infty} \sigma_0[\beta] = -\infty,$$

and

$$\lim_{\beta \downarrow -\infty} \sigma_n[\beta] = (n\pi)^2 \qquad \text{for all } n \geq 1$$

provide us with the eigenvalues of the limiting problem (8.115) (see Exercise 9.3 of J. López-Gómez [142]). Consequently, the second eigenvalue of (8.114), $\sigma_1[\beta]$, instead of $\sigma_0[\beta]$, approximates the principal eigenvalue of (8.115), π^2, as $\beta \downarrow -\infty$.

According to H. Berestycki, L. Nirenberg and S. R. S. Varadhan [27], the fact that

$$\sigma[\mathfrak{L}, \mathfrak{D}, \Omega] > 0,$$

or, equivalently, that $(\mathfrak{L}, \mathfrak{D}, \Omega)$ satisfies the strong maximum principle for sufficiently small $|\Omega|$, was first noted by I. J. Bakelman [21] and then used extensively by H. Berestycki and L. Nirenberg [26], though D. Gilbarg and N. Trudinger, after the proof of Lemma 8.4 of [79], had already established that $\mathfrak{L}$ is coercive for sufficiently small $|\Omega|$. Indeed, on pp. 52–53 of [27], H. Berestycki, L. Nirenberg and S. R. S. Varadhan claim that

This paper was motivated by the following observation which was first noted by Bakelman [2] and used extensively in [5].

PROPOSITION 1.1. Let L satisfy conditions (1.2) and (1.3) in Ω. Assume diam $\Omega \leq d$. There exists $\delta > 0$ depending only on n, c_0, b and d [the spatial dimension and the coefficients of L] such that the maximum principle holds for L in Ω if the measure of Ω, $|\Omega|$, satisfies $|\Omega| < \delta$.

Naturally, the low estimates (8.105) are attributable to C. Faber [61] and E. Krahn [117]. They were later refined to include some general classes of second order elliptic operators $\mathfrak{L}$, under Dirichlet boundary conditions, in Theorem 2.5 and Remark 2.2 of H. Berestycki, L. Nirenberg and S. R. S. Varadhan [27], and in Theorem 5.1 of J. López-Gómez [137]. In their greatest generality, Proposition 8.6 and Corollary 8.4 go back to Section 10 of S. Cano-Casanova and J. López-Gómez [39]. Essentially, they established that $(\mathfrak{L}, \mathfrak{B}[\beta], \Omega)$ satisfies the strong maximum principle if β is sufficiently large and $|\Omega|$ sufficiently small.

Chapter 9

Principal eigenvalues of linear weighted boundary value problems

This chapter studies the existence and multiplicity of principal eigenvalues for the linear weighted boundary value problem

$$
\begin{cases}
\mathfrak{L}\varphi = \lambda W(x)\varphi & \text{in } \Omega, \\
\mathfrak{B}\varphi = 0 & \text{on } \partial\Omega,
\end{cases}
\tag{9.1}
$$

where $\mathfrak{L}$ and $\mathfrak{B}$ satisfy the general requirements of Chapter 4, $\partial\Omega$ is of class $\mathcal{C}^2$, $W \in L^\infty(\Omega)$, and $\lambda \in \mathbb{R}$. Setting

$$
V := -W
\tag{9.2}
$$

and

$$
\Sigma(\lambda) := \sigma[\mathfrak{L} + \lambda V, \mathfrak{B}, \Omega], \qquad \lambda \in \mathbb{R},
\tag{9.3}
$$

a given value $\lambda^* \in \mathbb{R}$ is said to be a *principal eigenvalue* of (9.1) if

$$
\Sigma(\lambda^*) = 0.
$$

This chapter characterizes the existence of principal eigenvalues of (9.1) in all possible cases. The notations (9.2) and (9.3) will be maintained throughout it.

Subsequently, we will outline the distribution and main contents of this chapter. In Section 9.1 we study some general properties of the map $\Sigma(\lambda)$. Precisely, we show that it is real analytic and strictly concave if W changes sign, while it is strictly monotone if W does not change sign. As a result of these facts, collected in Theorem 9.1, the results of Section 9.2 characterize the existence of principal eigenvalues for (9.1). Precisely, in the special, but important, case when $W \in \mathcal{C}(\bar{\Omega})$, the next results hold:

- Suppose $W \geq 0$, or $W \leq 0$, and $W \neq 0$. Then, (9.1) has a principal eigenvalue if, and only if, $(\mathfrak{L} - \lambda_0 W, \mathfrak{B}, \Omega)$ satisfies the strong maximum principle for some $\lambda_0 \in \mathbb{R}$. In such case, the principal eigenvalue is unique and algebraically simple.

- Suppose $W \neq 0$ changes sign in Ω. Then, (9.1) admits a principal eigenvalue if, and only if, some of the following conditions hold:

  (a) $(\mathfrak{L} - \lambda_0 W, \mathfrak{B}, \Omega)$ satisfies the strong maximum principle for some $\lambda_0 \in \mathbb{R}$. In such case, (9.1) possesses two principal eigenvalues, which are algebraically simple.

  (b) For every $\lambda \in \mathbb{R}$, $(\mathfrak{L} - \lambda W, \mathfrak{B}, \Omega)$ does not satisfy the strong maximum principle, but there exists $\lambda_0 \in \mathbb{R}$ such that

  $$\sigma[\mathfrak{L} - \lambda_0 W, \mathfrak{B}, \Omega] = 0.$$

  In such case, λ_0 must be the unique principal eigenvalue of (9.1), and it is a double eigenvalue.

In Section 9.3, we introduce a general class of potentials $V \geq 0$ for which

$$\lim_{\lambda \uparrow \infty} \sigma[\mathfrak{L} + \lambda V, \mathfrak{B}, \Omega] = \sigma[\mathfrak{L}, \mathfrak{B}[\Omega_0], \Omega_0],$$

where

$$\Omega_0 := \operatorname{int} V^{-1}(0).$$

For that class of potentials, the next characterization holds:

- Suppose $W \geq 0$, or $W \leq 0$, and $W \neq 0$. Then, (9.1) has a principal eigenvalue if, and only if,

  $$\sigma[\mathfrak{L}, \mathfrak{B}[\Omega_0], \Omega_0] > 0.$$

According to Corollary 8.4, this holds if $|\Omega_0|$ is sufficiently small and β is sufficiently large on $\Gamma_1 \cap \partial\Omega_0$.

Finally, Section 9.4 uses these results to give a series of rather explicit sufficient conditions so that (9.1) possess two principal eigenvalues when W changes sign. Naturally, one has to obtain sufficient conditions so that $(\mathfrak{L} - \lambda_0 W, \mathfrak{B}, \Omega)$ satisfy the maximum principle for some $\lambda_0 \in \mathbb{R}$.

9.1 General properties of the map $\Sigma(\lambda)$

The next result collects some important properties of the map $\Sigma(\lambda)$.

Theorem 9.1. *Suppose $V \in L^\infty(\Omega)$. Then, the map $\Sigma(\lambda)$ defined by (9.3) satisfies the following properties:*

(a) *$\Sigma(\lambda)$ is real analytic and concave, in the sense that $\Sigma''(\lambda) \leq 0$ for all $\lambda \in \mathbb{R}$. Therefore, either $\Sigma'' = 0$ in $\mathbb{R}$, or there exists a discrete set $Z \subset \mathbb{R}$ such that $\Sigma''(\lambda) < 0$ for all $\lambda \in \mathbb{R} \setminus Z$. By discrete it means that $Z \cap K$ is finite for all compact subset $K \subset \mathbb{R}$.*

(b) *Assume there exist* $x_+ \in \Omega$ *and* $R > 0$ *such that*

$$B_+ := B_R(x_+) \subset \Omega \quad and \quad \inf_{B_+} V > 0. \tag{9.4}$$

Then,

$$\lim_{\lambda \downarrow -\infty} \Sigma(\lambda) = -\infty. \tag{9.5}$$

Suppose, in addition, that $V \geq 0$ *in* Ω. *Then,* $\Sigma'(\lambda) > 0$ *for all* $\lambda \in \mathbb{R}$.

(c) *Assume there exist* $x_- \in \Omega$ *and* $R > 0$ *such that*

$$B_- := B_R(x_-) \subset \Omega \quad and \quad \sup_{B_-} V < 0. \tag{9.6}$$

Then,

$$\lim_{\lambda \uparrow \infty} \Sigma(\lambda) = -\infty. \tag{9.7}$$

Suppose, in addition, that $V \leq 0$ *in* Ω. *Then,* $\Sigma'(\lambda) < 0$ *for all* $\lambda \in \mathbb{R}$.

(d) *Suppose there are* x_+, $x_- \in \Omega$ *and* $R > 0$ *satisfying* (9.4) *and* (9.6). *Then,* (9.5) *and* (9.7) *hold and, therefore, there exists* $\lambda_0 \in \mathbb{R}$ *such that*

$$\Sigma(\lambda_0) = \max_{\lambda \in \mathbb{R}} \Sigma(\lambda). \tag{9.8}$$

Moreover, $\Sigma'(\lambda_0) = 0$, $\Sigma'(\lambda) > 0$ *if* $\lambda < \lambda_0$, *and* $\Sigma'(\lambda) < 0$ *if* $\lambda > \lambda_0$. *Consequently,* λ_0 *must be unique.*

Proof. We will prove each of them separately.

Proof of Part (a): Subsequently, we set

$$\mathfrak{L}(\lambda) := \mathfrak{L} + \lambda V, \qquad \lambda \in \mathbb{R},$$

and regard $\mathfrak{L}(\lambda)$, $\lambda \in \mathbb{R}$, as a family of closed operators with common domain

$$D(\mathfrak{L}(\lambda)) = W_{\mathfrak{B}}^{2,\infty^-}(\Omega)$$

and values in $L^2(\Omega)$. Then, $\mathfrak{L}(\lambda)$ is a real holomorphic family of type (A) as discussed in Section VII.2 of T. Kato [112], because

$$\lambda \mapsto \int_\Omega v \mathfrak{L}(\lambda) u$$

is real analytic in λ for all $v \in L^2(\Omega)$ and $u \in W_{\mathfrak{B}}^{2,\infty^-}(\Omega)$. Therefore, by Theorems 1.7 and 1.8 in Section VII.1.3 of [112], $\Sigma(\lambda)$ is real analytic in λ. Moreover, if $\varphi(\lambda) \gg 0$ stands for the unique principal eigenfunction of $\Sigma(\lambda)$ such that

$$\int_\Omega \varphi^2(\lambda) = 1,$$

the map

$$\mathbb{R} \longrightarrow L^2(\Omega)$$
$$\lambda \mapsto \varphi(\lambda)$$

is also real analytic. Next, we will show that

$$\Sigma''(\lambda) \le 0 \qquad \text{for all} \quad \lambda \in \mathbb{R}. \tag{9.9}$$

According to Theorem 8.3, we have that

$$\begin{aligned}
\Sigma(t\lambda_1 + (1-t)\lambda_2) &= \sigma[\mathfrak{L} + t\lambda_1 V + (1-t)\lambda_2 V, \mathfrak{B}, \Omega] \\
&\ge t\sigma[\mathfrak{L} + \lambda_1 V, \mathfrak{B}, \Omega] + (1-t)\sigma[\mathfrak{L} + \lambda_2 V, \mathfrak{B}, \Omega] \\
&= t\Sigma(\lambda_1) + (1-t)\Sigma(\lambda_2)
\end{aligned}$$

for all $t \in [0,1]$ and $\lambda_1, \lambda_2 \in \mathbb{R}$. Thus,

$$\Sigma(\lambda_2 + t(\lambda_1 - \lambda_2)) \ge \Sigma(\lambda_2) + t\left(\Sigma(\lambda_1) - \Sigma(\lambda_2)\right)$$

and hence,

$$\frac{\Sigma(\lambda_2 + t(\lambda_1 - \lambda_2)) - \Sigma(\lambda_2)}{t} \ge \Sigma(\lambda_1) - \Sigma(\lambda_2)$$

for all $t \in (0,1]$ and $\lambda_1, \lambda_2 \in \mathbb{R}$. Consequently,

$$\frac{\Sigma(\lambda_2 + t(\lambda_1 - \lambda_2)) - \Sigma(\lambda_2)}{t(\lambda_1 - \lambda_2)} \ge \frac{\Sigma(\lambda_1) - \Sigma(\lambda_2)}{\lambda_1 - \lambda_2} \tag{9.10}$$

for all $t \in (0,1]$ and $\lambda_1, \lambda_2 \in \mathbb{R}$ with $\lambda_1 > \lambda_2$. Letting $t \downarrow 0$ in (9.10), it becomes apparent that

$$\Sigma'(\lambda_2) \ge \frac{\Sigma(\lambda_1) - \Sigma(\lambda_2)}{\lambda_1 - \lambda_2}$$

for all $\lambda_1 > \lambda_2$. Thus, by the mean value theorem, we find that for every $\lambda_1, \lambda_2 \in \mathbb{R}$ such that $\lambda_1 > \lambda_2$ there exists $\lambda \in (\lambda_2, \lambda_1)$ for which

$$\Sigma'(\lambda_2) \ge \Sigma'(\lambda). \tag{9.11}$$

Clearly, this is impossible if there exists $\lambda \in \mathbb{R}$ for which $\Sigma''(\lambda) > 0$, because Σ' should be increasing in a neighborhood of such λ. Therefore, (9.9) holds.

Finally, since Σ'' is real analytic, owing to the identity principle, it becomes apparent that either $\Sigma'' = 0$, or Σ'' vanishes, at most, in a discrete set, possibly empty. This concludes the proof of Part (a).

Proof of Part (b): According to Proposition 8.2,

$$\Sigma(\lambda) = \sigma[\mathfrak{L} + \lambda V, \mathfrak{B}, \Omega] < \sigma[\mathfrak{L} + \lambda V, \mathfrak{D}, B_+],$$

because $\mathfrak{B}[B_+] = \mathfrak{D}$. Thus, by Proposition 8.3, we have that

$$\Sigma(\lambda) < \sigma[\mathfrak{L}, \mathfrak{D}, B_+] + \lambda \inf_{B_+} V \qquad \text{for all} \quad \lambda < 0.$$

Clearly, (9.5) follows from (9.4) and this estimate by letting $\lambda \downarrow -\infty$.

Now, suppose $V \geq 0$. Then, by Proposition 8.3, $\lambda \mapsto \Sigma(\lambda)$ is increasing and hence $\Sigma'(\lambda) \geq 0$ for all $\lambda \in \mathbb{R}$. By analyticity, either $\Sigma' = 0$ in $\mathbb{R}$, or $\Sigma'(\lambda) > 0$ for all $\lambda \in \mathbb{R}$ except in a discrete set. According to (9.5), $\Sigma(\lambda)$ cannot be a constant. Therefore, the second option occurs. Suppose $\Sigma'(\lambda_0) = 0$ for some $\lambda_0 \in \mathbb{R}$. Then,

$$\Sigma'(\lambda) = \Sigma'(\lambda) - \Sigma'(\lambda_0) = \int_{\lambda_0}^{\lambda} \Sigma'' \leq 0 \qquad \text{for all} \quad \lambda \geq \lambda_0$$

and, consequently, $\Sigma' = 0$ in $[\lambda_0, \infty)$, which is impossible. Therefore, $\Sigma'(\lambda) > 0$ for all $\lambda \in \mathbb{R}$. This ends the proof of Part (b).

Proof of Part (c): As in the proof of Part (b), we have that

$$\Sigma(\lambda) = \sigma[\mathfrak{L} + \lambda V, \mathfrak{B}, \Omega] < \sigma[\mathfrak{L} + \lambda V, \mathfrak{D}, B_-],$$

for all $\lambda \in \mathbb{R}$ and hence,

$$\Sigma(\lambda) < \sigma[\mathfrak{L}, \mathfrak{D}, B_-] + \lambda \sup_{B_-} V \qquad \text{for all} \quad \lambda > 0.$$

Obviously, (9.7) follows from (9.6) and this estimate by letting $\lambda \uparrow \infty$.

Now, suppose $V \leq 0$. Then, by Proposition 8.3, $\lambda \mapsto \Sigma(\lambda)$ is decreasing and hence $\Sigma'(\lambda) \leq 0$ for all $\lambda \in \mathbb{R}$. By analyticity, either $\Sigma' = 0$ in $\mathbb{R}$, or $\Sigma'(\lambda) < 0$ for all $\lambda \in \mathbb{R}$ except in a discrete set. According to (9.7), $\Sigma(\lambda)$ cannot be a constant and, so, the second option occurs. Suppose $\Sigma'(\lambda_0) = 0$ for some $\lambda_0 \in \mathbb{R}$. Then,

$$-\Sigma'(\lambda) = \Sigma'(\lambda_0) - \Sigma'(\lambda) = \int_{\lambda}^{\lambda_0} \Sigma'' \leq 0 \qquad \text{for all} \quad \lambda \leq \lambda_0$$

and, consequently, $\Sigma' = 0$ in $(-\infty, \lambda_0]$, which is impossible. Therefore, $\Sigma'(\lambda) < 0$ for all $\lambda \in \mathbb{R}$. This shows Part (c).

Proof of Part (d): The existence of λ_0 is a direct consequence from the continuity of $\Sigma(\lambda)$ and of (9.5) and (9.7). Necessarily, $\Sigma'(\lambda_0) = 0$. Next, we will show that $\Sigma'(\lambda) > 0$ for all $\lambda < \lambda_0$. On the contrary, suppose there exists $\lambda_1 < \lambda_0$ such that

$$\Sigma'(\lambda_1) \leq 0.$$

Then,

$$0 \geq \Sigma'(\lambda_1) = \int_{\lambda_0}^{\lambda_1} \Sigma''(\lambda) \, d\lambda = -\int_{\lambda_1}^{\lambda_0} \Sigma'' \geq 0,$$

because $\Sigma'' \leq 0$. Hence,

$$\Sigma'(\lambda_1) = -\int_{\lambda_1}^{\lambda_0} \Sigma'' = 0$$

and, therefore, $\Sigma'' = 0$ in $[\lambda_1, \lambda_0]$. By Part (a), $\Sigma'' = 0$ in $\mathbb{R}$ and, so, there exist a, $b \in \mathbb{R}$ such that

$$\Sigma(\lambda) = a\lambda + b \qquad \text{for all} \quad \lambda \in \mathbb{R}.$$

By (9.5) and (9.7), this cannot occur. Thus, $\Sigma'(\lambda) > 0$ for all $\lambda < \lambda_0$.

Finally, we will prove that $\Sigma'(\lambda) < 0$ for all $\lambda > \lambda_0$. On the contrary, suppose that there exists $\lambda_2 > \lambda_0$ such that

$$\Sigma'(\lambda_2) \geq 0.$$

Then,

$$0 \leq \Sigma'(\lambda_2) = \int_{\lambda_0}^{\lambda_2} \Sigma'' \leq 0$$

and so,

$$\Sigma'(\lambda_2) = \int_{\lambda_0}^{\lambda_2} \Sigma'' = 0$$

which implies $\Sigma'' = 0$ in $[\lambda_0, \lambda_2]$ and, consequently, by analyticity, $\Sigma'' = 0$ in $\mathbb{R}$, which is impossible. The proof is complete. $\square$

9.2   Characterizing the existence of a principal eigenvalue

This section characterizes the existence and multiplicity of principal eigenvalues for the linear weighted boundary value problem (9.1). According to the sign of the weight function W, we will distinguish three different cases. When $W \geq 0$, the main result reads as follows. Throughout this section, it should not be forgotten that

$$V := -W.$$

Theorem 9.2. *Suppose $W \geq 0$ and there are $x_- \in \Omega$ and $R > 0$ such that*

$$B_- := B_R(x_-) \subset \Omega \quad \text{and} \quad \inf_{B_-} W > 0. \tag{9.12}$$

Then, (9.1) possesses a principal eigenvalue if and only if

$$\Sigma(-\infty) := \lim_{\lambda \downarrow -\infty} \Sigma(\lambda) > 0. \tag{9.13}$$

Moreover, it is unique if it exists and if we denote it by λ^*, then, λ^* is a simple eigenvalue of $(\mathfrak{L} - \lambda W, W)$ as discussed by M. G. Crandall and P. H. Rabinowitz [45], i.e.,

$$W\varphi^* \notin R[\mathfrak{L} - \lambda^* W] \tag{9.14}$$

for any principal eigenfunction $\varphi^* \gg 0$ of (9.1) associated to λ^*.

Proof. By (9.12), the function $V := -W \leq 0$ satisfies (9.6) and hence, according to Theorem 9.1(c), (9.7) holds. Moreover,

$$\Sigma'(\lambda) < 0 \quad \text{for all} \quad \lambda \in \mathbb{R} \tag{9.15}$$

and, therefore, the limit (9.13) is well defined. Note that $\Sigma(-\infty)$ might be finite or infinity. Indeed,

$$\Sigma(-\infty) = \infty \quad \text{if} \quad \inf_{\Omega} W > 0,$$

because, in such case, for every $\lambda < 0$ we have that

$$\Sigma(\lambda) = \sigma[\mathfrak{L} - \lambda W, \mathfrak{B}, \Omega] \geq \sigma[\mathfrak{L}, \mathfrak{B}, \Omega] - \lambda \inf_{\Omega} W,$$

while, due to Proposition 8.2, when $W = 0$ in some open set $\Omega_0 \subset \Omega$ with $\bar{\Omega}_0 \subset \Omega$, then,

$$\Sigma(\lambda) = \sigma[\mathfrak{L} - \lambda W, \mathfrak{B}, \Omega] \leq \sigma[\mathfrak{L}, \mathfrak{D}, \Omega_0] \quad \text{for all} \quad \lambda \in \mathbb{R}$$

and, consequently,

$$\Sigma(-\infty) \leq \sigma[\mathfrak{L}, \mathfrak{D}, \Omega_0].$$

In (9.13) we use the natural convention that $\infty > 0$ if $\Sigma(-\infty) = \infty$.

Suppose $\Sigma(-\infty) > 0$. Then, $\Sigma(\lambda_1) > 0$ for some $\lambda_1 \in \mathbb{R}$ and hence, by (9.7), there exists $\lambda^* > \lambda_1$ such that $\Sigma(\lambda^*) = 0$. Conversely, if there exists $\lambda^* \in \mathbb{R}$ such that $\Sigma(\lambda^*) = 0$, then, by (9.15), $\Sigma'(\lambda^*) < 0$ and hence $\Sigma(\lambda) > 0$ for all $\lambda < \lambda^*$. Therefore, $\Sigma(-\infty) > 0$. This shows that λ^* exists if and only if (9.13) holds.

The uniqueness of λ^* can be obtained, for example, with the following argument. Suppose

$$\Sigma(\lambda_1^*) = \Sigma(\lambda_2^*) = 0$$

for some $\lambda_1^* < \lambda_2^*$. Then, by the mean value theorem, there exists $\lambda \in (\lambda_1^*, \lambda_2^*)$ such that $\Sigma'(\lambda) = 0$, which contradicts (9.15). Therefore, λ^* is unique.

It remains to prove (9.14). Let $\varphi(\lambda)$ be the principal eigenfunction associated to $\Sigma(\lambda)$ normalized so that

$$\int_{\Omega} \varphi^2(\lambda) = 1. \tag{9.16}$$

According to the proof of Theorem 9.1(a), the map $\lambda \mapsto \varphi(\lambda)$ is real analytic. Thus, differentiating the identity

$$(\mathfrak{L} - \lambda W)\varphi(\lambda) = \Sigma(\lambda)\varphi(\lambda), \qquad \lambda \in \mathbb{R},$$

with respect to λ yields

$$(\mathfrak{L} - \lambda W)\varphi'(\lambda) - W\varphi(\lambda) = \Sigma'(\lambda)\varphi(\lambda) + \Sigma(\lambda)\varphi'(\lambda), \qquad \lambda \in \mathbb{R},$$

and, consequently, particularizing at $\lambda = \lambda^*$ shows that

$$(\mathfrak{L} - \lambda^* W)\varphi'(\lambda^*) = W\varphi(\lambda^*) + \Sigma'(\lambda^*)\varphi(\lambda^*). \tag{9.17}$$

Suppose

$$W\varphi(\lambda^*) \in R[\mathfrak{L} - \lambda^* W].$$

Then, since $\Sigma'(\lambda^*) < 0$, (9.17) implies that

$$\varphi(\lambda^*) \in R[\mathfrak{L} - \lambda^* W],$$

which is impossible, because

$$N[\mathfrak{L} - \lambda^* W] = \operatorname{span}[\varphi(\lambda^*)]$$

and, according to Theorem 7.8, $\Sigma(\lambda^*) = 0$ is a simple eigenvalue of $(\mathfrak{L} - \lambda^* W, \mathfrak{B}, \Omega)$. The proof is complete.   □

Figure 9.1 shows a genuine graph of the map $\lambda \mapsto \Sigma(\lambda)$ in the special case when $W \geq 0$ and $\Sigma_\alpha := \Sigma(-\infty) > 0$. Note that, under the assump-

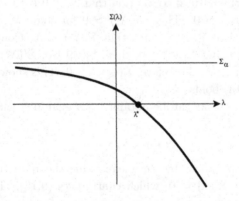

Fig. 9.1   A genuine graph of $\lambda \mapsto \Sigma(\lambda)$ when $W \geq 0$

tions of Theorem 9.2, $\lambda^* > 0$ if

$$\Sigma(0) = \sigma[\mathfrak{L}, \mathfrak{B}, \Omega] > 0,$$

$\lambda^* = 0$ if $\sigma[\mathfrak{L}, \mathfrak{B}, \Omega] = 0$, and $\lambda^* < 0$ if $\sigma[\mathfrak{L}, \mathfrak{B}, \Omega] < 0$.

Remark 9.1. The proof of (9.14) is based on the fact that $\Sigma'(\lambda^*) \neq 0$, rather than on the sign of W. Therefore, the last assertion of Theorem 9.2 holds true as soon as

$$\Sigma(\lambda^*) = 0 \quad \text{and} \quad \Sigma'(\lambda^*) \neq 0.$$

Similarly, the next result holds in the special case when $W \leq 0$.

Theorem 9.3. *Suppose $W \leq 0$ and there are $x_+ \in \Omega$ and $R > 0$ such that*

$$B_+ := B_R(x_+) \subset \Omega \quad \text{and} \quad \sup_{B_+} W < 0. \tag{9.18}$$

Then, (9.1) possesses a principal eigenvalue if and only if

$$\Sigma(\infty) := \lim_{\lambda \uparrow \infty} \Sigma(\lambda) > 0. \tag{9.19}$$

Moreover, it is unique if it exists and, if we denote it by λ^*, then, λ^* is a simple eigenvalue of $(\mathfrak{L} - \lambda W, W)$ as discussed by M. G. Crandall and P. H. Rabinowitz [45].

Proof. By (9.18), the function $V := -W \leq 0$ satisfies (9.4) and hence, according to Theorem 9.1(b), (9.5) holds. Moreover,

$$\Sigma'(\lambda) > 0 \quad \text{for all} \quad \lambda \in \mathbb{R} \tag{9.20}$$

and, consequently, the limit (9.19) is well defined. It might be finite or infinite. Indeed,

$$\Sigma(\infty) = \infty \quad \text{if} \quad \sup_{\Omega} W < 0,$$

because, in such case, for every $\lambda > 0$ we have that

$$\Sigma(\lambda) = \sigma[\mathfrak{L} - \lambda W, \mathfrak{B}, \Omega] \geq \sigma[\mathfrak{L}, \mathfrak{B}, \Omega] - \lambda \sup_{\Omega} W,$$

while, owing to Proposition 8.2, if $W = 0$ in some open subset $\Omega_0 \subset \Omega$ with $\bar{\Omega}_0 \subset \Omega$, then,

$$\Sigma(\lambda) = \sigma[\mathfrak{L} - \lambda W, \mathfrak{B}, \Omega] \leq \sigma[\mathfrak{L}, \mathfrak{D}, \Omega_0] \quad \text{for all} \quad \lambda \in \mathbb{R},$$

and, consequently,

$$\Sigma(\infty) \leq \sigma[\mathfrak{L}, \mathfrak{D}, \Omega_0].$$

Naturally, in (9.19) we use the convention $\infty > 0$ if $\Sigma(\infty) = \infty$.

Suppose $\Sigma(\infty) > 0$. Then, $\Sigma(\lambda_1) > 0$ for some $\lambda_1 \in \mathbb{R}$ and hence, by (9.5), there exists $\lambda^* < \lambda_1$ such that $\Sigma(\lambda^*) = 0$. Conversely, if there is $\lambda^* \in \mathbb{R}$ such that $\Sigma(\lambda^*) = 0$, then, by (9.20), $\Sigma'(\lambda^*) > 0$ and hence $\Sigma(\lambda) > 0$ for all $\lambda > \lambda^*$. Therefore, $\Sigma(\infty) > 0$. This shows that λ^* exists if and only if (9.19) holds.

The uniqueness of λ^* follows as in the proof of Theorem 9.2 and, so, we omit the details here. The last assertion follows from Remark 9.1, by (9.20). The proof is complete.   □

Figure 9.2 shows a genuine graph of the map $\lambda \mapsto \Sigma(\lambda)$ in the special case when $W \leq 0$ and $\Sigma_\omega := \Sigma(\infty) > 0$. Under the assumptions of Theorem 9.3, $\lambda^* > 0$ if

$$\Sigma(0) = \sigma[\mathfrak{L}, \mathfrak{B}, \Omega] < 0,$$

$\lambda^* = 0$ if $\Sigma(0) = 0$, and $\lambda^* < 0$ if $\Sigma(0) > 0$. Figure 9.2 shows an admissible case when $\lambda^* < 0$.

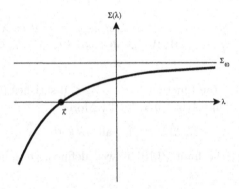

Fig. 9.2   A genuine graph of $\lambda \mapsto \Sigma(\lambda)$ when $W \leq 0$

Finally, the next result holds when the weight function W changes of sign in Ω.

Theorem 9.4. *Suppose W changes sign in Ω, in the sense that there are x_+, $x_- \in \Omega$ and $R > 0$ for which (9.12) and (9.18) hold. Then,*

$$\lim_{\lambda \downarrow -\infty} \Sigma(\lambda) = \lim_{\lambda \uparrow \infty} \Sigma(\lambda) = -\infty, \tag{9.21}$$

there exists a unique $\lambda_0 \in \mathbb{R}$ such that

$$\Sigma(\lambda_0) = \max_{\lambda \in \mathbb{R}} \Sigma(\lambda), \tag{9.22}$$

and $\Sigma'(\lambda_0) = 0$, $\Sigma'(\lambda) > 0$ if $\lambda < \lambda_0$, and $\Sigma'(\lambda) < 0$ for all $\lambda > \lambda_0$. Therefore, (9.1) possesses a principal eigenvalue if, and only if, $\Sigma(\lambda_0) \geq 0$.

Moreover, λ_0 provides us with the unique principal eigenvalue of (9.1) *if $\Sigma(\lambda_0) = 0$, while* (9.1) *possesses two principal eigenvalues, say $\lambda_-^* < \lambda_+^*$ if $\Sigma(\lambda_0) > 0$. Furthermore, in such case,*

$$\lambda_-^* < \lambda_0 < \lambda_+^*$$

and λ_-^*, λ_+^* are simple eigenvalues of $(\mathfrak{L} - \lambda W, W)$ as discussed by M. G. Crandall and P. H. Rabinowitz [45].

Proof.   This result is an immediate consequence from Theorem 9.1(d) and Remark 9.1. It should be noted that $\Sigma'(\lambda_0) = 0$ and, therefore, the principal eigenvalue cannot be a simple eigenvalue of $(\mathfrak{L} - \lambda W, W)$ if $\Sigma(\lambda_0) = 0$. $\quad\square$

Figure 9.3 shows a genuine graph of the map $\lambda \mapsto \Sigma(\lambda)$ in the general case when W changes sign in Ω and $\Sigma(\lambda_0) > 0$.

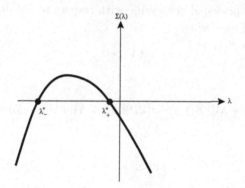

Fig. 9.3   A genuine graph of $\lambda \mapsto \Sigma(\lambda)$ when W changes sign in Ω

Under the assumptions of Theorem 9.4, suppose $\Sigma(\lambda_0) > 0$. Then, $\lambda_-^* < 0 < \lambda_+^*$ if

$$\Sigma(0) = \sigma[\mathfrak{L}, \mathfrak{B}, \Omega] > 0, \tag{9.23}$$

$0 = \lambda_-^* < \lambda_+^*$ if

$$\Sigma(0) = \sigma[\mathfrak{L}, \mathfrak{B}, \Omega] = 0 \quad \text{and} \quad \Sigma'(0) > 0,$$

$\lambda_-^* < \lambda_+^* = 0$ if

$$\Sigma(0) = \sigma[\mathfrak{L}, \mathfrak{B}, \Omega] = 0 \quad \text{and} \quad \Sigma'(0) < 0,$$

$0 < \lambda_-^* < \lambda_+^*$ if

$$\Sigma(0) = \sigma[\mathfrak{L}, \mathfrak{B}, \Omega] < 0 \quad \text{and} \quad \Sigma'(0) > 0,$$

and, finally, $\lambda_-^* < \lambda_+^* < 0$ if

$$\Sigma(0) = \sigma[\mathfrak{L}, \mathfrak{B}, \Omega] < 0 \quad \text{and} \quad \Sigma'(0) < 0. \tag{9.24}$$

The very special case when condition (9.23) holds is the classical situation covered by the theorem of P. Hess and T. Kato [97] (see P. Hess [96]). In Figure 9.3, (9.24) holds and hence $\lambda_-^* < \lambda_+^* < 0$. By changing the sign of the weight function W, i.e., working with $-W$, instead of W, the corresponding $\Sigma(\lambda)$ has two positive zeros. Namely, $-\lambda_-^*$ and $-\lambda_+^*$.

9.3   Ascertaining $\lim_{\lambda \to \infty} \sigma[\mathfrak{L} + \lambda V, \mathfrak{B}, \Omega]$ when $V \geq 0$

Throughout this section we will assume (8.26) and $V \geq 0$ with $V \neq 0$. Imposing (8.26) is imperative for applying the results of Section 8.5 on continuity of the principal eigenvalue with respect to the domain.

In the special case when

$$\inf_{\Omega} V > 0,$$

we have that

$$\Sigma(\lambda) = \sigma[\mathfrak{L} + \lambda V, \mathfrak{B}, \Omega] > \sigma[\mathfrak{L}, \mathfrak{B}, \Omega] + \lambda \inf_{\Omega} V \quad \text{for all} \quad \lambda > 0$$

and, therefore,

$$\lim_{\lambda \uparrow \infty} \Sigma(\lambda) = \infty. \tag{9.25}$$

Consequently, throughout this section, we will assume that

$$\inf_{\Omega} V = 0. \tag{9.26}$$

This section consists of three parts. In the first one, we will ascertain

$$\lim_{\lambda \uparrow \infty} \sigma[\mathfrak{L} + \lambda V, \mathfrak{D}, \Omega]$$

for a special class of potentials V for which characterizing this limit does not require so many technicalities as the proof of the most general case covered by the main theorem. Then, we will introduce the most general class of potentials V for which we will characterize this limit. Finally, we will establish and prove the main result, through a rather natural but extremely technical and lengthy proof.

Naturally, all results of this section can be easily adapted to cover the case when $V \leq 0$, but the corresponding details are not given here.

9.3.1 The simplest case

The main result of this section reads as follows.

Theorem 9.5. *Suppose* (8.26) *and the open set*

$$\Omega_0 := \operatorname{int} V^{-1}(0) \tag{9.27}$$

is connected and of class $\mathcal{C}^2$, and it satisfies $\bar{\Omega}_0 \subset \Omega$.

For sufficiently small $\delta > 0$, let Ω_δ be the open δ-neighborhood of Ω_0

$$\Omega_\delta := \{\, x \in \Omega \,:\, \operatorname{dist}(x, \Omega_0) < \delta \,\}, \tag{9.28}$$

and suppose, in addition, that

$$\inf_{\bar{\Omega} \backslash \Omega_\delta} V > 0 \quad \text{for sufficiently small } \ \delta > 0. \tag{9.29}$$

Then,

$$\lim_{\lambda \uparrow \infty} \sigma[\mathfrak{L} + \lambda V, \mathfrak{D}, \Omega] = \sigma[\mathfrak{L}, \mathfrak{D}, \Omega_0]. \tag{9.30}$$

As Ω_0 is of class $\mathcal{C}^2$, $\partial \Omega_0$ necessarily has finitely many components. Note that a sufficient condition for (9.29) is

$$V \in \mathcal{C}(\bar{\Omega}) \quad \text{and} \quad V(x) > 0 \quad \forall\, x \in \bar{\Omega} \backslash \bar{\Omega}_0,$$

because $\bar{\Omega} \backslash \Omega_\delta$ is a compact subset of $\bar{\Omega} \backslash \bar{\Omega}_0$, for $\delta \sim 0$, and V is continuous and positive therein. Figure 9.4 shows an admissible situation where $\partial \Omega$ has one component and $\partial \Omega_0$ two. The darker area is the region where $V > 0$. Its complement is Ω_0, the region where $V = 0$.

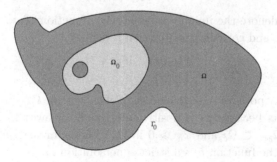

Fig. 9.4   A nodal configuration within the setting of Theorem 9.5

Proof.   According to Proposition 8.2,

$$\sigma[\mathfrak{L} + \lambda V, \mathfrak{D}, \Omega] < \sigma[\mathfrak{L} + \lambda V, \mathfrak{D}, \Omega_0] = \sigma[\mathfrak{L}, \mathfrak{D}, \Omega_0]$$

for all $\lambda \in \mathbb{R}$ and hence,

$$\Sigma(\infty) := \lim_{\lambda \uparrow \infty} \sigma[\mathfrak{L} + \lambda V, \mathfrak{D}, \Omega] \leq \sigma[\mathfrak{L}, \mathfrak{D}, \Omega_0].$$

Thus, to complete the proof of (9.30) we should prove that for every $\epsilon > 0$ there exists $\lambda_1 = \lambda_1(\epsilon)$ such that

$$\sigma[\mathfrak{L} + \lambda V, \mathfrak{D}, \Omega] > \sigma[\mathfrak{L}, \mathfrak{D}, \Omega_0] - \epsilon \qquad \forall\, \lambda \geq \lambda_1. \tag{9.31}$$

Subsequently, we fix $\epsilon > 0$ and consider the δ-neighborhoods of Ω_0 defined by (9.28). According to Definition 8.1(E), Ω_δ converges to Ω_0 from its exterior as $\delta \downarrow 0$. Thus, thanks to Theorems 8.4 and 8.5, we find that

$$\lim_{\delta \downarrow 0} \sigma[\mathfrak{L}, \mathfrak{D}, \Omega_\delta] = \sigma[\mathfrak{L}, \mathfrak{D}, \Omega_0].$$

Therefore, for sufficiently small $\delta > 0$,

$$\sigma[\mathfrak{L}, \mathfrak{D}, \Omega_0] - \epsilon < \sigma[\mathfrak{L}, \mathfrak{D}, \Omega_\delta] < \sigma[\mathfrak{L}, \mathfrak{D}, \Omega_0].$$

Fix one of these δ's. Then, the estimate

$$\sigma[\mathfrak{L} + \lambda V, \mathfrak{D}, \Omega] > \sigma[\mathfrak{L}, \mathfrak{D}, \Omega_\delta] \qquad \forall\, \lambda \geq \lambda_1, \tag{9.32}$$

implies (9.31) and hence, to complete the proof, it suffices to show (9.32), or, equivalently,

$$\sigma\left[\mathfrak{L} + \lambda V - \sigma[\mathfrak{L}, \mathfrak{D}, \Omega_\delta], \mathfrak{D}, \Omega\right] > 0 \qquad \forall\, \lambda \geq \lambda_1. \tag{9.33}$$

This will be a consequence from Theorem 7.10 through the construction of a positive strict supersolution for

$$\left(\mathfrak{L} + \lambda V - \sigma[\mathfrak{L}, \mathfrak{D}, \Omega_\delta], \mathfrak{D}, \Omega\right).$$

Let $\varphi_\delta \gg 0$ denote the unique principal eigenfunction of $\sigma[\mathfrak{L}, \mathfrak{D}, \Omega_\delta]$ with $\max_\Omega \varphi_\delta = 1$, and consider the function h defined by

$$h(x) := \begin{cases} \varphi_\delta(x) & \text{if } x \in \Omega_{\delta/2}, \\ \psi(x) & \text{if } x \in \bar{\Omega} \setminus \Omega_{\delta/2}, \end{cases} \tag{9.34}$$

where ψ is any positive smooth extension of φ_δ outside $\Omega_{\delta/2}$ separated from zero. It exists because φ_δ is positive and bounded away from zero along $\partial\Omega_{\delta/2}$, as $\partial\Omega_{\delta/2} \subset \Omega_\delta$ and $\varphi_\delta \gg 0$ in Ω_δ. We claim that, for sufficiently large $\lambda > 0$, the function h is a strict supersolution of

$$\mathfrak{L}_\lambda := \mathfrak{L} + \lambda V - \sigma[\mathfrak{L}, \mathfrak{D}, \Omega_\delta]$$

in Ω under homogeneous Dirichlet boundary conditions. Indeed, by construction, $h(x) > 0$ for all $x \in \bar{\Omega}$, and, in $\Omega_{\delta/2}$, we have that

$$\mathfrak{L}_\lambda h = \mathfrak{L}_\lambda \varphi_\delta = (\mathfrak{L} - \sigma[\mathfrak{L}, \mathfrak{D}, \Omega_\delta])\varphi_\delta + \lambda V \varphi_\delta = \lambda V \varphi_\delta \geq 0,$$

while, in the complement of $\Omega_{\delta/2}$, for sufficiently large $\lambda > 0$,

$$\mathfrak{L}_\lambda h = \mathfrak{L}_\lambda \psi = (\mathfrak{L} - \sigma[\mathfrak{L}, \mathfrak{D}, \Omega_\delta])\psi + \lambda V \psi$$
$$\geq (\mathfrak{L} - \sigma[\mathfrak{L}, \mathfrak{D}, \Omega_\delta])\psi + \lambda \inf_{\bar{\Omega}\backslash\Omega_{\delta/2}} V \inf_{\bar{\Omega}\backslash\Omega_{\delta/2}} \psi > 0$$

because, thanks to (9.29) and the construction of ψ,

$$\inf_{\bar{\Omega}\backslash\Omega_{\delta/2}} V > 0 \quad \text{and} \quad \inf_{\bar{\Omega}\backslash\Omega_{\delta/2}} \psi > 0.$$

This shows the claim above. Theorem 7.10 ends the proof. $\qquad\square$

As a consequence from Theorem 9.3, Proposition 8.6 and (9.30), the next result holds.

Corollary 9.1. *Suppose V satisfies the requirements of Theorem 9.5. Then, (9.1) admits a principal eigenvalue if, and only if,*

$$\sigma[\mathfrak{L}, \mathfrak{D}, \Omega_0] > 0.$$

Moreover, this holds for sufficiently small $|\Omega_0|$.

9.3.2 The admissible V's satisfying the main theorem

In this section we introduce a class of non-negative potentials $V \in L^\infty(\Omega)$, denoted by $\mathcal{A}$, for which

$$\lim_{\lambda\uparrow\infty} \sigma[\mathfrak{L} + \lambda V, \mathfrak{B}, \Omega] = \sigma[\mathfrak{L}, \mathfrak{B}[\Omega_0], \Omega_0], \qquad (9.35)$$

where Ω_0 is the maximal open subset of Ω where $V = 0$ and $\mathfrak{B}[\Omega_0]$ stands for the boundary operator introduced in (8.2). The class $\mathcal{A}$ is introduced through the next definition.

Definition 9.1. *Suppose $V \in L^\infty(\Omega)$ and $V \geq 0$. It is said that V is an admissible potential for (9.35), or, shortly, that $V \in \mathcal{A}$, if there exist an open subset Ω_0 of Ω and a compact subset K of $\bar{\Omega}$ with Lebesgue measure zero such that*

$$K \cap (\bar{\Omega}_0 \cup \Gamma_1) = \emptyset, \qquad (9.36)$$

$$\Omega_+ := \{\, x \in \Omega \ : \ V(x) > 0 \,\} = \Omega \setminus (\bar{\Omega}_0 \cup K), \qquad (9.37)$$

for which the following conditions are satisfied:

(a) Ω_0 *possesses finitely many components of class* $\mathcal{C}^2$, *say* Ω_0^j, $1 \le j \le m$, *such that*

$$\bar{\Omega}_0^i \cap \bar{\Omega}_0^j = \emptyset \quad if \quad i \ne j,$$

and

$$\text{dist}\,(\Gamma_1, \partial\Omega_0 \cap \Omega) > 0. \tag{9.38}$$

Thus, if we denote by Γ_1^i, $1 \le i \le n_1$, *the components of* Γ_1, *then, for each* $1 \le i \le n_1$, *either* Γ_1^i *is a component of* $\partial\Omega_0$, *or* $\Gamma_1^i \cap \partial\Omega_0 = \emptyset$. *Indeed, if* $\Gamma_1^i \cap \partial\Omega_0 \ne \emptyset$ *but* Γ_1^i *is not a component of* $\partial\Omega_0$, *then*

$$\text{dist}(\Gamma_1^i, \partial\Omega_0 \cap \Omega) = 0,$$

which contradicts (9.38).

(b) *Let* $\{i_1, ..., i_p\}$ *denote the subset of* $\{1, ..., n_1\}$ *for which*

$$\Gamma_1^i \cap \partial\Omega_0 = \emptyset \quad \Longleftrightarrow \quad i \in \{i_1, ..., i_p\}.$$

Then, V *is bounded away from zero in any compact subset of*

$$\Omega_+ \cup \bigcup_{j=1}^{p} \Gamma_1^{i_j}.$$

When $\Gamma_1 \subset \partial\Omega_0$, *i.e.,* $\{i_1, ..., i_p\} = \emptyset$, *then we are only imposing that* V *is bounded away from zero in any compact subset of* Ω_+.

Now, we will explain the meaning of this condition in the special, but important, case when $V \in \mathcal{C}(\bar{\Omega})$. In such case, by continuity, V is bounded away from zero in any compact subset of Ω_+ and, therefore, condition (b) holds if either $\Gamma_1 \subset \partial\Omega_0$, or $\{i_1, ..., i_p\} \ne \emptyset$ and

$$V(x) > 0 \quad \text{for all} \quad x \in \Gamma_1^+ := \bigcup_{j=1}^{p} \Gamma_1^{i_j}, \tag{9.39}$$

because Γ_1^+ is compact and V continuous. According to condition (a), Γ_1^i must be a component of $\partial\Omega_0$ for all

$$i \in \{1, ..., n_1\} \setminus \{i_1, ..., i_p\}$$

and hence $V = 0$ on all these components.

(c) *Let* Γ_0^i, $1 \le i \le n_0$, *denote the components of* Γ_0, *and let* $\{i_1, ..., i_q\}$ *be the subset of* $\{1, ..., n_0\}$ *for which*

$$(\partial\Omega_0 \cup K) \cap \Gamma_0^i \ne \emptyset \quad \Longleftrightarrow \quad i \in \{i_1, ..., i_q\}.$$

Then, V is bounded away from zero in any compact subset of

$$\Omega_+ \cup \bigcup_{j=1}^{q} \Gamma_0^{i_j} \setminus (\partial\Omega_0 \cup K).$$

When $(\partial\Omega_0 \cup K) \cap \Gamma_0 = \emptyset$, we are only imposing that V is bounded away from zero in any compact subset of Ω_+.

Now, we will explain this condition in case $V \in \mathcal{C}(\bar{\Omega})$. In such situation, by continuity, V is bounded away from zero in Ω_+, and, therefore, condition (c) holds if $(\partial\Omega_0 \cup K) \cap \Gamma_0 = \emptyset$. So, suppose

$$(\partial\Omega_0 \cup K) \cap \Gamma_0 \neq \emptyset.$$

Then, according to condition (a), $V = 0$ in $\partial\Omega_0 \cup K$ and hence,

$$V = 0 \quad \text{in} \quad \bigcup_{j=1}^{q} \Gamma_0^{i_j} \cap (\partial\Omega_0 \cup K).$$

Consequently, the condition (c) holds if and only if

$$V(x) > 0 \qquad \text{for all} \quad x \in \bigcup_{j=1}^{q} \Gamma_0^{i_j} \setminus (\partial\Omega_0 \cup K). \tag{9.40}$$

(d) *For every $\eta > 0$ there exist a natural number $\ell(\eta) \geq 1$ and $\ell(\eta)$ open subsets of $\mathbb{R}^N$, G_j^η, $1 \leq j \leq \ell(\eta)$, with $|G_j^\eta| < \eta$, $1 \leq j \leq \ell(\eta)$, such that*

$$\bar{G}_i^\eta \cap \bar{G}_j^\eta = \emptyset \quad \text{if} \quad i \neq j, \qquad K \subset \bigcup_{j=1}^{\ell(\eta)} G_j^\eta,$$

and $G_j^\eta \cap \Omega$ is connected and of class $\mathcal{C}^2$ for all $1 \leq j \leq \ell(\eta)$.

When Ω_0 satisfies all the requirements of Definition 9.1(a), one can introduce the following concept.

Definition 9.2. *Let Ω_0 be an open subset of Ω satisfying the requirements of Definition 9.1(a). Then, the principal eigenvalue of $(\mathfrak{L}, \mathfrak{B}[\Omega_0], \Omega_0)$ is defined through*

$$\sigma[\mathfrak{L}, \mathfrak{B}[\Omega_0], \Omega_0] := \min_{1 \leq j \leq m} \sigma[\mathfrak{L}, \mathfrak{B}[\Omega_0^j], \Omega_0^j].$$

Remark 9.2. As (9.38) is (8.1) and Ω_0 is of class $\mathcal{C}^2$, the principal eigenvalues

$$\sigma[\mathfrak{L}, \mathfrak{B}[\Omega_0], \Omega_0] \quad \text{and} \quad \sigma[\mathfrak{L}, \mathfrak{B}[\Omega_0^j], \Omega_0^j], \qquad 1 \leq j \leq m,$$

are well defined.

9.3.3 The main theorem

Now, we are ready to state and prove the main result of this section.

Theorem 9.6. *Suppose* (8.26) *and* $V \in \mathcal{A}$ *satisfies* $\operatorname{int} \Omega_+ \neq \emptyset$. *Then,* (9.35) *holds.*

Proof. Thanks to Proposition 8.2, since $V = 0$ in Ω_0, we have that

$$\sigma[\mathfrak{L} + \lambda V, \mathfrak{B}, \Omega] < \sigma[\mathfrak{L}, \mathfrak{B}[\Omega_0^j], \Omega_0^j]$$

for all $1 \leq j \leq m$ and $\lambda \in \mathbb{R}$. Thus, according to Definition 9.2,

$$\sigma[\mathfrak{L} + \lambda V, \mathfrak{B}, \Omega] < \sigma[\mathfrak{L}, \mathfrak{B}[\Omega_0], \Omega_0] \qquad \forall \, \lambda \in \mathbb{R}. \tag{9.41}$$

Note that V satisfies (9.4) if $V \in \mathcal{A}$ with $\operatorname{int} \Omega_+ \neq \emptyset$. Hence, by Theorem 9.1(a), the map

$$\lambda \mapsto \Sigma(\lambda) = \sigma[\mathfrak{L} + \lambda V, \mathfrak{B}, \Omega]$$

is increasing and, consequently,

$$\Sigma(\infty) = \lim_{\lambda \uparrow \infty} \Sigma(\lambda) \leq \sigma[\mathfrak{L}, \mathfrak{B}[\Omega_0], \Omega_0].$$

Therefore, to complete the proof of (9.35) it remains to show that for every $\epsilon > 0$ there exists $\lambda_1 = \lambda_1(\epsilon) \in \mathbb{R}$ such that

$$\Sigma(\lambda) > \sigma[\mathfrak{L}, \mathfrak{B}[\Omega_0], \Omega_0] - \epsilon \qquad \forall \, \lambda \geq \lambda_1,$$

or, equivalently,

$$\sigma[\mathfrak{L} + \lambda V - \sigma[\mathfrak{L}, \mathfrak{B}[\Omega_0], \Omega_0] + \epsilon, \mathfrak{B}, \Omega] > 0 \qquad \forall \, \lambda \geq \lambda_1. \tag{9.42}$$

According to Theorem 7.10, (9.42) holds if and only if

$$(\mathfrak{L} + \lambda V - \sigma[\mathfrak{L}, \mathfrak{B}[\Omega_0], \Omega_0] + \epsilon, \mathfrak{B}, \Omega) \tag{9.43}$$

possesses a positive strict supersolution for all $\lambda > \lambda_1$. Therefore, much like in the proof of Theorem 9.5, the rest of the proof is devoted to the construction of a positive strict supersolution of (9.43) for sufficiently large λ. In the construction of the supersolution we will distinguish several different cases according to the structure of Ω_0. First, we will consider the simplest cases when Ω_0 is connected and $K = \emptyset$. Then, we shall consider the most general cases.

Step 1: Suppose

$$m = 1, \qquad K = \emptyset, \qquad \text{and} \quad \Gamma_0 \cap \partial\Omega_0 = \emptyset. \tag{9.44}$$

Then, Definition 9.1(c) only requires V to be bounded away from zero in any compact subset of Ω_+, which had already been imposed by Definition 9.1(b).

For each $k \in \{0,1\}$, let Γ_k^j, $1 \le j \le n_k$, denote the components of Γ_k, and let $\{i_1, ..., i_p\}$ be the subset of $\{1, ..., n_1\}$ for which

$$\Gamma_1^j \cap \partial\Omega_0 = \emptyset \quad \Longleftrightarrow \quad j \in \{i_1, ..., i_p\}. \tag{9.45}$$

According to Definition 9.1(a), Γ_1^i is a component of $\partial\Omega_0$ for all $i \in \{1, ..., n_1\} \setminus \{i_1, ..., i_p\}$.

Figure 9.5 shows an admissible example satisfying (9.44). In this case, Γ_0 consists of two components, Γ_0^1 and Γ_0^2, as well as Γ_1, whose components have been named by Γ_1^1 and Γ_1^2. The boundary of Ω_0, $\partial\Omega_0$, has two components too. One of them is Γ_1^2. The other one lies within Ω.

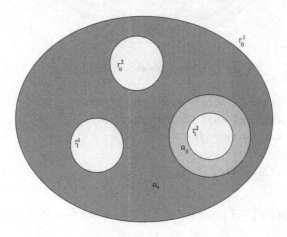

Fig. 9.5   A nodal configuration satisfying (9.44)

In Figure 9.5, Ω_+ is the darker subdomain of Ω, while Ω_0 is its complement, so that $\Omega_+ = \Omega \setminus \bar{\Omega}_0$. Any potential $V \in \mathcal{C}(\bar{\Omega})$ such that $V = 0$ in Ω_0 and $V(x) > 0$ for all $x \in \Omega_+$ satisfies Definition 9.1(a), (c) and (d), and it satisfies Definition 9.1(b) if, and only if, $V(x) > 0$ for all $x \in \Gamma_1^1$.

Subsequently, we fix $\epsilon > 0$ and, for sufficiently small $\delta > 0$, consider the open δ-neighborhoods

$$\Omega_\delta := (\Omega_0 + B_\delta) \cap \Omega,$$
$$\mathcal{N}_\delta^{0,j} := (\Gamma_0^j + B_\delta) \cap \Omega, \qquad 1 \le j \le n_0, \tag{9.46}$$
$$\mathcal{N}_\delta^{1,j} := (\Gamma_1^j + B_\delta) \cap \Omega, \qquad j \in \{i_1, ..., i_p\}.$$

Recall that $B_\delta \subset \mathbb{R}^N$ is the ball of radius δ centered at the origin. Figure 9.6 shows these neighborhoods, for sufficiently small $\delta > 0$, for the example of Figure 9.5.

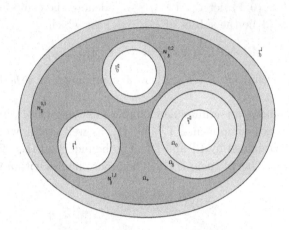

Fig. 9.6   The δ-neighborhoods defined in (9.46)

As Ω_0 is of class $\mathcal{C}^2$, $\partial\Omega_0$ possesses finitely many components and, therefore, Ω_0 possesses, at most, finitely many holes. By (9.44), there exists $\delta_0 > 0$ such that, for every $0 < \delta < \delta_0$,

$$\partial\Omega_\delta \setminus (\Gamma_1 \cap \partial\Omega_0) \subset \Omega_+, \qquad \bar{\Omega}_\delta \cap \bigcup_{j=1}^{n_0} \mathcal{N}_\delta^{0,j} = \emptyset, \qquad \bigcup_{j=1}^{n_0} \mathcal{N}_\delta^{0,j} \setminus \Gamma_0 \subset \Omega_+.$$

Moreover, since $\Gamma_k^j \cap \Gamma_\ell^i = \emptyset$ if $i \neq j$, there exists $\delta_1 \in (0, \delta_0)$ such that, for each $0 < \delta < \delta_1$,

$$\bar{\mathcal{N}}_\delta^{k,j} \cap \bar{\mathcal{N}}_\delta^{\ell,i} = \emptyset \qquad \text{if} \quad (i, \ell) \neq (j, k), \quad k, \ell \in \{0, 1\}.$$

Also, according to (9.45), we have that

$$\partial\Omega_0 \cap \bigcup_{j=1}^{p} \Gamma_1^{i_j} = \emptyset$$

and hence, there exists $\delta_2 \in (0, \delta_1)$ such that

$$\bar{\Omega}_\delta \cap \bigcup_{j=1}^{p} \bar{\mathcal{N}}_\delta^{1,i_j} = \emptyset$$

for all $0 < \delta < \delta_2$.

By construction, Ω_0 is a proper subdomain of Ω_δ such that

$$\lim_{\delta\downarrow 0}\Omega_\delta = \Omega_0$$

in the sense of Definition 8.1(E). Thus, owing to (9.38), Proposition 8.2 implies that

$$\sigma[\mathfrak{L},\mathfrak{B}[\Omega_\delta],\Omega_\delta] < \sigma[\mathfrak{L},\mathfrak{B}[\Omega_0],\Omega_0], \quad 0 < \delta < \delta_2,$$

and, thanks to Theorems 8.4 and 8.5,

$$\lim_{\delta\downarrow 0}\sigma[\mathfrak{L},\mathfrak{B}[\Omega_\delta],\Omega_\delta] = \sigma[\mathfrak{L},\mathfrak{B}[\Omega_0],\Omega_0].$$

Therefore, there exists $\delta_3 \in (0,\delta_2)$ such that

$$\sigma[\mathfrak{L},\mathfrak{B}[\Omega_\delta],\Omega_\delta] < \sigma[\mathfrak{L},\mathfrak{B}[\Omega_0],\Omega_0] < \sigma[\mathfrak{L},\mathfrak{B}[\Omega_\delta],\Omega_\delta] + \epsilon \qquad (9.47)$$

for all $0 < \delta < \delta_3$.

On the other hand, since

$$\lim_{\delta\downarrow 0}|\mathcal{N}_\delta^{0,j}| = 0, \qquad 1 \le j \le n_0,$$

it follows from Proposition 8.6 that

$$\lim_{\delta\downarrow 0}\sigma[\mathfrak{L},\mathfrak{D},\mathcal{N}_\delta^{0,j}] = \infty, \qquad 1 \le j \le n_0,$$

and, consequently, there exists $\delta_4 \in (0,\delta_3)$ such that

$$\sigma[\mathfrak{L},\mathfrak{D},\mathcal{N}_\delta^{0,j}] > \sigma[\mathfrak{L},\mathfrak{B}[\Omega_0],\Omega_0], \qquad 1 \le j \le n_0, \qquad (9.48)$$

for all $0 < \delta < \delta_4$.

Subsequently, we fix $\delta \in (0,\delta_4)$. Let φ_δ, ψ_δ^i, $i \in \{i_1,..,i_p\}$, and ξ_δ^j, $j \in \{1,...,n_0\}$, denote three arbitrary principal eigenfunctions associated with

$$\sigma[\mathfrak{L},\mathfrak{B}[\Omega_\delta],\Omega_\delta], \qquad \sigma[\mathfrak{L},\mathfrak{B}[\mathcal{N}_\delta^{1,i}],\mathcal{N}_\delta^{1,i}], \quad i \in \{i_1,...,i_p\},$$

and

$$\sigma[\mathfrak{L},\mathfrak{D},\mathcal{N}_\delta^{0,j}], \quad j \in \{1,...,n_0\},$$

respectively, and consider the function $\Phi : \bar{\Omega} \to [0,\infty)$ defined through

$$\Phi := \begin{cases} \varphi_\delta & \text{in } \bar{\Omega}_{\delta/2}, \\[2mm] \psi_\delta^{i_j} & \text{in } \bar{\mathcal{N}}_{\delta/2}^{1,i_j}, \quad 1 \le j \le p, \\[2mm] \xi_\delta^j & \text{in } \bar{\mathcal{N}}_{\delta/2}^{0,j}, \quad 1 \le j \le n_0, \\[2mm] \zeta_\delta & \text{in } \bar{\Omega} \setminus (\bar{\Omega}_{\delta/2} \cup \bigcup_{j=1}^{p}\bar{\mathcal{N}}_{\delta/2}^{1,i_j} \cup \bigcup_{j=1}^{n_0}\bar{\mathcal{N}}_{\delta/2}^{0,j}), \end{cases} \qquad (9.49)$$

where ζ_δ is any smooth positive extension of φ_δ, $\psi_\delta^{i_j}$, $1 \le j \le p$, and ξ_δ^j, $1 \le j \le n_0$, from

$$\mathcal{K}_\delta := \bar{\Omega}_{\delta/2} \cup \bigcup_{j=1}^{p} \bar{\mathcal{N}}_{\delta/2}^{1,i_j} \cup \bigcup_{j=1}^{n_0} \bar{\mathcal{N}}_{\delta/2}^{0,j}$$

to $\bar{\Omega}$ which is bounded away from zero in $\bar{\Omega} \setminus \mathcal{K}_\delta$. The function ζ_δ does exist because the restrictions

$$\varphi_\delta|_{\partial\Omega_{\delta/2} \cap \Omega}, \qquad \psi_\delta^{i_j}|_{\partial\mathcal{N}_{\delta/2}^{1,i_j} \setminus \Gamma_1^{i_j}}, \quad 1 \le j \le p,$$

and $\xi_\delta^j|_{\partial\mathcal{N}_{\delta/2}^{0,j} \setminus \Gamma_0^j}$, $1 \le j \le n_0$, are positive and bounded away from zero. By construction,

$$\Phi(x) > 0 \qquad \text{for each} \quad x \in \Omega.$$

Naturally, if $\Gamma_1 \subset \partial\Omega_0$, i.e., $\{i_1, ..., i_p\} = \emptyset$, then, in the definition of Φ the $\psi_\delta^{i_j}$'s should be deleted.

Next, we will show that there exists $\lambda_1 = \lambda_1(\epsilon)$ such that Φ provides us with a strict supersolution of (9.43) for all $\lambda \ge \lambda_1$. By Theorem 7.10, this completes the proof under condition (9.44).

By (9.47), the following estimate holds in $\Omega_{\delta/2}$

$$(\mathfrak{L} + \lambda V - \sigma[\mathfrak{L}, \mathfrak{B}[\Omega_0], \Omega_0] + \epsilon)\Phi = (\mathfrak{L} + \lambda V - \sigma[\mathfrak{L}, \mathfrak{B}[\Omega_0], \Omega_0] + \epsilon)\varphi_\delta$$

$$= (\sigma[\mathfrak{L}, \mathfrak{B}[\Omega_\delta], \Omega_\delta] - \sigma[\mathfrak{L}, \mathfrak{B}[\Omega_0], \Omega_0] + \epsilon)\varphi_\delta > 0$$

for all $\lambda \in \mathbb{R}$. Similarly, by (9.48), the next estimate holds in $\mathcal{N}_{\delta/2}^{0,j}$

$$(\mathfrak{L} + \lambda V - \sigma[\mathfrak{L}, \mathfrak{B}[\Omega_0], \Omega_0] + \epsilon)\Phi = (\mathfrak{L} + \lambda V - \sigma[\mathfrak{L}, \mathfrak{B}[\Omega_0], \Omega_0] + \epsilon)\xi_\delta^j$$

$$> (\sigma[\mathfrak{L}, \mathfrak{D}, \mathcal{N}_\delta^{0,j}] - \sigma[\mathfrak{L}, \mathfrak{B}[\Omega_0], \Omega_0] + \epsilon)\xi_\delta^j > 0$$

for all $1 \le j \le n_0$. Now, note that

$$\bar{\Omega} \setminus \left(\Omega_{\delta/2} \cup \bigcup_{j=1}^{n_0} \mathcal{N}_{\delta/2}^{0,j} \right)$$

is a compact subset of

$$\Omega_+ \cup \bigcup_{j=1}^{p} \Gamma_1^{i_j}$$

and hence, according to Definition 9.1(b), there exists a constant $\omega > 0$ such that

$$V \ge \omega > 0 \qquad \text{in} \quad \bar{\Omega} \setminus \left(\bar{\Omega}_{\delta/2} \cup \bigcup_{j=1}^{n_0} \mathcal{N}_{\delta/2}^{0,j} \right). \tag{9.50}$$

Thanks to (9.50), for every $1 \le j \le p$, in $\bar{\mathcal{N}}_{\delta/2}^{1,i_j}$ we have that

$$(\mathfrak{L} + \lambda V - \sigma[\mathfrak{L}, \mathfrak{B}[\Omega_0], \Omega_0] + \epsilon)\Phi = (\mathfrak{L} + \lambda V - \sigma[\mathfrak{L}, \mathfrak{B}[\Omega_0], \Omega_0] + \epsilon)\psi_\delta^{i_j}$$
$$\ge (\sigma[\mathfrak{L}, \mathfrak{B}[\mathcal{N}_\delta^{1,i_j}], \mathcal{N}_\delta^{1,i_j}] - \sigma[\mathfrak{L}, \mathfrak{B}[\Omega_0], \Omega_0] + \epsilon + \lambda\omega)\psi_\delta^{i_j} > 0$$

provided

$$\lambda > \max\left\{ \omega^{-1}\left(\sigma[\mathfrak{L}, \mathfrak{B}[\Omega_0], \Omega_0] - \epsilon - \sigma[\mathfrak{L}, \mathfrak{B}[\mathcal{N}_\delta^{1,i_j}], \mathcal{N}_\delta^{1,i_j}]\right), 0 \right\},$$

whereas in

$$\bar{\Omega} \setminus \left(\bar{\Omega}_{\delta/2} \cup \bigcup_{j=1}^{p} \bar{\mathcal{N}}_{\delta/2}^{1,i_j} \cup \bigcup_{j=1}^{n_0} \bar{\mathcal{N}}_{\delta/2}^{0,j} \right)$$

we have that

$$(\mathfrak{L} + \lambda V - \sigma[\mathfrak{L}, \mathfrak{B}[\Omega_0], \Omega_0] + \epsilon)\Phi = (\mathfrak{L} + \lambda V - \sigma[\mathfrak{L}, \mathfrak{B}[\Omega_0], \Omega_0] + \epsilon)\zeta_\delta$$
$$\ge (\mathfrak{L} - \sigma[\mathfrak{L}, \mathfrak{B}[\Omega_0], \Omega_0] + \epsilon)\zeta_\delta + \lambda\omega\zeta_\delta > 0$$

for sufficiently large $\lambda > 0$, because the function

$$(\mathfrak{L} - \sigma[\mathfrak{L}, \mathfrak{B}[\Omega_0], \Omega_0] + \epsilon)\zeta_\delta$$

does not depend on λ, and ζ_δ is positive and bounded away from zero.

Finally, by construction,

$$\mathfrak{B}\Phi = \mathfrak{D}\xi_\delta^j = 0 \quad \text{on} \quad \Gamma_0^j, \quad 1 \le j \le n_0,$$

$$\mathfrak{B}\Phi = \mathfrak{B}\psi_\delta^{i_j} = 0 \quad \text{on} \quad \Gamma_1^{i_j}, \quad 1 \le j \le p,$$

and

$$\mathfrak{B}\Phi = \mathfrak{B}\varphi_\delta = 0 \quad \text{on} \quad \partial\Omega_0 \cap \Gamma_1.$$

This completes the proof of the theorem under condition (9.44).

Step 2: Suppose

$$m = 1, \quad K = \emptyset, \quad \text{and} \quad \Gamma_0 \cap \partial\Omega_0 \ne \emptyset, \tag{9.51}$$

instead of (9.44). Let Γ_0^i, $1 \le i \le n_0$, denote the components of Γ_0, and $\{i_1, ..., i_q\}$ be the subset of $\{1, ..., n_0\}$ for which

$$\partial\Omega_0 \cap \Gamma_0^j \ne \emptyset \quad \Longleftrightarrow \quad j \in \{i_1, ..., i_q\}.$$

Figure 9.7 illustrates a possible nodal configuration of V satisfying these requirements with $q = 1$ and $i_1 = 2$.

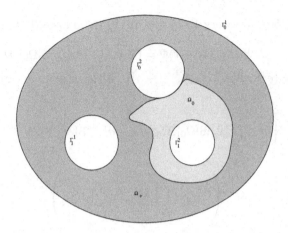

Fig. 9.7   A nodal configuration satisfying (9.51)

Fix $\epsilon > 0$ and, for sufficiently small $\eta > 0$, consider the auxiliary domain

$$G_\eta := \Omega \cup \left(\bigcup_{j=1}^{q} \Gamma_0^{ij} + B_\eta \right).$$

In the special case illustrated by Figure 9.7, it becomes apparent that

$$G_\eta = \Omega \cup (\Gamma_0^2 + B_\eta);$$

equivalently, G_η consists of Ω plus the set of points $x \in \mathbb{R}^N$ such that $\operatorname{dist}(x, \Gamma_0^2) < \eta$.

By (8.26), we have that

$$a_{ij} \in \mathcal{C}^1(\bar{\Omega}), \quad b_i \in \mathcal{C}(\bar{\Omega}), \qquad 1 \le i, j \le N.$$

Fix $\eta > 0$ and let

$$\tilde{a}_{ij} = \tilde{a}_{ji} \in \mathcal{C}^1(\bar{G}_\eta), \quad \tilde{b}_i \in \mathcal{C}(\bar{G}_\eta), \quad \tilde{c} \in L^\infty(G_\eta), \qquad 1 \le i, j \le N,$$

be regular extensions from $\bar{\Omega}$ to $\bar{G}_\eta$ of the coefficients $a_{ij} = a_{ji}$, b_i, and c, $1 \le i, j \le N$, respectively.

Now, consider the differential operator

$$\tilde{\mathfrak{L}} := -\operatorname{div}(\tilde{A}\nabla \cdot) + \langle \tilde{b}, \nabla \cdot \rangle + \tilde{c}$$

in G_η, where we have denoted

$$\tilde{A} := (\tilde{a}_{ij})_{1 \le i, j \le N}, \qquad \tilde{b} = (\tilde{b}_1, ..., \tilde{b}_N).$$

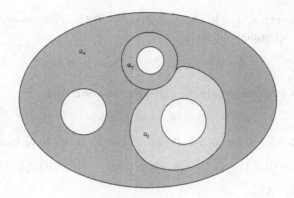

Fig. 9.8   Enlarging Ω through Γ_0^2

As $\mathfrak{L}$ is uniformly elliptic in Ω with constant $\mu > 0$, there exists $\tilde{\eta} \in (0, \eta)$ such that $\tilde{\mathfrak{L}}$ is uniformly elliptic in $G_{\tilde{\eta}}$ with constant $\mu/2$. Subsequently, we set

$$\tilde{\Omega} := G_{\tilde{\eta}}$$

and consider the extended potential

$$\tilde{V} := \begin{cases} 1 & \text{in} \quad \tilde{\Omega} \setminus \Omega, \\ V & \text{in} \quad \Omega, \end{cases}$$

and the boundary operator

$$\tilde{\mathfrak{B}} := \begin{cases} \mathfrak{D} & \text{on} \quad \partial\tilde{\Omega} \setminus \Gamma_1, \\ \mathfrak{B}, & \text{on} \quad \Gamma_1. \end{cases}$$

By construction, it is easy to check that $\tilde{V} \in \tilde{\mathcal{A}}$, where $\tilde{\mathcal{A}}$ stands for the class of admissible potentials in the open set $\tilde{\Omega}$. Moreover,

$$\tilde{\Omega}_0 = \Omega_0, \qquad \tilde{\Gamma}_1 = \Gamma_1, \qquad (\partial\tilde{\Omega} \setminus \tilde{\Gamma}_1) \cap \partial\tilde{\Omega}_0 = \emptyset.$$

Thus, the extended problem in $\tilde{\Omega}$ satisfies (9.44) and, consequently, by Step 1, there exist $\tilde{\lambda}_1 > 0$ and a function

$$\tilde{\Phi} : \bar{\tilde{\Omega}} \to [0, \infty)$$

such that $\tilde{\Phi}(x) > 0$ for all $x \in \tilde{\Omega}$, and it is a strict supersolution of

$$\left(\tilde{\mathfrak{L}} + \lambda\tilde{V} - \sigma[\tilde{\mathfrak{L}}, \tilde{\mathfrak{B}}[\tilde{\Omega}_0], \tilde{\Omega}_0] + \epsilon, \tilde{\mathfrak{B}}, \tilde{\Omega} \right)$$

for all $\lambda > \tilde{\lambda}_1$. Note that

$$\sigma[\tilde{\mathfrak{L}}, \tilde{\mathfrak{B}}[\tilde{\Omega}_0], \tilde{\Omega}_0] = \sigma[\tilde{\mathfrak{L}}, \tilde{\mathfrak{B}}[\Omega_0], \Omega_0] = \sigma[\mathfrak{L}, \mathfrak{B}[\Omega_0], \Omega_0],$$

because $\tilde{\Omega}_0 = \Omega_0$, $\tilde{\mathfrak{L}} = \mathfrak{L}$ in $\bar{\Omega}_0$, and $\tilde{\mathfrak{B}}[\Omega_0] = \mathfrak{B}[\Omega_0]$. Therefore, $\tilde{\Phi}$ provides us with a strict supersolution of

$$\left(\tilde{\mathfrak{L}} + \lambda \tilde{V} - \sigma[\mathfrak{L}, \mathfrak{B}[\Omega_0], \Omega_0] + \epsilon, \tilde{\mathfrak{B}}, \tilde{\Omega} \right)$$

for all $\lambda > \tilde{\lambda}_1$. Subsequently, we consider the restriction

$$\Phi := \tilde{\Phi}|_{\bar{\Omega}}.$$

Obviously, $\Phi(x) > 0$ for all $x \in \Omega$. Moreover, in Ω, we have that

$$(\mathfrak{L} + \lambda V - \sigma[\mathfrak{L}, \mathfrak{B}[\Omega_0], \Omega_0] + \epsilon)\Phi = (\tilde{\mathfrak{L}} + \lambda \tilde{V} - \sigma[\mathfrak{L}, \mathfrak{B}[\Omega_0], \Omega_0] + \epsilon)\tilde{\Phi} \geq 0$$

for all $\lambda > \tilde{\lambda}_1$. Also,

$$\Phi(x) = \tilde{\Phi}(x) > 0 \qquad \text{for all } x \in \bigcup_{j=1}^{q} \Gamma_0^{i_j},$$

because, by construction,

$$\bigcup_{j=1}^{q} \Gamma_0^{i_j} \subset \tilde{\Omega} \quad \text{and} \quad \tilde{\Phi}(x) > 0 \quad \text{for all } x \in \tilde{\Omega}.$$

In addition,

$$\Phi = \tilde{\Phi} = 0 \qquad \text{on } \Gamma_0 \setminus \bigcup_{j=1}^{q} \Gamma_0^{i_j}$$

and

$$\mathfrak{B}\Phi = (\partial_\nu + \beta)\tilde{\Phi} \geq 0 \qquad \text{on } \Gamma_1,$$

by the construction of $\tilde{\Phi}$. It should be noted that, in a neighborhood of Γ_1, $\tilde{\Phi} = \Phi$ and, hence, not only $\beta\Phi = \beta\tilde{\Phi}$, but also $\partial_\nu \Phi = \partial_\nu \tilde{\Phi}$.

Consequently, $\mathfrak{B}\Phi > 0$ on $\partial\Omega$, and, therefore, Φ provides us with a positive strict supersolution of (9.43) for all $\lambda > \tilde{\lambda}_1$. This completes the proof of Step 2.

Step 3: Now, suppose

$$m \geq 1, \qquad K \neq \emptyset, \quad \text{and} \quad \Gamma_0 \cap (\partial\Omega_0 \cup K) = \emptyset. \qquad (9.52)$$

Figure 9.9 shows an admissible situation where (9.52) holds. In this example, Γ_0 consists of two components, Γ_0^1 and Γ_0^2, as well as Γ_1, whose components have been named Γ_1^1 and Γ_1^2, and Ω_0, whose components are Ω_0^1 and Ω_0^2. So, in this example, $m = 2$. In Figure 9.9, $V = 0$ in

$$\Omega_0 := \Omega_0^1 \cup \Omega_0^2,$$

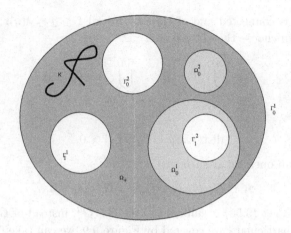

Fig. 9.9 $V(x) > 0$ for all $x \in \Omega_+ \cup \Gamma_1^1$ and $V = 0$ in $\Omega_0^1 \cup \Omega_0^2 \cup K$

and in K, which is a negligible set. As due to (9.36),

$$K \cap (\bar{\Omega}_0 \cup \Gamma_1) = \emptyset, \tag{9.53}$$

all structural requirements of Definition 9.1 are fulfilled, as well as (9.52), as soon as, for instance, $V \in \mathcal{C}(\bar{\Omega})$ satisfies $V(x) > 0$ for all $x \in \Omega_+ \cup \Gamma_1^1$ and $V(x) = 0$ for all $x \in \Omega_0^1 \cup \Omega_0^2 \cup K$.

By (9.52) and (9.53), we have that $K \cap \Gamma_1 = \emptyset$ and $K \cap \Gamma_0 = \emptyset$, respectively. Thus, $K \cap \partial\Omega = \emptyset$ and hence, $K \subset \Omega$. Moreover, by (9.52), $\bar{\Omega}_0 \subset \Omega \cup \Gamma_1$, and, due to (9.53), $K \cap \bar{\Omega}_0 = \emptyset$. Summarizing,

$$K \subset \Omega, \qquad \bar{\Omega}_0 \subset \Omega \cup \Gamma_1, \qquad K \cap \bar{\Omega}_0 = \emptyset, \tag{9.54}$$

and, therefore,

$$\text{dist}(\Gamma_0, \bar{\Omega}_0 \cup K) > 0, \quad \text{dist}(\Gamma_1, K) > 0, \quad \text{dist}(K, \bar{\Omega}_0) > 0, \tag{9.55}$$

because each of the pairs consists of two disjoint compact subsets.

Subsequently, we denote by Ω_0^i, $1 \le i \le m$, the components of Ω_0. Without loss of generality, they can be relabeled so that

$$\sigma[\mathcal{L}, \mathcal{B}, \Omega_0^i] \le \sigma[\mathcal{L}, \mathcal{B}, \Omega_0^{i+1}], \qquad 1 \le i \le m - 1,$$

and, consequently, according to Definition 9.2, we have that

$$\sigma[\mathcal{L}, \mathcal{B}[\Omega_0], \Omega_0] = \sigma[\mathcal{L}, \mathcal{B}[\Omega_0^1], \Omega_0^1].$$

Fix $\eta > 0$. Thanks to Definition 9.1(d), there exist a natural number $\ell(\eta) \ge 1$ and $\ell(\eta)$ open sets $G_j^\eta \subset \mathbb{R}^N$, $1 \le j \le \ell(\eta)$, with $|G_j^\eta| < \eta$, $1 \le j \le \ell(\eta)$, such that

$$K \subset \bigcup_{j=1}^{\ell(\eta)} (G_j^\eta \cap \Omega), \qquad \bar{G}_i^\eta \cap \bar{G}_j^\eta = \emptyset \quad \text{if } i \ne j,$$

and $G_j^\eta \cap \Omega$ is connected and of class $\mathcal{C}^2$ for all $1 \leq j \leq \ell(\eta)$. Thanks to (9.54), we can choose the G_j^η's so that

$$K \subset \bigcup_{j=1}^{\ell(\eta)} \bar{G}_j^\eta \subset \Omega, \qquad \bigcup_{j=1}^{\ell(\eta)} \bar{G}_j^\eta \cap \bar{\Omega}_0 = \emptyset. \qquad (9.56)$$

Indeed, since

$$\mathrm{dist}(K, \bar{\Omega}_0 \cup \Gamma_0 \cup \Gamma_1) > 0,$$

there exists an open set G such that

$$K \subset G, \qquad \bar{G} \subset \Omega, \qquad \bar{G} \cap \bar{\Omega}_0 = \emptyset,$$

and hence, to get (9.56) it suffices to take $G \cap G_j^\eta$, instead of G_j^η, $1 \leq j \leq \ell(\eta)$. In the particular case covered by Figure 9.9, we can take $\ell(\eta) = 1$ for all $\eta > 0$, because K is connected, and, actually,

$$G_1^\eta = K + B_\delta$$

for an appropriate sufficiently small $\delta > 0$.

Owing to Proposition 8.6, there exists $\eta_0 > 0$ such that, for every $\eta \in (0, \eta_0)$ and $1 \leq j \leq \ell(\eta)$,

$$\sigma[\mathfrak{L}, \mathfrak{D}, G_j^\eta] \geq \mu \Sigma_1 |B_1|^{\frac{2}{N}} \eta^{-\frac{2}{N}} - \|\mathbf{b}\|_\infty \sqrt{\Sigma} |B_1|^{\frac{1}{N}} \eta^{-\frac{1}{N}} + \inf_\Omega c.$$

Therefore, there exists $\eta_1 \in (0, \eta_0)$ such that

$$\sigma[\mathfrak{L}, \mathfrak{B}[\Omega_0^m], \Omega_0^m] < \min_{1 \leq j \leq \ell(\eta)} \sigma[\mathfrak{L}, \mathfrak{D}, G_j^\eta] \quad \text{for all} \quad \eta \in (0, \eta_1). \qquad (9.57)$$

Without loss of generality, by rearranging the G_j^η's, if necessary, we can assume that

$$\sigma[\mathfrak{L}, \mathfrak{D}, G_j^\eta] \leq \sigma[\mathfrak{L}, \mathfrak{D}, G_{j+1}^\eta], \qquad 1 \leq j \leq \ell(\eta) - 1.$$

Fix $\eta \in (0, \eta_1)$ and consider the δ-neighborhoods of the components Ω_0^i's

$$\Omega_\delta^i := (\Omega_0^i + B_\delta) \cap \Omega$$

for all $1 \leq i \leq m$ and sufficiently small $\delta > 0$. As $\bar{\Omega}_0^i \cap \bar{\Omega}_0^j = \emptyset$ if $i \neq j$, there exists $\delta_0 > 0$ such that

$$\bar{\Omega}_\delta^i \cap \bar{\Omega}_\delta^j = \emptyset \qquad \text{if} \quad i \neq j \qquad (9.58)$$

for all $0 < \delta < \delta_0$. Moreover, by (9.56), there exists $\delta_1 \in (0, \delta_0)$ such that

$$\left(\bigcup_{j=1}^{\ell(\eta)} \bar{G}_j^\eta \right) \cap \left(\bigcup_{i=1}^{m} \bar{\Omega}_\delta^i \right) = \emptyset \qquad \text{for all} \quad 0 < \delta < \delta_0. \qquad (9.59)$$

Subsequently, we consider the m potentials

$$V_i := \begin{cases} V & \text{in } \bar{\Omega}_\delta^i \\ 1 & \text{in } \bar{\Omega} \setminus \bar{\Omega}_\delta^i, \end{cases} \qquad 1 \leq i \leq m. \qquad (9.60)$$

We claim that

$$V_i \in \mathcal{A}, \qquad 1 \leq i \leq m. \qquad (9.61)$$

Note that, for each $1 \leq i \leq m$, the vanishing set associated to V_i is Ω_0^i, which is connected, and that the corresponding K, say K_i, is the empty set, since (9.56) and (9.59) imply that

$$K \cap (\bar{\Omega}_\delta^i \cup \Gamma_1) = \emptyset, \qquad 1 \leq i \leq m,$$

and $V_i = 1$ in the complement of $\bar{\Omega}_\delta^i$. Therefore, if (9.61) holds, necessarily $m_i = 1$, $K_i = \emptyset$ and, due to (9.52), $\partial\Omega_0^i \cap \Gamma_0 = \emptyset$, where m_i stands for the number of components of Ω_0^i. Consequently, V_i would be an admissible potential to apply the previous Step 1. To prove (9.61) we proceed as follows. Thanks to (9.37), (9.56), (9.58) and (9.59), we have that

$$\Omega \cap \bigcup_{i=1}^m (\bar{\Omega}_\delta^i \setminus \bar{\Omega}_0^i) \subset \Omega_+$$

and hence, for every $1 \leq i \leq m$, V_i is bounded away from zero in any compact subset of Ω_+, as V is bounded away from zero in any compact subset of Ω_+. Let Γ_1^j, $1 \leq j \leq n_1$, denote the components of Γ_1 and, for every $1 \leq i \leq m$, let $\{j_1, ..., j_{p_i}\}$ denote the subset of $\{1, ..., n_1\}$ for which

$$\Gamma_1^j \cap \partial\Omega_0^i = \emptyset \quad \Longleftrightarrow \quad j \in \{j_1, ..., j_{p_i}\}.$$

Then, for each $1 \leq i \leq m$ we have that

$$\partial\Omega_0^i \cap \bigcup_{k=1}^{p_i} \Gamma_1^{j_k} = \emptyset$$

and hence,

$$\text{dist}(\partial\Omega_0^i, \bigcup_{k=1}^{p_i} \Gamma_1^{j_k}) > 0, \qquad 1 \leq i \leq m.$$

Thus, there exists $\delta_2 \in (0, \delta_1)$ such that

$$\left(\bigcup_{k=1}^{p_i} \Gamma_1^{j_k} + B_\delta \right) \cap \bar{\Omega}_\delta^i = \emptyset \qquad (9.62)$$

for all $1 \leq i \leq m$ and $0 < \delta < \delta_2$. Fix $\delta \in (0, \delta_2)$. Then, it follows from (9.60) and (9.62) that

$$V_i = 1 \quad \text{in} \quad \left(\bigcup_{k=1}^{p_i} \Gamma_1^{j_k} + B_\delta \right) \cap \Omega$$

for all $1 \leq i \leq m$. Consequently, for every $1 \leq i \leq m$, V_i is bounded away from zero in any compact subset of

$$\Omega_+ \cup \bigcup_{k=1}^{p_i} \Gamma_1^{j_k}$$

and, therefore, (9.61) indeed holds.

By the result of Step 1, for every $\epsilon > 0$ there exist $\lambda_1 = \lambda_1(\epsilon) > 0$ and m smooth functions

$$\Phi_i : \bar{\Omega} \to [0, \infty), \qquad 1 \leq i \leq m,$$

such that

$$\Phi_i(x) > 0 \quad \text{for all} \quad x \in \Omega \quad \text{and} \quad 1 \leq i \leq m, \qquad (9.63)$$

and Φ_i is a strict supersolution of

$$(\mathfrak{L} + \lambda V_i - \sigma[\mathfrak{L}, \mathfrak{B}[\Omega_0^i], \Omega_0^i] + \epsilon, \mathfrak{B}, \Omega)$$

for all $1 \leq i \leq m$ and $\lambda > \lambda_1$.

Similarly, we will consider the $\ell(\eta)$ potentials

$$\hat{V}_j := \begin{cases} 0 & \text{in} \quad G_j^\eta, \\ 1 & \text{in} \quad \bar{\Omega} \setminus G_j^\eta, \end{cases} \qquad 1 \leq j \leq \ell(\eta). \qquad (9.64)$$

By definition, for every $1 \leq j \leq \ell(\eta)$, the vanishing set of $\hat{V}_j$ equals G_j^η, which is connected and of class $\mathcal{C}^2$. Moreover, by (9.56), $\bar{G}_j^\eta \subset \Omega$. Thus, there exists $\rho > 0$ such that

$$\hat{V}_j = 1 \quad \text{in} \quad (\Gamma_1 + B_\rho) \cap \bar{\Omega},$$

and hence,

$$\hat{V}_j \in \mathcal{A}, \qquad 1 \leq j \leq \ell(\eta).$$

Actually, since

$$\Gamma_0 \cap G_j^\eta = \emptyset, \qquad 1 \leq j \leq \ell(\eta),$$

each of these potentials fits into the abstract framework of Step 1. Consequently, there are $\lambda_2 = \lambda_2(\epsilon) > \lambda_1(\epsilon)$ and $\ell(\eta)$ smooth functions

$$\hat{\Phi}_j : \bar{\Omega} \to [0, \infty), \qquad 1 \leq j \leq \ell(\eta),$$

such that

$$\hat{\Phi}_j(x) > 0 \quad \text{for all} \quad x \in \Omega, \quad 1 \leq j \leq \ell(\eta), \tag{9.65}$$

and $\hat{\Phi}_j$ is a strict supersolution of

$$(\mathfrak{L} + \lambda \hat{V}_j - \sigma[\mathfrak{L}, \mathfrak{B}[G_j^\eta], G_j^\eta] + \epsilon, \mathfrak{B}, \Omega)$$

for all $1 \leq j \leq \ell(\eta)$ and $\lambda > \lambda_2$. Note that $\bar{G}_j^\eta \subset \Omega$, $1 \leq j \leq \ell(\eta)$, implies

$$\mathfrak{B}[G_j^\eta] = \mathfrak{D}, \quad 1 \leq j \leq \ell(\eta),$$

and, therefore, $\hat{\Phi}_j$ is a strict supersolution of

$$(\mathfrak{L} + \lambda \hat{V}_j - \sigma[\mathfrak{L}, \mathfrak{D}, G_j^\eta] + \epsilon, \mathfrak{B}, \Omega)$$

for all $1 \leq j \leq \ell(\eta)$ and $\lambda > \lambda_2$.

Let Γ_1^j, $1 \leq j \leq n_1$, be the components of Γ_1 and let $\{i_1, ..., i_p\}$ denote the subset of $\{1, ..., n_1\}$ for which

$$\Gamma_1^j \cap \partial\Omega_0 = \emptyset \quad \Longleftrightarrow \quad j \in \{i_1, ..., i_p\}.$$

According to Definition 9.1(a), Γ_1^j is a component of $\partial\Omega_0$ for all $j \in \{1, ..., n_1\} \setminus \{i_1, ..., i_p\}$. Moreover,

$$\bigcup_{j=1}^p \Gamma_1^{i_j} \cap \partial\Omega_0 = \emptyset$$

and hence,

$$\text{dist}\left(\bigcup_{j=1}^p \Gamma_1^{i_j}, \partial\Omega_0\right) > 0. \tag{9.66}$$

Now, consider the δ-neighborhoods $\mathcal{N}_\delta^{0,j}$ and $\mathcal{N}_\delta^{1,j}$ defined in (9.46). Thanks to (9.52), (9.56) and (9.66), there exists $\delta_3 \in (0, \delta_2)$ such that

$$\left(\bigcup_{j=1}^p \bar{\mathcal{N}}_\delta^{1,i_j} \cup \bigcup_{j=1}^{n_0} \bar{\mathcal{N}}_\delta^{0,j}\right) \cap \left(\bigcup_{j=1}^m \bar{\Omega}_\delta^j \cup \bigcup_{j=1}^{\ell(\eta)} \bar{G}_j^\eta\right) = \emptyset \tag{9.67}$$

for all $0 < \delta < \delta_3$. Moreover, since

$$\Gamma_k^j \cap \Gamma_\ell^i = \emptyset \quad \text{if} \quad (i, \ell) \neq (j, k),$$

there exists $\delta_4 \in (0, \delta_3)$ such that

$$\bar{\mathcal{N}}_\delta^{k,j} \cap \bar{\mathcal{N}}_\delta^{\ell,i} = \emptyset \quad \text{if} \quad (i, \ell) \neq (j, k), \quad k, \ell \in \{0, 1\}, \tag{9.68}$$

for each $0 < \delta < \delta_4$. Furthermore, since

$$\lim_{\delta \downarrow 0} |\mathcal{N}_\delta^{0,j}| = 0 \quad \text{for each} \quad 1 \leq j \leq n_0,$$

by Proposition 8.6, there exists $\delta_5 \in (0, \delta_4)$ such that

$$\sigma[\mathfrak{L}, \mathfrak{D}, \mathcal{N}_\delta^{0,j}] > \sigma[\mathfrak{L}, \mathfrak{B}[\Omega_0], \Omega_0], \qquad 1 \le j \le n_0, \qquad (9.69)$$

for all $0 < \delta < \delta_5$. Finally, let ψ_δ^i, $i \in \{i_1, ..., i_p\}$, and ξ_δ^j, $1 \le j \le n_0$, denote the principal eigenfunctions associated to

$$\sigma[\mathfrak{L}, \mathfrak{B}[\mathcal{N}_\delta^{1,i}], \mathcal{N}_\delta^{1,i}], \quad i \in \{i_1, ..., i_p\}, \quad \text{and} \quad \sigma[\mathfrak{L}, \mathfrak{D}, \mathcal{N}_\delta^{0,j}], \quad 1 \le j \le n_0,$$

respectively.

Thanks to (9.59), (9.67) and (9.68), the next function is well defined

$$\Phi := \begin{cases} \Phi_i & \text{in } \bar{\Omega}_\delta^i, \quad 1 \le i \le m, \\[2mm] \hat{\Phi}_j & \text{in } \bar{G}_j^\eta, \quad 1 \le j \le \ell(\eta), \\[2mm] \psi_\delta^{i_j} & \text{in } \bar{\mathcal{N}}_{\delta/2}^{1,i_j}, \quad 1 \le j \le p, \\[2mm] \xi_\delta^j & \text{in } \bar{\mathcal{N}}_{\delta/2}^{0,j}, \quad 1 \le j \le n_0, \\[2mm] \zeta_\delta & \text{in } \bar{\Omega} \setminus \mathcal{K}_{\delta,\eta}, \end{cases} \qquad (9.70)$$

where we have denoted

$$\mathcal{K}_{\delta,\eta} := \bigcup_{i=1}^m \bar{\Omega}_\delta^i \cup \bigcup_{j=1}^{\ell(\eta)} \bar{G}_j^\eta \cup \bigcup_{j=1}^p \bar{\mathcal{N}}_{\delta/2}^{1,i_j} \cup \bigcup_{j=1}^{n_0} \bar{\mathcal{N}}_{\delta/2}^{0,j}$$

and ζ_δ is any positive regular extension of Φ_i, $1 \le i \le m$, $\hat{\Phi}_j$, $1 \le j \le \ell(\eta)$, $\psi_\delta^{i_j}$, $1 \le j \le p$, and ξ_δ^j, $1 \le j \le n_0$, from $\mathcal{K}_{\delta,\eta}$ to $\bar{\Omega}$ which is bounded away from zero in $\bar{\Omega} \setminus \mathcal{K}_{\delta,\eta}$. The function ζ_δ does exit because, thanks to (9.56), (9.63) and (9.65), the restrictions

$$\Phi_i|_{\partial\Omega_\delta^i \setminus \Gamma_1}, \qquad \hat{\Phi}_j|_{\partial G_j^\eta}, \qquad 1 \le i \le m, \quad 1 \le j \le \ell(\eta),$$

$$\psi_\delta^{i_j}|_{\partial\mathcal{N}_{\delta/2}^{1,i_j} \setminus \Gamma_1}, \qquad \xi_\delta^i|_{\partial\mathcal{N}_{\delta/2}^{0,i} \setminus \Gamma_0}, \qquad 1 \le j \le p, \quad 1 \le i \le n_0,$$

are positive and bounded away from zero. As in Step 1, the functions $\psi_\delta^{i_j}$'s should not appear in (9.70) if $\Gamma_1 \subset \partial\Omega_0$.

According to the definition of Φ_i, for each $1 \le i \le m$, the following estimates hold in Ω_δ^i

$$\begin{aligned} (\mathfrak{L} + \lambda V - \sigma[\mathfrak{L}, \mathfrak{B}[\Omega_0], \Omega_0] + \epsilon)\, \Phi &= (\mathfrak{L} + \lambda V_i - \sigma[\mathfrak{L}, \mathfrak{B}[\Omega_0^1], \Omega_0^1] + \epsilon)\, \Phi_i \\ &\ge (\mathfrak{L} + \lambda V_i - \sigma[\mathfrak{L}, \mathfrak{B}[\Omega_0^i], \Omega_0^i] + \epsilon)\, \Phi_i \\ &\ge 0 \end{aligned}$$

for all $\lambda > \lambda_2$, because $V = V_i$ in Ω_δ^i. Moreover, thanks to (9.57), for every $1 \leq j \leq \ell(\eta)$, it follows from the definition of $\hat{\Phi}_j$ that the following estimates are satisfied in G_j^η

$$(\mathcal{L}+\lambda V-\sigma[\mathcal{L},\mathfrak{B}[\Omega_0],\Omega_0]+\epsilon)\,\Phi = (\mathcal{L}+\lambda\hat{V}_j-\sigma[\mathcal{L},\mathfrak{B}[\Omega_0^1],\Omega_0^1]+\epsilon)\hat{\Phi}_j$$
$$> (\mathcal{L}+\lambda\hat{V}_j-\sigma[\mathcal{L},\mathfrak{D},G_j^\eta]+\epsilon)\hat{\Phi}_j$$
$$\geq 0$$

for all $\lambda > \lambda_2$, because $V \geq \hat{V}_j = 0$ in G_j^η and $\lambda > 0$.

On the other hand, by Definition 9.1(b), V is positive and bounded away from zero in any compact subset of

$$\Omega_+ \cup \bigcup_{j=1}^{p} \Gamma_1^{i_j}$$

and hence there exists $\omega > 0$ such that

$$V \geq \omega > 0 \quad \text{in} \quad \bigcup_{j=1}^{p} \mathcal{N}_{\delta/2}^{1,i_j}.$$

Thus, for every $1 \leq j \leq p$, the following estimates hold in $\mathcal{N}_{\delta/2}^{1,i_j}$

$$(\mathcal{L}+\lambda V-\sigma[\mathcal{L},\mathfrak{B}[\Omega_0],\Omega_0]+\epsilon)\Phi = (\mathcal{L}+\lambda V-\sigma[\mathcal{L},\mathfrak{B}[\Omega_0],\Omega_0]+\epsilon)\psi_\delta^{i_j}$$
$$= (\sigma[\mathcal{L},\mathfrak{B}[\mathcal{N}_\delta^{1,i_j}],\mathcal{N}_\delta^{1,i_j}] + \lambda V - \sigma[\mathcal{L},\mathfrak{B}[\Omega_0],\Omega_0] + \epsilon)\psi_\delta^{i_j}$$
$$> (\sigma[\mathcal{L},\mathfrak{B}[\mathcal{N}_\delta^{1,i_j}],\mathcal{N}_\delta^{1,i_j}] + \lambda\omega - \sigma[\mathcal{L},\mathfrak{B}[\Omega_0],\Omega_0] + \epsilon)\psi_\delta^{i_j}$$
$$> 0$$

for sufficiently large $\lambda > 0$. Also, due (9.69), for each $1 \leq j \leq n_0$, we find from the definitions of Φ and ξ_δ^j that the next estimates hold in $\mathcal{N}_{\delta/2}^{0,j}$

$$(\mathcal{L} + \lambda V - \sigma[\mathcal{L},\mathfrak{B}[\Omega_0],\Omega_0] + \epsilon)\Phi = (\mathcal{L} + \lambda V - \sigma[\mathcal{L},\mathfrak{B}[\Omega_0],\Omega_0] + \epsilon)\xi_\delta^j$$
$$= (\sigma[\mathcal{L},\mathfrak{D},\mathcal{N}_\delta^{0,j}] + \lambda V - \sigma[\mathcal{L},\mathfrak{B}[\Omega_0^1],\Omega_0^1] + \epsilon)\xi_\delta^j$$
$$> (\sigma[\mathcal{L},\mathfrak{D},\mathcal{N}_\delta^{0,j}] - \sigma[\mathcal{L},\mathfrak{B}[\Omega_0^1],\Omega_0^1] + \epsilon)\xi_\delta^j > 0$$

for all $\lambda > 0$, while, in $\bar{\Omega} \setminus \mathcal{K}_{\delta,\eta}$, we have that

$$(\mathcal{L} + \lambda V - \sigma[\mathcal{L},\mathfrak{B}[\Omega_0],\Omega_0] + \epsilon)\Phi = (\mathcal{L} + \lambda V - \sigma[\mathcal{L},\mathfrak{B}[\Omega_0],\Omega_0] + \epsilon)\zeta_\delta$$
$$\geq (\mathcal{L} - \sigma[\mathcal{L},\mathfrak{B}[\Omega_0],\Omega_0] + \epsilon)\zeta_\delta + \lambda V\zeta_\delta > 0$$

for sufficiently large $\lambda > 0$, because $(\mathcal{L}-\sigma[\mathcal{L},\mathfrak{B}[\Omega_0],\Omega_0]+\epsilon)\zeta_\delta$ is independent of λ and $V\zeta_\delta$ is separated away from zero in $\bar{\Omega} \setminus \mathcal{K}_{\delta,\eta}$, since the closure of this set is a compact subset of Ω_+.

Finally, by construction,

$$\mathfrak{B}\Phi = \mathfrak{D}\Phi = \xi_\delta^j = 0 \quad \text{on} \quad \Gamma_0^j, \quad 1 \le j \le n_0,$$

$$\mathfrak{B}\Phi = (\partial_\nu + b)\Phi = (\partial_\nu + b)\psi_\delta^{i_j} = 0 \quad \text{on} \quad \Gamma_1^{i_j}, \quad 1 \le j \le p,$$

and, for every $1 \le j \le m$ such that $\partial\Omega_0^j \cap \Gamma_1 \ne \emptyset$,

$$\mathfrak{B}\Phi = (\partial_\nu + b)\Phi = (\partial_\nu + b)\Phi_j \ge 0 \quad \text{on} \quad \partial\Omega_\delta^j \cap \Gamma_1 = \partial\Omega_0^j \cap \Gamma_1.$$

Therefore,

$$\mathfrak{B}\Phi \ge 0 \quad \text{on} \quad \partial\Omega$$

and, consequently, the function Φ defined by (9.70) provides us with a positive strict supersolution of (9.43) for sufficiently large $\lambda > 0$. This completes the proof of Step 3.

Step 4: Finally, suppose

$$m \ge 1, \quad K \ne \emptyset, \quad \text{and} \quad \Gamma_0 \cap (\partial\Omega_0 \cup K) \ne \emptyset. \tag{9.71}$$

Let Γ_0^j, $1 \le j \le n_0$, be the components of Γ_0, and let $\{i_1, ..., i_q\}$ denote the subset of $\{1, ..., n_0\}$ for which

$$\Gamma_0^j \cap (\partial\Omega_0 \cup K) \ne \emptyset \quad \Longleftrightarrow \quad j \in \{i_1, ..., i_q\}.$$

As in the proof of Step 2, for sufficiently small $\eta > 0$, we consider the extended open set

$$\tilde{\Omega} := G_\eta := \Omega \cup \left(\bigcup_{j=1}^{q} \Gamma_0^{i_j} + B_\eta \right).$$

The rest of the proof consists in constructing $\tilde{\mathfrak{L}}$, $\tilde{V}$ and $\tilde{\mathfrak{B}}$, as in the proof of Step 2, satisfying

$$\tilde{m} \ge 1, \quad \tilde{K} \ne \emptyset, \quad \text{and} \quad \tilde{\Gamma}_0 \cap (\partial\tilde{\Omega}_0 \cup \tilde{K}) = \emptyset.$$

Reasoning as in the proof of the second part of Step 2, but this time using the result of Step 3, instead of the result of Step 1, ends the proof. $\square$

9.4   Characterizing the existence of principal eigenvalues for admissible potentials

The main result for sign definite potentials W reads as follows.

Theorem 9.7. *Suppose* $W \in L^{\infty}(\Omega)$ *and* $W \geq 0$. *Then, the following assertions are true:*

(a) *If*

$$\inf_{\Omega} W > 0, \tag{9.72}$$

then (9.1) possesses a unique principal eigenvalue.

(b) *Suppose*

$$\inf_{\Omega} W = 0, \tag{9.73}$$

$W \in \mathcal{A}$ *in the sense of Definition 9.1 with* $\text{int } \Omega_+ \neq \emptyset$, *and the coefficients of* $\mathfrak{L}$ *satisfy (8.26). Then, (9.1) admits a principal eigenvalue if, and only if,*

$$\sigma[\mathfrak{L}, \mathfrak{B}[\Omega_0], \Omega_0] > 0, \tag{9.74}$$

and it is unique if it exists.

Similarly, (9.1) possesses a (unique) principal eigenvalue if

$$\sup_{\Omega} W < 0, \tag{9.75}$$

whereas, if

$$\sup_{\Omega} W = 0, \tag{9.76}$$

$-W \in \mathcal{A}$ *with* $\Omega_+ \neq \emptyset$ *and* $\mathfrak{L}$ *satisfies (8.26), then (9.1) possesses a principal eigenvalue if, and only if, (9.74) holds, and it is unique if it exists.*

Proof.   Suppose (9.72). Then, for every $\lambda < 0$,

$$\Sigma(\lambda) = \sigma[\mathfrak{L} - \lambda W, \mathfrak{B}, \Omega] \geq \sigma[\mathfrak{L}, \mathfrak{B}, \Omega] - \lambda \inf_{\Omega} W,$$

and hence,

$$\lim_{\lambda \downarrow -\infty} \Sigma(\lambda) = \infty.$$

Thus, (9.13) holds and, therefore, Theorem 9.2 ends the proof of Part (a).

Now, suppose W satisfies (9.73), $W \in \mathcal{A}$, and $\mathfrak{L}$ satisfies (8.26). Then, according to Theorem 9.6, we have that

$$\lim_{\lambda \downarrow -\infty} \sigma[\mathfrak{L} - \lambda W, \mathfrak{B}, \Omega] = \sigma[\mathfrak{L}, \mathfrak{B}[\Omega_0], \Omega_0]$$

and, consequently, Part (b) again is a corollary from Theorem 9.2.

Subsequently, we suppose $W \leq 0$, instead of $W \geq 0$. If, in addition, (9.75) is satisfied, then,

$$\Sigma(\lambda) = \sigma[\mathfrak{L} - \lambda W, \mathfrak{B}, \Omega] \geq \sigma[\mathfrak{L}, \mathfrak{B}, \Omega] - \lambda \sup_{\Omega} W$$

for all $\lambda > 0$ and hence,

$$\lim_{\lambda \uparrow \infty} \Sigma(\lambda) = \infty.$$

Consequently, (9.19) holds and Theorem 9.3 ends the proof.

Finally, suppose W satisfies (9.73), $-W \in \mathcal{A}$ with $\Omega_+ \neq \emptyset$ and $\mathfrak{L}$ satisfies (8.26). Then, thanks to Theorem 9.6,

$$\lim_{\lambda \uparrow \infty} \sigma[\mathfrak{L} - \lambda W, \mathfrak{B}, \Omega] = \sigma[\mathfrak{L}, \mathfrak{B}[\Omega_0], \Omega_0]$$

and, consequently, in such case, the result also follows from Theorem 9.3. The proof is complete.   □

When W changes sign in Ω, i.e., there are x_+, $x_- \in \Omega$ and $R > 0$ for which (9.12) and (9.18) are satisfied, and

$$\sigma[\mathfrak{L}, \mathfrak{B}, \Omega] > 0,$$

then, it follows from Theorem 9.4 that (9.1) possesses two principal eigenvalues $\lambda_-^* < 0 < \lambda_+^*$. If $(\mathfrak{L}, \mathfrak{B}, \Omega)$ does not satisfy the maximum principle, then, the next result holds.

Theorem 9.8. *Suppose $\mathfrak{L}$ satisfies (8.26) and*

$$\sigma[\mathfrak{L}, \mathfrak{B}, \Omega] < 0. \tag{9.77}$$

Let $W \in L^\infty(\Omega)$ satisfy (9.12) and (9.18) for some x_+, $x_- \in \Omega$ and $R > 0$. Set

$$W^+ := \max\{W, 0\}, \qquad W^- := W^+ - W, \tag{9.78}$$

and assume one of the following two conditions holds:

(a) *$W^+ \in \mathcal{A}$ and*

$$\sigma[\mathfrak{L}, \mathfrak{B}[\Omega_0^+], \Omega_0^+] > 0, \quad \|W^-\|_{L^\infty(\Omega)} < \max_{\lambda < \lambda^+} \frac{\Sigma_+(\lambda)}{-\lambda}, \tag{9.79}$$

where Ω_0^+ is the set Ω_0 associated to W^+ through Definition 9.1(a),

$$\Sigma_+(\lambda) := \sigma[\mathfrak{L} - \lambda W^+, \mathfrak{B}, \Omega], \qquad \lambda \in \mathbb{R}, \tag{9.80}$$

and $\lambda^+ < 0$ is the unique zero of Σ_+.

(b) $W^- \in \mathcal{A}$ and

$$\sigma[\mathfrak{L}, \mathfrak{B}[\Omega_0^-], \Omega_0^-] > 0, \quad \|W^+\|_{L^\infty(\Omega)} < \max_{\lambda > \lambda^-} \frac{\Sigma_-(\lambda)}{\lambda}, \qquad (9.81)$$

where Ω_0^- is the set Ω_0 associated to W^- through Definition 9.1(a),

$$\Sigma_-(\lambda) := \sigma[\mathfrak{L} + \lambda W^-, \mathfrak{B}, \Omega], \qquad \lambda \in \mathbb{R}, \qquad (9.82)$$

and $\lambda^- > 0$ is the unique zero of Σ_-.

Then, (9.1) possesses exactly two principal eigenvalues. Moreover, in case (a) both are negative, while both are positive in case (b).

Owing to Corollary 8.4, condition (9.79) holds if $|\Omega_0^+|$ and $\|W^-\|_\infty$ are sufficiently small and β is sufficiently large on $\Gamma_1 \cap \partial\Omega_0^+$. Similarly, (9.81) holds if $|\Omega_0^-|$ and $\|W^+\|_\infty$ are sufficiently small and β is sufficiently large on $\Gamma_1 \cap \partial\Omega_0^-$. Naturally, the restrictions on the size of β are unnecessary if $\Gamma_1 \cap \partial\Omega_0^+ = \emptyset$, or $\Gamma_1 \cap \partial\Omega_0^- = \emptyset$.

Proof. First, we will make sure that (9.79) and (9.81) make sense. As W satisfies (9.12), it becomes apparent that

$$\text{int}\, \Omega_+^+ \neq \emptyset$$

if $W^+ \in \mathcal{A}$, where Ω_+^+ is the set Ω_+ associated to W^+ through Definition 9.1(a). Consequently, if $W^+ \in \mathcal{A}$ satisfies

$$\sigma[\mathfrak{L}, \mathfrak{B}[\Omega_0^+], \Omega_0^+] > 0,$$

then, according to Theorems 9.1 and 9.6,

$$\lim_{\lambda \downarrow -\infty} \Sigma_+(\lambda) = \sigma[\mathfrak{L}, \mathfrak{B}[\Omega_0^+], \Omega_0^+] > 0, \qquad \lim_{\lambda \uparrow \infty} \Sigma_+(\lambda) = -\infty,$$

and, therefore, there exists a unique value of λ, denoted by λ^+, such that $\Sigma_+(\lambda^+) = 0$. By (9.77), $\Sigma(0) < 0$. Thus, since Σ_+ is decreasing, necessarily $\lambda^+ < 0$. Subsequently, we set

$$q_+(\lambda) := \frac{\Sigma_+(\lambda)}{-\lambda}, \qquad \lambda \leq \lambda^+.$$

By construction, $q_+(\lambda^+) = 0$. Moreover,

$$\Sigma_+(\lambda) > \Sigma_+(\lambda^+) = 0$$

for all $\lambda < \lambda^+ < 0$ and hence, $q_+(\lambda) > 0$ for all $\lambda < \lambda^+$. As

$$\lim_{\lambda \downarrow -\infty} \Sigma_+(\lambda) = \sigma[\mathfrak{L}, \mathfrak{B}[\Omega_0^+], \Omega_0^+],$$

we have that

$$\lim_{\lambda \downarrow -\infty} q_+(\lambda) = 0$$

and, consequently,

$$\max_{\lambda < \lambda^+} \frac{\Sigma_+(\lambda)}{-\lambda} = \max_{\lambda < \lambda^+} q_+(\lambda) \in (0, \infty)$$

is well defined and condition (9.79) makes sense.

Now, suppose $W^- \in \mathcal{A}$ satisfies

$$\sigma[\mathfrak{L}, \mathfrak{B}[\Omega_0^-], \Omega_0^-] > 0.$$

According to (9.18), we have that

$$\text{int} \, \Omega_+^- \neq \emptyset$$

where Ω_+^- is the set Ω_+ associated to W^- through Definition 9.1(a). Moreover, arguing as above, we have that

$$\lim_{\lambda \downarrow -\infty} \Sigma_-(\lambda) = -\infty, \qquad \lim_{\lambda \uparrow \infty} \Sigma_-(\lambda) = \sigma[\mathfrak{L}, \mathfrak{B}[\Omega_0^-], \Omega_0^-] > 0,$$

and hence, there exists a unique λ_- such that $\Sigma_-(\lambda_-) = 0$. By (9.77), $\Sigma(0) < 0$. Thus, since Σ_- is increasing, necessarily $\lambda^- > 0$. From these features, it becomes apparent that the auxiliary function defined by

$$q_-(\lambda) := \frac{\Sigma_-(\lambda)}{\lambda}, \qquad \lambda \geq \lambda^-,$$

satisfies

$$q_-(\lambda_-) = 0, \qquad \lim_{\lambda \uparrow \infty} q_-(\lambda) = 0,$$

and $q_-(\lambda) > 0$ for all $\lambda > \lambda_-$. Therefore, (9.81) is consistent too.

Suppose $W^+ \in \mathcal{A}$ satisfies (9.79). Then, there exists $\tilde{\lambda} < \lambda^+ < 0$ such that

$$\|W^-\|_{L^\infty(\Omega)} < \frac{\Sigma_+(\tilde{\lambda})}{-\tilde{\lambda}}$$

and hence, by (9.80),

$$\Sigma_+(\tilde{\lambda}) := \sigma[\mathfrak{L} - \tilde{\lambda} W^+, \mathfrak{B}, \Omega] > -\tilde{\lambda} \|W^-\|_{L^\infty(\Omega)}.$$

Therefore, since $\tilde{\lambda} < 0$,

$$\Sigma(\tilde{\lambda}) = \sigma[\mathfrak{L} - \tilde{\lambda} W, \mathfrak{B}, \Omega] = \sigma[\mathfrak{L} - \tilde{\lambda} W^+ + \tilde{\lambda} W^-, \mathfrak{B}, \Omega]$$
$$\geq \sigma[\mathfrak{L} - \tilde{\lambda} W^+, \mathfrak{B}, \Omega] + \tilde{\lambda} \|W^-\|_{L^\infty(\Omega)} > 0$$

and, consequently, the conclusions hold from Theorem 9.4.

Similarly, if $W^- \in \mathcal{A}$ satisfies (9.81), there is $\tilde{\lambda} > \lambda^- > 0$ such that

$$\|W^+\|_{L^\infty(\Omega)} < \frac{\Sigma_-(\tilde{\lambda})}{\tilde{\lambda}}$$

and hence,

$$\Sigma_-(\tilde{\lambda}) := \sigma[\mathfrak{L} + \tilde{\lambda}W^-, \mathfrak{B}, \Omega] > \tilde{\lambda}\|W^+\|_{L^\infty(\Omega)}.$$

Therefore,

$$\Sigma(\tilde{\lambda}) = \sigma[\mathfrak{L} - \tilde{\lambda}W, \mathfrak{B}, \Omega] = \sigma[\mathfrak{L} - \tilde{\lambda}W^+ + \tilde{\lambda}W^-, \mathfrak{B}, \Omega]$$
$$\geq \sigma[\mathfrak{L} + \tilde{\lambda}W^-, \mathfrak{B}, \Omega] - \tilde{\lambda}\|W^+\|_{L^\infty(\Omega)} > 0$$

and Theorem 9.4 ends the proof. □

9.5   Comments on Chapter 9

The analysis of the classical case when the potential W has definite sign and it is bounded away from zero does not entail any special difficulty and it goes back, at least, to R. Courant and D. Hilbert [44]. Seemingly, the analysis of (9.1) in the most general and interesting situation when the potential W changes sign goes back to the works of A. Manes and A. M. Micheletti [157], P. Hess and T. Kato [97], and K. J. Brown and S. S. Lin [33].

In the special case when $\Gamma_1 = \emptyset$ ($\mathfrak{B} = \mathfrak{D}$) and $\mathfrak{L}$ is self-adjoint with coercive associated bilinear form, A. Manes and A. M. Micheletti [157] established the existence of two principal eigenvalues, $\lambda_- < 0 < \lambda_+$, from a strong maximum principle of G. Stampacchia [214]. This result was later extended by P. Hess and T. Kato [97] to cover a more general class of operators $\mathfrak{L}$, not necessarily self-adjoint, with $c \geq 0$, by establishing that if W is continuous and $W(x_+) > 0$ for some $x_+ \in \Omega$, then (9.1) admits a unique positive principal eigenvalue. Almost simultaneously, K. J. Brown and S. S. Lin [33] extended these results to cover the special case when $\mathfrak{L} = -\Delta$, $\Gamma_0 = \emptyset$ and $\beta = 0$ ($\mathfrak{B} = \mathfrak{N}$). In such case, their main result established that (9.1) possesses a positive principal eigenvalue, $\lambda_+ > 0$, if, and only if,

$$\int_\Omega W < 0. \tag{9.83}$$

Moreover, λ_+ is unique and algebraically simple if it exists.

In the settings of A. Manes and A. M. Micheletti [157] and P. Hess and T. Kato [97], $\Gamma_1 = \emptyset$ and

$$\sigma[\mathfrak{L}, \mathfrak{B}, \Omega] = \sigma[\mathfrak{L}, \mathfrak{D}, \Omega] > 0. \tag{9.84}$$

Therefore, the existence and the uniqueness of a positive and a negative eigenvalue is a consequence from Theorem 9.4.

Incidentally, in the framework of K. J. Brown and S. S. Lin [33],

$$\sigma[\mathfrak{L}, \mathfrak{B}, \Omega] = \sigma[-\Delta, \mathfrak{N}, \Omega] = 0, \tag{9.85}$$

since 0 is the principal eigenvalue of $-\Delta$ under Neumann boundary conditions in Ω. Therefore, 0 must be a principal eigenvalue of (9.1) and, according to Theorem 9.4, (9.1) admits a positive eigenvalue if, and only if, $\Sigma'(0) > 0$, where

$$\Sigma(\lambda) := \sigma[-\Delta - \lambda W, \mathfrak{N}, \Omega], \qquad \lambda \in \mathbb{R}.$$

Indeed, 0 must be the unique eigenvalue of (9.1) if $\Sigma'(0) = 0$, while the second eigenvalue of (9.1) must be negative if $\Sigma'(0) < 0$. Note that

$$\Sigma(0) = 0 \quad \text{and} \quad \Sigma'(0) \neq 0 \quad \text{imply} \quad \max_{\lambda \in \mathbb{R}} \Sigma(\lambda) > 0.$$

Moreover, adapting the last part of the proof of Theorem 9.2, with the choice $\varphi(0) = 1$, it is easy to check that

$$\Sigma'(0) = -\frac{1}{|\Omega|} \int_\Omega W$$

and, consequently,

$$\Sigma'(0) > 0 \quad \Longleftrightarrow \quad \int_\Omega W < 0,$$

which provides us with the main theorem of K. J. Brown and S. S. Lin [33].

Summarizing, the most pioneering results established that if either

$$\sigma[\mathfrak{L}, \mathfrak{B}, \Omega] > 0, \tag{9.86}$$

or

$$\sigma[\mathfrak{L}, \mathfrak{B}, \Omega] = 0 \quad \text{and} \quad \Sigma'(0) > 0, \tag{9.87}$$

then (9.1) has two principal eigenvalues

$$\lambda_- \leq 0 < \lambda_+.$$

According to Theorem 7.10, we already know that (9.86) occurs if, and only if, $(\mathfrak{L}, \mathfrak{B}, \Omega)$ satisfies the strong maximum principle, which is a very severe restriction. Actually, this chapter has developed a general abstract theory to overcome it.

Even in the special case when W has definite sign, the results of this chapter provide with some substantial and very deep extensions of all classical available results. Indeed, according to Theorem 9.6, if $W \geq 0$ and

$$\Omega_0 := \text{int} \, W^{-1}(0)$$

is a nice open subset of Ω, then (9.1) admits a principle eigenvalue, necessarily unique, if and only if

$$\sigma[\mathfrak{L}, \mathfrak{B}[\Omega_0], \Omega_0] > 0. \tag{9.88}$$

Moreover, by Theorem 7.10, this occurs if, and only if, $(\mathfrak{L}, \mathfrak{B}[\Omega_0], \Omega_0)$ satisfies the strong maximum principle, and, due to Corollary 8.4, this holds true whenever $|\Omega_0|$ is sufficiently small and β is sufficiently large on $\Gamma_1 \cap \partial\Omega_0$.

Theorem 9.6 goes back to J. López-Gómez [135, 137] when $\Gamma_1 = \emptyset$, and to S. Cano-Casanova and J. López-Gómez [39] in the general setting covered in this chapter, as well as the general abstract theory of this chapter in its full generality. Essentially, this chapter polishes the contents of [39].

Besides Theorem 9.6 is a pivotal result for characterizing the existence of principal eigenvalues of (9.1), it has a number of fundamental applications in a variety of scientific fields.

In the context of the *semiclassical analysis* of Schrödinger operators, one is interested in the problem of analyzing the behavior as $h \downarrow 0$ of the principal eigenvalue and associated eigenfunctions of the linear boundary value problem

$$\begin{cases} h^2\mathfrak{L}\varphi + V(x)\varphi = \sigma\varphi & \text{in} \quad \Omega, \\ \varphi = 0 & \text{on} \quad \partial\Omega. \end{cases} \tag{9.89}$$

In quantum physics,

$$h \sim 6.624 \times 10^{-27} \text{erg/sec}$$

is Planck's constant. Except at atomic scales, h is very small; the length of the scales over which quantum effects are important being dependent on it. This fact provides us with a way of transition from classical to quantum mechanics, by comparing the classical limit when $h = 0$, where all quantum effects are neglected, with the original quantum system through the semiclassical regime, where h is assumed to be arbitrarily small.

When $V \in \mathcal{C}^\infty$ is a non-negative potential bounded away from zero at infinity and having a finite number of non-degenerate zeros, it is a rather classical result that the *fundamental energy* of $-h^2\Delta + V(x)$, subsequently denoted by $E(h)$, satisfies

$$E(h) := \sigma[-h^2\Delta + V, \mathfrak{D}, \Omega] = E_1 h + O(h^2) \qquad \text{as} \quad h \downarrow 0, \qquad (9.90)$$

where E_1 stands for the fundamental energy of the associated harmonic oscillator localized at the wells of $V(x)$. Moreover, the associated principal eigenfunctions, or *ground states*, concentrate in $V^{-1}(0)$, and either there is a rapid eigenvalue degeneracy, or the limiting ground states reside asymptotically in a single well (see Theorem 1.2 of B. Simon [204]).

It was the intention of B. Simon on p. 22 of [203] to study the case when $V^{-1}(0)$ is a manifold, instead of a discrete set, but he only observed that if some of the zeros of $V(x)$ are degenerate, then, the fundamental energy $E(h)$ might decay to zero faster than linearly, and that, under the appropriate hypothesis on the shape of $V(x)$ at the degenerate minima, it should be possible to get lower bounds for that decay. Theorem 9.6 provides us with an extremely satisfactory answer to these open questions, as it reveals that

$$\lim_{h \downarrow 0} h^{-2}\sigma[-h^2\Delta + V, \mathfrak{D}, \Omega] = \sigma[-\Delta, \mathfrak{D}, \Omega_0] \qquad (9.91)$$

if $\Omega_0 := \operatorname{int} V^{-1}(0)$ is a nice open subdomain of Ω and, therefore,

$$E(h) = \sigma[-\Delta, \mathfrak{D}, \Omega_0] h^2 + o(h^2) \qquad \text{as} \quad h \downarrow 0. \qquad (9.92)$$

Consequently, the fundamental energy decays quadratically in these degenerate situations. The interested reader is sent to E. N. Dancer and J. López-Gómez [47] for further details.

In the context of population dynamics, Theorem 9.6 has tremendously facilitated some recent studies about the range of validity of the *principle of competitive exclusion*. Actually, Theorem 9.6 has revealed that, in the presence of refuge areas, the most paradigmatic spatial competing species models predict *permanence* even when the level of the aggressions between the antagonist species grows up. Basically, because in the presence of refuges the species can segregate into them to avoid the aggressions from competitors. Naturally, as soon as the refuges can maintain the corresponding species in isolation (see J. López-Gómez [136], J. López-Gómez and J. C. Sabina [155], S. Cano-Casanova and J. López-Gómez [40], and J. López-Gómez and M. Molina-Meyer [150]). Consequently, in spatial models, the principle of competitive exclusion fails in the presence of refuges. To prove this result, we take as a model for competing species the next spatially heterogeneous evolutionary model of Lotka–Volterra type, which incorporates diffusion and transport effects,

$$\begin{cases} \partial_t u + \mathfrak{L}_1 u = \lambda u - A(x)u^2 - b\, B(x)uv & x \in \Omega, \ t > 0, \\ \partial_t v + \mathfrak{L}_2 v = \mu v - c\, C(x)uv - D(x)v^2 & x \in \Omega, \ t > 0, \\ \mathfrak{B}_1 u(x,t) = \mathfrak{B}_2 v(x,t) = 0 & x \in \partial\Omega, \ t > 0, \\ (u(x,0), v(x,0)) = (u_0(x), v_0(x)) & x \in \Omega, \end{cases} \qquad (9.93)$$

where:

i) Ω is a bounded domain of $\mathbb{R}^N$, $N \geq 1$, of class $\mathcal{C}^2$.

ii) λ, $\mu \in \mathbb{R}$, $b, c > 0$, and $\mathfrak{L}_1$, $\mathfrak{L}_2$ are second order uniformly elliptic operators in Ω

$$\mathfrak{L}_j := -\operatorname{div}(A_j \nabla \cdot) + \langle \mathbf{b}_j, \nabla \cdot \rangle + c_j, \qquad j \in \{1, 2\}.$$

iii) For each $j \in \{1, 2\}$, $\mathfrak{B}_j$ stands for the boundary operator

$$\mathfrak{B}_j := \begin{cases} \mathfrak{D} & \text{on } \Gamma_0^j \\ \partial_{\nu_j} + \beta_j & \text{on } \Gamma_1^j \end{cases}$$

where Γ_0^j and Γ_1^j are two disjoint open and closed subsets of $\partial\Omega$ with $\Gamma_0^j \cup \Gamma_1^j = \partial\Omega$, $\beta_j \in \mathcal{C}(\Gamma_1^j)$, and $\nu_j = A_j \mathbf{n}$ is the co-normal vector-field.

iv) $A, B, C, D \in \mathcal{C}(\bar{\Omega})$ are non-negative functions such that

$$A(x) > 0 \quad \text{and} \quad D(x) > 0 \quad \text{for all } x \in \bar{\Omega},$$

$B, C \in \mathcal{A}$, with respect to the partitions of $\partial\Omega$ induced by Γ_0^j and Γ_1^j, $j \in \{1, 2\}$, respectively, and

$$\operatorname{int} \Omega_+^B \neq \emptyset, \qquad \operatorname{int} \Omega_+^C \neq \emptyset,$$

where Ω_+^B and Ω_+^C stand for the respective regions Ω_+ associated to $B(x)$ and $C(x)$ through Definition 9.1(a).

In population dynamics, (9.93) models the evolution of two competing species u and v in the inhabiting area Ω when the individuals disperse randomly within Ω according to the patterns

$$\mathfrak{L}_j := -\operatorname{div}(A_j \nabla \cdot) + \langle \mathbf{b}_j, \nabla \cdot \rangle, \qquad j \in \{1, 2\},$$

where the $\mathbf{b}_j$'s represent the transport effects of each of the species. Typically, for every $x \in \Omega$ and $t > 0$, $u(x,t)$ and $v(x,t)$ measure the densities of the populations at the point $x \in \Omega$ at time $t > 0$, and $\lambda - c_1(x)$ and $\mu - c_2(x)$ measure the intrinsic growth (or death) rates of u and v, respectively, while the function coefficients $A(x)$ and $D(x)$ stand for the normalized carrying capacities of the species, and $B(x)$ and $C(x)$ fix the nature of the competition between u and v. Precisely, the region where $B > 0$ (resp. $C > 0$)

provides us with the patches of the territory where u (resp. v) receives aggressions from v (resp. u), while the region where $B = 0$ (resp. $C = 0$), denoted by Ω_0^B (resp. Ω_0^C), is the spatial refuge of the species u (resp. v). In model (9.93), the parameters $b > 0$ and $c > 0$ measure the intensity of the aggressions between u and v, and u_0 and v_0 stand for the initial population densities. Actually, in the most classical non-spatial models where $\mathfrak{L}_1 = \mathfrak{L}_2 = 0$ and

$$A = B = C = D = 1,$$

the competition is considered to be of low intensity if $bc < 1$, whereas it is of high intensity if $bc > 1$. The book of R. S. Cantrell and C. Cosner [41] is an excellent monograph on spatial ecology through reaction diffusion systems.

For every initial data $u_0, v_0 \in \mathcal{C}(\bar{\Omega})$, (9.93) has a unique global solution $(u(x, t; u_0, v_0), v(x, t; u_0, v_0))$ (see, e.g., H. Amann [10]), and, due to the parabolic maximum principle,

$$0 \leq u(\cdot, t; u_0, v_0) \leq T_1(t)u_0 \quad \text{and} \quad 0 \leq v(\cdot, t; u_0, v_0) \leq T_2(t)v_0$$

for all $t > 0$, where

$$T_1(t) = e^{t(\lambda - \mathfrak{L}_1)}, \qquad T_2(t) = e^{t(\mu - \mathfrak{L}_2)}.$$

According to the pioneering results of M. W. Hirsch [100] and H. Matano [158], and the synthesis of H. L. Smith and H. R. Thieme [208], the limiting profiles as $t \uparrow \infty$ of the positive solutions of (9.93) are, generically, non-negative steady-states (see the monograph of H. L. Smith [207]), which are the non-negative solutions of

$$\begin{cases} \mathfrak{L}_1 u = \lambda u - A(x)u^2 - b\,B(x)uv & \text{in } \Omega, \\ \mathfrak{L}_2 v = \mu v - c\,C(x)uv - D(x)v^2 & \text{in } \Omega, \\ \mathfrak{B}_1 u = \mathfrak{B}_2 v = 0 & \text{on } \partial\Omega. \end{cases} \qquad (9.94)$$

Besides $(0, 0)$, the problem (9.94) admits three types of component-wise non-negative solutions. Namely, the solutions having one component vanishing, $(u, 0)$ or $(0, v)$, known as the *semi-trivial positive solutions*, and the solutions having both component positive, known as the *coexistence states*.

According to Theorem 2.14 of [40], (9.94) possesses a semi-trivial positive solution of the form $(u, 0)$ if, and only if,

$$\lambda > \sigma[\mathfrak{L}_1, \mathfrak{B}_1, \Omega]$$

and, in such case, $(U_\lambda, 0)$ is the unique one, where U_λ is the unique positive solution of

$$\begin{cases} \mathfrak{L}_1 u = \lambda u - A(x)u^2 & \text{in } \Omega, \\ \mathfrak{B}_1 u = 0 & \text{on } \partial\Omega. \end{cases}$$

Similarly, (9.94) possesses a semi-trivial positive solution of the form $(0, v)$ if, and only if,

$$\mu > \sigma[\mathfrak{L}_2, \mathfrak{B}_2, \Omega]$$

and, in such case, $(0, V_\mu)$ is the unique one, where V_μ is the unique positive solution of

$$\begin{cases} \mathfrak{L}_2 v = \mu v - D(x) v^2 & \text{in } \Omega, \\ \mathfrak{B}_2 v = 0 & \text{on } \partial\Omega. \end{cases}$$

Moreover, according to Proposition 5.1 [40], for every $\lambda > \sigma[\mathfrak{L}_1, \mathfrak{B}_1, \Omega]$, $(U_\lambda, 0)$ is linearly unstable if and only if

$$\mu > \sigma[\mathfrak{L}_2 + cC(x) U_\lambda, \mathfrak{B}_2, \Omega],$$

and, for any $\mu > \sigma[\mathfrak{L}_2, \mathfrak{B}_2, \Omega]$, $(0, V_\mu)$ is linearly unstable if and only if

$$\lambda > \sigma[\mathfrak{L}_1 + bB(x) V_\mu, \mathfrak{B}_1, \Omega].$$

Therefore, under the estimates

$$\begin{cases} \mu > \sigma[\mathfrak{L}_2 + cC(x) U_\lambda, \mathfrak{B}_2, \Omega], \\ \lambda > \sigma[\mathfrak{L}_1 + bB(x) V_\mu, \mathfrak{B}_1, \Omega], \end{cases} \tag{9.95}$$

both semi-trivial positive solutions, $(U_\lambda, 0)$ and $(0, V_\mu)$, are linearly unstable. It turns out that, in such case, the model is *permament* in the sense that no species is driven to extinction by the other (see, e.g., Theorem 5.3 of [40]). Moreover, (9.94) has a coexistence state.

On the other hand, according to Theorem 9.6, we have that

$$\lim_{c\uparrow\infty} \sigma[\mathfrak{L}_2 + cC(x) U_\lambda, \mathfrak{B}_2, \Omega] = \sigma[\mathfrak{L}_2, \mathfrak{B}_2[\Omega_0^C], \Omega_0^C],$$

$$\lim_{b\uparrow\infty} \sigma[\mathfrak{L}_1 + bB(x) V_\mu, \mathfrak{B}_1, \Omega] = \sigma[\mathfrak{L}_1, \mathfrak{B}_1[\Omega_0^B], \Omega_0^B],$$

and, therefore, the estimates

$$\mu > \sigma[\mathfrak{L}_2, \mathfrak{B}_2[\Omega_0^C], \Omega_0^C], \qquad \lambda > \sigma[\mathfrak{L}_1, \mathfrak{B}_1[\Omega_0^B], \Omega_0^B], \tag{9.96}$$

guarantee the permanence of u and v for all $b > 0$ and $c > 0$. Consequently, under condition (9.96), the species u and v are permanent independently of the intensity of their mutual aggressions; measured by b and c.

As conditions (9.96) utterly mean that the refuge areas of u and v, Ω_0^B and Ω_0^C, respectively, can maintain the corresponding species in the absence of antagonists, Theorem 9.6 establishes how the principle of competitive exclusion fails in spatial models in the presence of refuges for each of the competitors.

The previous ideas are on the foundations of the mathematical analysis of the effects of facilitation in competitive environments, as well as in the analysis of the effects of strategic alliances in management (see J. López-Gómez and M. Molina-Meyer [149, 151, 152]).

Bibliography

[1] Adams, R. A. (1975). *Sobolev Spaces* (Academic Press, New York).

[2] Adams, D. R. and Hedberg, L. I. (1996). *Function Spaces and Potential Theory*, Grundlehren der Mathematischen Wissenschaften 314 (Springer, Berlin).

[3] Agmon, S. (1983). On positivity and decay of solutions of second order elliptic equations on Riemanian manifolds, in D. Greco (ed.) *Methods of Functional Analysis and Theory of Elliptic Equations* (Liguori, Napples), pp. 187–204.

[4] Agmon, S., Douglis, A. and Nirenberg, L. (1959). Estimates near the boundary for solutions of elliptic partial differential equations satisfying general boundary value conditions I, *Comm. Pure Appl. Math.* **12**, pp. 623–727.

[5] Alexandroff, A. D. (1962). A characterization property of the spheres, *Ann. Mat. Pura Appl.* **58**, pp. 303–354.

[6] Alikakos, N. D. and Fusco, G. (1991). A dynamical systems proof of the Krein–Rutman theorem and an extension of the Perron theorem, *Proc. Roy. Soc. Edinburgh* **117A**, pp. 209–214.

[7] Amann, H. (1972). On the number of solutions of nonlinear equations in ordered Banach spaces, *J. Funct. Anal.* **11**, pp. 346–384.

[8] Amann, H. (1976). Fixed point equations and nonlinear eigenvalue problems in ordered Banach spaces, *SIAM Rev.* **18**, pp. 620–709.

[9] Amann, H. (1983). Dual semigroups and second order linear elliptic boundary value problems, *Israel J. Math.* **45**, pp. 225–254.

[10] Amann, H. (1995). *Linear and Quasilinear Parabolic Problems, Vol. I: Abstract Linear Theory* (Birkhäuser, Bassel).

[11] Amann, H. (2005). Maximum principles and principal eigenvalues, in J. Ferrera, J. López-Gómez and F. R. Ruiz del Portal (eds.) *10 Mathematical Essays on Approximation in Analysis and Topology* (Elsevier, Amsterdam), pp. 1–60.

[12] Amann, H. (In preparation). *Linear and Quasilinear Parabolic Problems, Volume II: Function Spaces and Linear Differential Operators*.

[13] Amann, H. and López-Gómez, J. (1998). A priori bounds and multiple solutions for superlinear indefinite elliptic problems, *J. Diff. Eqns.* **146**, pp.

336–374.

[14] Aronszajn, N. (1965). Potentiels Besseliens, *Ann. Inst. Fourier (Grenoble)* **15**, pp. 43–58.

[15] Aronszajn, N. and Smith, K. T. (1961). Theory of Bessel potentials I, *Ann. Inst. Fourier (Grenoble)* **11**, pp. 385–475.

[16] Aronszajn, N., Mulla, F. and Szeptycki, P. (1963). On spaces of potentials connected with L^p-classes, *Ann. Inst. Fourier (Grenoble)* **13**, pp. 211–306.

[17] Arzela, C. (1889). Funzioni di linee, *Atti R. Accad. Lincei Rend.* **4**, pp. 342–348.

[18] Ascoli, G. (1883). Le curve limiti di una varietà data di curve, *Mem. R. Accad. Lincei* **18**, pp. 521–586.

[19] Babuška, I. (1961). Stability of the domain with respect to the fundamental problem in the theory of partial differential equations, mainly in connection with the theory of elasticity, I, II, *Czechoslovak Math. J.* **11**, pp. 76–105, 165–203.

[20] Babuška, I. and Vyborny, R. (1965). Continuous dependence of eigenvalues on the domain, *Czechoslovak Math. J.* **15**, pp. 169–178.

[21] Bakelman, I. J. (1994). In D. Taliaferro (ed.) *Convex Analysis and Nonlinear Geometric Elliptic Equations*. With an obituary for the author by William Rundell (Springer, New York).

[22] Banach, S. (1932). *Théorie des opérations linéaires* (Monografje Matematyczne, Warsaw).

[23] Barta, J. (1937). Sur la vibration fondamentale d'une membrane, *C. R. Acad. Sci. Paris* **204**, pp. 472–473.

[24] Beauzamy, B. (1985). *Introduction to Banach Spaces and their Geometry*, 2nd edn. North-Holland Mathematic Studies 68 (North-Holland, Amsterdam).

[25] Beltramo, A. and Hess, P. (1984). On the principal eigenvalue of a periodic-parabolic operator, *Comm. Part. Diff. Eqns.* **9**, pp. 919–941.

[26] Berestycki, H. and Nirenberg, L. (1991). On the method of moving planes and the sliding method, *Boll. Soc. Brasil Mat. (Nova Ser.)* **22**, pp. 1–37.

[27] Berestycki, H., Nirenberg, L. and Varadhan, S. R. S. (1994). The principal eigenvalue and maximum principle for second order elliptic operators in general domains, *Comm. Pure Appl. Math.* **XLVII**, 1, pp. 47–92.

[28] Bony, J. M. (1967). Principe du maximum dans les espaces de Sobolev, *C. R. Acad. Sci. Paris* **265**, pp. 333–336.

[29] Brézis, H. (1983). *Analyse Fontionnelle* (Masson, Paris).

[30] Browder, F. E. (1960). A priori estimates for solutions of elliptic boundary value problems I, *Neder. Akad. Wetensch. Indag. Math.* **22**, pp. 149–159.

[31] Browder, F. E. (1960). A priori estimates for solutions of elliptic boundary value problems II, *Neder. Akad. Wetensch. Indag. Math.* **22**, pp. 160–169.

[32] Browder, F. E. (1961). A priori estimates for solutions of elliptic boundary value problems III, *Neder. Akad. Wetensch. Indag. Math.* **23**, pp. 404–410.

[33] Brown, K. J. and Lin, C. C. (1980). On the existence of positive eigenfunctions for an eigenvalue problem with indefinite weight function, *J. Math. Anal. Appl.* **75**, pp. 112–120.

[34] Butzer, P. L. and Berens, H. (1967). *Semi-Groups of Operators and Approximation* (Springer, Berlin).

[35] Calderón, A. P. (1961). Lebesgue spaces of differentiable functions and distributions, in *Partial Differential Equations, Proc. Sympos. Pure Math.* **4**, (Amer. Math. Soc., Providence), pp. 33–49.

[36] Calderón, A. P. and Zygmund, A. (1952). On the existence of certain singular integrals, *Acta Math.* **88**, pp. 85–139.

[37] Calderón, A. P. and Zygmund, A. (1961). Local properties of solutions of elliptic partial differential equations, *Studia Math.* **20**, pp. 171–225.

[38] Cano-Casanova, S. (1998). *Caracterización del principio del máximo para problemas elípticos con condiciones de contorno generales*, Master's thesis, Complutense University of Madrid, Spain.

[39] Cano-Casanova, S. and López-Gómez, J. (2002). Properties of the principal eigenvalues of a general class of non-classical mixed boundary value problems, *J. Diff. Eqns.* **178**, pp. 123–211.

[40] Cano-Casanova, S. and López-Gómez, J. (2003). Permanence under strong aggressions is possible, *Ann. Inst. H. Poincaré Anal. Non Linéaire* **20**, 6, pp. 999–1041.

[41] Cantrell, R. S. and Cosner, C. (2003). *Spatial Ecology via Reaction-Diffusion Equations*, Wiley Series in Mathematical and Computational Biology (John Wiley and Sons, Chichester).

[42] Clarkson, J. A. (1936). Uniformly convex spaces, *Trans. Amer. Math. Soc.* **40**, pp. 396–414.

[43] Clément, Ph. and Peletier, L. A. (1979). An anti-maximum principle for second-order elliptic operators, *J. Diff. Eqns.* **34**, pp. 218–229.

[44] Courant, R. and Hilbert, D. (1962). *Methods of Mathematical Physics I-II* (Wiley-Interscience, New York).

[45] Crandall, M. G. and Rabinowitz, P. H. (1971). Bifurcation from simple eigenvalues, *J. Funct. Anal.* **8**, pp. 321–340.

[46] Dancer, E. N. (1990). The effect of domain shape on the number of positive solutions of certain nonlinear equations II, *J. Diff. Eqs.* **87**, pp. 316–339.

[47] Dancer, E. N. and López-Gómez, J. (2000). Semiclassical analysis of general second order elliptic operators on bounded domains, *Trans. Amer. Math. Soc.* **352**, pp. 3723–3742.

[48] Day, M. M. (1941). Reflexive Banach spaces not isomorphic to uniformly convex spaces, *Bull. Amer. Math. Soc.* **47**, pp. 313–317.

[49] Delgado, M., López-Gómez, J. and Suárez, A. (2000). On the symbiotic Lotka–Volterra model with diffusion and transport effects, *J. Diff. Eqns.* **160**, pp. 175–262.

[50] Denk, R., Hieber, M. and Prüss, J. (2003), $\mathcal{R}$-boundedness. Fourier multipliers and problems of elliptic and parabolic type, *Mem. Amer. Math. Soc.* **166**, no. 788.

[51] Denk, R., Hieber, M. and Prüss, J. (2007). Optimal $L^p - L^q$ estimates for parabolic boundary value problems with inhomogeneous data, *Math. Z.* **257**, pp. 193–224.

[52] Deny, J. and Lions, J. L. (1953). Les espaces de type Beppo Levi, *Ann. Inst.*

Fourier, Grenoble **5**, pp. 305–370.

[53] Diestel, J. (1975). *Geometry of Banach Spaces: Selected Topics*, Lecture Notes in Mathematics 485 (Springer, Berlin).

[54] Donsker, M. D. and Varadhan, S. R. S. (1975). On a variational formula for the principal eigenvalue for operators with maximum principle, *Proc. Nat. Acad. USA* **72**, pp. 780–783.

[55] Donsker, M. D. and Varadhan, S. R. S. (1976). On the principal eigenvalue of second order elliptic differential operators, *Comm. Pure Appl. Math.* **29**, pp. 595–621.

[56] Du, Y. (2006). *Order Structure and Topological Methods in Nonlinear Partial Differential Equations*, Series on Partial Differential Equations and Applications 2 (World Scientific, Singapore).

[57] Duffin, R. J. (1947). Lower bounds for eigenvalues, *Phys. Rev.* **71**, pp. 827–828.

[58] Earnshaw, S. (1839). On the nature of the molecular forces which regulate the constitution of the luminiferous ether, *Cambridge Phil. Soc. Trans.* **7**, pp. 97–112.

[59] Eberlein, W. F. (1947). Weak compactness in Banach spaces, *Proc. Nat. Acad. Sci. USA* **38**, pp. 51–53.

[60] Evans, L. C. (1998). *Partial Differential Equations*, Graduate Studies in Mathematics 19 (Amer. Math. Soc., Providence, RI).

[61] Faber, C. (1923). Beweis das unter allen homogenen Membranen von gleicher Fläche und gleicher Spannung die kreisdörmige den tiefsten Grundton gibt, *Sitzungsber. Bayer. Akad. der Wiss. Math. Phys.* pp. 169–171.

[62] Figueiredo, D. G. and Mitidieri, E. (1990), Maximum principles for cooperative elliptic systems, *C. R. Acad. Sci. Paris* **310**, pp. 49–52.

[63] Figueiredo, D. G. and Mitidieri, E. (1992). Maximum principles for linear elliptic systems, *Rend. Istit. Mat. Univ. Trieste* **22**, pp. 36–66.

[64] Fleckinger, J. Hernández, J. and Thélin, F. (1993). A maximum principle for linear cooperative elliptic systems, in *Differential Equations with Applications to Mathematical Physics*, Math. Sci. Eng. 192, (Academic Press, Boston, MA), pp. 79–86.

[65] Fourier, J. B. J. (1888-90). *Oeuvres I-II* (Gauthier-Villars, Paris).

[66] Fraile, J. M., Koch-Medina, P., López-Gómez, J. and Merino, S. (1996). Elliptic eigenvalue problems and unbounded continua of positive solutions of a semilinear elliptic equation, *J. Diff. Eqns.* **127**, pp. 295–319.

[67] Friedman, A. (1969). *Partial Differential Equations* (Holt, Rinehart and Winston, New York).

[68] Friedrichs, K. O. (1928). Die Randwert und Eigenwertprobleme aus der Theorie des elastichen Platten, *Math. Ann.* **98**, pp. 205–247.

[69] Friedrichs, K. O. (1944). The identity of weak and strong extensions of differential operators, *Trans. Amer. Math. Soc.* **55**, pp. 131–151.

[70] Friedrichs, K. O. (1953). Differentiability of solutions of elliptic partial differential equations, *Comm. Pure and Appl. Maths.* **5**, pp. 299–326.

[71] Frobenius, G. (1908). Über matrizen aus positiven elementen I, *S. B. Preuss. Akad. Wiss. Berlin*, pp. 471–476.

[72] Frobenius, G. (1909). Über matrizen aus positiven elementen II, *S. B. Preuss. Akad. Wiss. Berlin*, pp. 514–518.

[73] Frobenius, G. (1912). Über matrizen aus nicht negativen elementen, *S. B. Preuss. Akad. Wiss. Berlin*, pp. 456–477.

[74] Gagliardo, E. (1958). Proprieta di alcune classi di funzioni in piu variabili, *Ric. Mat.* **7**, pp. 102–137.

[75] Gauss, C. F. (1839). Allgemeine Theorie des Erdmagnetismus, *Beobachtungen des magnetischen Vereins im Jahre 1838*, (Leipzig).

[76] García-Melián, J., Gómez-Reñasco, R., López-Gómez, J. and Sabina, J. C. (1998). Point-wise growth and uniqueness of positive solutions for a class of sublinear elliptic problems where bifurcation from infinity occurs, *Arch. Rat. Mech. Anal.* **145**, pp. 261–289.

[77] Gårding, L. (1953). Dirichlet's problem for linear elliptic partial differential equations, *Math. Scand.* **1**, pp. 55–72.

[78] Gidas, B. Ni, W. M. and Nirenberg, L. (1979). Symmetry and related properties via the maximum principle, *Comm. Math. Phys.* **68**, pp. 209–243.

[79] Gilbarg, D. and Trudinger, N. (2001). *Elliptic Partial Differential Equations of Second Order*, Classics in Mathematics (Springer, Berlin and Heildelberg).

[80] Giraud, G. (1932). Généralizations des problèmes sur les opérations du type elliptiques, *Bull. des Sciences Math.* **56**, pp. 248–272, 281–312, 316–352.

[81] Giraud, G. (1933). Problèmes de valeurs á la frontière relatifs á certainnes données discontinues, *Bull. de la Soc. Math. de France* **61**, pp. 1–54.

[82] Göhberg, I. C., Goldberg, S. and Kaashoek, M. A. (1990). *Classes of Linear Operators*, Operator Theory: Advances and Applications 49 (Birkhäuser, Bassel).

[83] Göhberg, I. C. and Krein, M. G. (1957). The basis properties on defect numbers, root numbers and indices of linear operators, *Usp. Mat. Nauk.* **12**, pp. 43–118.

[84] Göhberg, I. C., Lancaster, P. and Rodman, L. (1982). *Matrix Polynomials*, Comp. Sci. Appl. Mathematics (Academic Press, New York).

[85] Gómez-Reñasco, R. (1999). *The effect of varying coefficients in semilinear elliptic boundary value problems. From classical solutions to metasolutions*, Ph.D. thesis, University of La Laguna, Tenerife, Spain.

[86] Gómez-Reñasco, R. and López-Gómez, J. (2002). On the existence and numerical computation of classical and non-classical solutions for a family of elliptic boundary value problems, *Nonl. Anal. TMA* **48**, pp. 567–605.

[87] Gossez, J. P. and Lami-Dozo, E. (1985). On the principal eigenvalue of a second order linear elliptic problem, *Arch. Rat. Mech. Anal.* **89**, pp. 169–175.

[88] Greiner, G. (1981). Zur Perron-Frobenius Theorie stark stetiger Halbgruppen, *Math. Z.* **177**, pp. 401–423.

[89] Grosberg, J. and Krein, M. G. (1939). Sur la Décomposition des fonctionnelles en composants positives, *C.R. (Doklady) Acad. Sci. CCCR (N.S.)* **25**, pp. 723–726.

[90] Grosswald, E. (1974). *Collected Papers of Hans Rademacher*, (MIT Press,

Cambridge).

[91] Guzmán, M. and Rubio, B. (1979). _Integración: Teoría y Técnicas_ (Alhambra, Madrid).

[92] Hale, J. K. (2005). Eigenvalues and perturbed domains, in J. Ferrera, J. López-Gómez and F. R. Ruiz del Portal (eds.) _10 Mathematical Essays on Approximation in Analysis and Topology_ (Elsevier, Amsterdam), pp. 95–123.

[93] Henry, D. (1981). _Geometric Theory of Semi-linear Parabolic Equations_, Lecture Notes in Mathematics 840 (Springer, Berlin).

[94] Hersch, J. (1960). Sur la fréquence fondamentale d'une membrane vibrante: évaluations par déefaut et principe de maximum, _Z. Angew. Math. Phys._ **11**, pp. 387–413.

[95] Hess, P. (1986). On the spectrum of elliptic operators with respect to indefinite weights, _Lin. Alg. and its Appns._ **84**, pp. 99–109.

[96] Hess, P. (1991). _Periodic-Parabolic Boundary Value Problems and Positivity_, Pitman Research Notes in Mathematics 247 (Longman, London).

[97] Hess, P. and Kato, T. (1980). On some linear and nonlinear eigenvalue problems with an indefinite weight function, _Comm. Partial Diff. Eqns._ **5**, pp. 999—1030.

[98] Hilbert, D. (1900). Über das Dirichletsche Prinzip, _Jber. Deutsch. Math. Verein_ **8**, pp. 184–188.

[99] Hilbert, D. (1909). Wesen und Ziele einer Analysis der unendlich vielen unabhängigen variablen, _Rend. Circ. Mat. Palermo_ **27**, pp. 59–74.

[100] Hirsch, M. W. (1988). Stability and convergence in strongly monotone dynamical systems, _J. Reine Angew. Math._ **383**, pp. 1–53.

[101] Hölder, O. (1889). Über einen Mittelwerthsatz, _Göttingen Nachr._ pp. 38–47.

[102] Hooker, W. W. (1960). Lower bounds for the first eigenvalue of elliptic equations of order two and four, _Tech. Rep. Univ. California at Berkeley_ **10**, AF49(638) p. 398.

[103] Hopf, E. (1927). Elementare Bemerkungen über die Lösungen partieller Differentialgleichungen zweiter Ordnung vom elliptischen Typus, _Sitzungsber. Preuss. Akad. Wiss._ **19**, pp. 147–152.

[104] Hopf, E. (1952). A remark on linear elliptic differential equations of the second order, _Proc. of the Amer. Math. Soc._ **3**, pp. 791–793.

[105] Hörmander, L. (1955). On the theory of general partial differential operators, _Acta Math._ **94**, pp. 161–248.

[106] Hörmander, L. (1983). _The Analysis of Linear Partial Differential Operators I-IV_ (Springer, Berlin and New York).

[107] Jameson, G. (1970). _Ordered Linear Spaces_, Lectures Notes in Mathematics 141 (Springer, New York).

[108] John, F. (1978). _Partial Differential Equations_, Applied Mathematical Sciences 1 (Springer, New York).

[109] Jordan, P. and von Neumann, J. (1935). On inner products in linear, metric spaces, _Ann. of Math._ **36**, pp. 719–723.

[110] Kakutani, S. (1939). Weak topologies and regularity of Banach spaces,

Proc. Imp. Acad. Tokyo **15**, pp. 169–173.

[111] Kato, T. (1982). Superconvexity of the spectral radius and convexity of the spectral bound and type, *Math. Z.* **180**, pp. 265–273.

[112] Kato, T. (1995). *Perturbation Theory for Linear Operators*, Classics in Mathematics (Springer, Berlin).

[113] Kinderlehrer, D. and Stampacchia, G. (1980). *An Introduction to Variational Inequalities and their Applications* (Academic Press, London).

[114] Klee, V. L. (1955). Boundness and continuity of linear functionals, *Duke Math. J.* **22**, pp. 263–270.

[115] Kondrachov, V. I. (1945). On some properties of functions from the space L_p, *Dokl. Akad. Nauk SSSR* **48**, pp. 533–538.

[116] Korn, A. (1909). *Über Minimalflächen deren Randkurven wenig von ebenen Kurven abweichen*, Adhandl. Königl. Preuss. Akad. Wiss. 2, (Preuss. Akad. Press, Berlin).

[117] Krahn, E. (1925). Über eine von Rayleigh formulierte Minimaleigenschaft des Kreises, *Math. Ann.* **91**, pp. 97–100.

[118] Krasnoselskij, M. A. (1964). *Positive Solutions of Operator Equations* (Noordhoff, Groninggen).

[119] Krein, M. G. (1940). Propriétés fondamentales des ensembles coniques normaux dans l'espace de Banach, *C. R. (Doklady) Acad. Sci. CCCR (N.S.)* **28**, pp. 13–17.

[120] Krein, M. G. and Rutman. M. A. (1948). Linear operators leaving invariant a cone in a Banach space (in Russian), *Usp. Mat. Nauk. (N.S.)* **3**, pp. 3–95.

[121] Kresin, G. and Maz'ja, V. G. (2013). *Maximum Principles and Sharp Constants for Solutions of Elliptic and Parabolic Systems* (Amer. Math. Soc., Providence).

[122] Lax, P. D. and Milgram, A. N. (1954). Contributions to the theory of partial differential equations, *Ann. Maths. Studies* **33**, pp. 167–190.

[123] Lebesgue, H. (1907). Sur le problème de Dirichlet, *Rend. Circ. Mat. Palermo* **24**, pp. 371–404.

[124] Lebesgue, H. (1910). Sur l'integration des fonctions discontinues, *Ann. Ec. Norm. Sup.* **XXVII**, pp. 361–450.

[125] Lichtenstein, L. (1912). Beiträge zur Theorie der linearen partiellen Differentialgleichungen zweiter Ordnung vom elliptischen Typus, *Rend. Circ. Mat. Palermo* **33**, pp. 201–211.

[126] Lichtenstein, L. (1924). Neue Beiträge zur Theorie der linearen partiellen Differentialgleichungen zweiter Ordnung vom elliptischen Typus, *Math. Z.* **20**, pp. 194–212.

[127] Lindenstrauss, J. and Tzafriri, L. (1971). On complemented subspace problem, *Israel J. Math.* **9**, pp. 263–269.

[128] Lindenstrauss, J. and Tzafriri, L. (1973). *Classical Banach Spaces*, Lectures Notes in Mathematics 338 (Springer, Berlin).

[129] Lions, J. L. (1963). Théorèmes de traces et d'interpolation IV, *Math. Ann.* **151**, pp. 42–56.

[130] Lions, J. L. and Magenes, E. (1968). *Problèmes aux Limites non Homogènes I-III* (Dunod, Paris).

[131] Lions, J. L. and Stampacchia, G. (1967). Variational inequalities, *Comm. Pure Appl. Maths.* **20**, pp 493–519.

[132] Lions, P. L. (1982). A remark on Bony maximum principle, *Proc. Amer. Math. Soc.* **88**, pp. 503–508.

[133] Lipschitz, R. (1876). Sur la possibilité d'intégrer complétement un système donné d'équations différentielles, *Bull. Sci. Math. Astro.* **10**, pp. 149–159.

[134] López-Gómez, J. (1981). *Problemas de contorno y valor inicial para ecuaciones parabólicas lineales y semilineales*, Master's thesis, Complutense University of Madrid, Spain.

[135] López-Gómez, J. (1994). On linear weighted boundary value problems, in *Partial Differential Equations. Models in Physics and Biology*, Mathematical Research 82 (Akademie-Verlag, Berlin), pp. 188–203.

[136] López-Gómez, J. (1995). Permanence under strong competition, Dynamical systems and applications, *World Sci. Ser. Appl. Anal.* **4**, pp. 473–488.

[137] López-Gómez, J. (1996). The maximum principle and the existence of principal eigenvalues for some linear weighted boundary value problems, *J. Diff. Eqns.* **127**, pp. 263–294.

[138] López-Gómez, J. (2000). Large solutions, metasolutions, and asymptotic behaviour of the regular positive solutions of sublinear parabolic problems, *El. J. Diff. Eqs.* **Conf. 05**, pp. 135–171.

[139] López-Gómez, J. (2001). Approaching metasolutions by classical solutions, *Diff. Int. Eqs.* **14**, pp. 739–750.

[140] López-Gómez, J. (2001). *Spectral Theory and Nonlinear Functional Analysis*, Research Notes in Mathematics Vol. 426 (Chapman and Hall/CRC Press, Boca Raton).

[141] López-Gómez, J. (2001). *Ecuaciones Diferenciales y Variable Compleja* (Prentice-Hall, Madrid).

[142] López-Gómez, J. (2002). *Ecuaciones Diferenciales y Variable Compleja, Problemas y Ejercicios Resueltos* (Prentice-Práctica, Madrid).

[143] López-Gómez, J. (2003). Coexistence and metacoexistence states in competing species models, *Houston J. Math.* **29**, pp. 485–538.

[144] López-Gómez, J. (2003). Classifying smooth supersolutions for a general class of elliptic boundary value problems, *Adv. Diff. Eqns.* **8**, pp. 1025–1042.

[145] López-Gómez, J. (2005). Metasolutions: Malthus versus Verhulst in population dynamics. A dream of Volterra, in M. Chipot and P. Quittner (eds.) *Stationary Partial Differential Equations II* (Elsevier, Amsterdam), pp. 211–309.

[146] López-Gómez, J. (2009). The strong maximum principle, *RIMS Kôkyûroku Bessatsu* **B15**, pp. 113–123.

[147] López-Gómez, J. (2011). The existence of weak solutions for a general class of mixed boundary value problems, *Disc. and Cont. Dyn. Syst.* **Supp. 2011**, pp. 1015–1024.

[148] López-Gómez, J. and Molina-Meyer, M. (1994). The maximum principle for cooperative weakly elliptic systems and some applications, *Diff. Int. Eqs.* **7**, pp. 383–398.

[149] López-Gómez, J. and Molina-Meyer, M. (2004). Singular perturbations in

Economy and Ecology. The effect of strategic symbiosis in random competitive environments, *Adv. Math. Sci. Appns.* **14**, pp. 87–107.

[150] López-Gómez, J. and Molina-Meyer, M. (2006). The competitive exclusion principle versus biodiversity through segregation and further adaptation to spatial heterogeneities, *Theor. Population Biol.* **69**, pp. 94–109.

[151] López-Gómez, J. and Molina-Meyer, M. (2006). Superlinear indefinite systems: Beyond Lotka-Volterra models, *J. Diff. Eqns.* **221**, pp. 343–411.

[152] López-Gómez, J. and Molina-Meyer, M. (2007). Biodiversity through coopetition, *Discrete Contin. Dyn. Syst. Ser. B* **8**, pp. 187–205.

[153] López-Gómez, J. and Mora-Corral, C. (2007). *Algebraic Multiplicities of Eigenvalues for Linear Operators*, Operator Theory: Advances and Applications Vol. 177 (Birkhäuser, Bassel).

[154] López-Gómez, J. and Pardo, R. M. (1993). The existence and the uniqueness for the predator-prey model with diffusion, *Diff. Int. Eqs.* **6**, pp. 1025–1031.

[155] López-Gómez, J. and Sabina, J. C. (1995). Coexistence states and global attractivity for some convective diffusive competing species models, *Trans. Amer. Math. Soc.* **347**, 3797–3833.

[156] Malliavin, P. (1982). *Intégration et Probabilitiés, Analyse de Fourier et Analyse Spectrale* (Masson, Paris).

[157] Manes, A. and Micheletti, A. M. (1973). Un'estensiones della teoria variazionale classica degli autovalori per operatori ellittici del secondo ordine, *Boll. Un. Mat. Ital.* **7**, pp. 285–301.

[158] Matano, H. (1984). Existence of nontrivial unstable sets for equilibriums of strongly order-preserving systems, *J. Fac. Sci. Univ. Tokyo Sect. IA Math.* **30**, pp. 645–673.

[159] Maz'ja, V. G. (1985). *Sobolev Spaces* (Springer, Berlin).

[160] McNabb, A. (1961). Strong comparison theorems for elliptic equations of second order, *J. Math. Mech.* **10**, pp. 431–440.

[161] Meyers, N. G. and Serrin, J. (1964). $H = W$, *Proc. Nat. Acad. Sci. USA* **51**, pp. 1055–1056.

[162] Milman, D. P. (1938). On some criteria for the regularity of spaces of type (B), *Doklady Akad. Nauk. SSSR* **20**, pp. 234.

[163] Molina-Meyer, M. (1995). Existence and uniqueness of coexistence states for some nonlinear elliptic systems, *Nonl. Anal. T.M.A.* **25**, pp. 279–296.

[164] Molina-Meyer, M. (1996). Global attractivity and singular perturbation for a class of nonlinear cooperative systems, *J. Diff. Eqns.* **128**, pp. 347–378.

[165] Molina-Meyer, M. (1997). Uniqueness and existence of positive solutions for weakly coupled general sublinear systems, *Nonl. Anal. T.M.A.* **30**, pp. 5375–5380.

[166] Morrey, C. B. (1938). On the solutions of quasi-linear elliptic partial differential equations, *Trans. Amer. Math. Soc.* **43**, pp. 126–166.

[167] Morrey, C. B. (1966). *Multiple Integrals in the Calculus of Variations* (Springer, Berlin).

[168] Motzkin, Th. (1935). Sur quelques propriétés charactéristiques des ensembles convexes, *Atti Acad. Naz. Lincei Rend.* **6**, pp. 562–567.

[169] Moutard, T. (1894). Notes sur les équations aux dérivées partielles, *J. de l'École Polytechnique* **64**, pp. 35–69.

[170] Nečas, J. (1967). *Les méthodes directes en théorie des équations elliptiques* (Éditeurs Academia, Prague).

[171] von Neumann, J. (1929). Allgemeine Eigenwerttheorie Hermitescher Funktionaloperatoren, *Math. Ann.* **102**, pp. 49–131.

[172] Oleinik, O. A. (1952). On properties of some boundary problems for equations of elliptic type, *Math. Sbornik, N. S.* **30**, pp. 695–702.

[173] Pagter, B. (1986). Irreducible compact operators, *Math. Z.* **192**, pp. 149–153.

[174] Paraf, A. (1892). Sur le problème de Dirichlet at son extension au cas de l'équation linéaire générale du second ordre, *Ann. Fac. Sci. Toulouse Ser. I*, **6**, pp. 1–75.

[175] Perron, O. (1907). Zur Theorie der matrices, *Math. Ann.* **64**, pp. 248–263.

[176] Pettis, B. J. (1939). A proof that every uniformly convex space is reflexive, *Duke Math. J.* **5**, pp. 249–253.

[177] Phillips, R. S. (1955). The adjoint semi-group, *Pacific J. Math.* **5**, pp. 269–283.

[178] Picard, E. (1905). *Traité d'Analyse Vol. 2* (Gauthier Villars, Paris).

[179] Picone, M. (1927). Maggiorazione degli integrali di equazioni lineari ellitico-paraboliche alle derivate parziali del secondo ordine, *Atti Accad. Naz. dei Lincei* **5**, pp. 138–143.

[180] Poincaré, H. (1916-1956). *Oeuvres Vol. 1-11* (Gauthier-Villars, Paris).

[181] Protter, M. H. (1960). Lower bounds for the first eigenvalue of elliptic equations, *Ann. of Math.* **71**, pp. 423–444.

[182] Protter, M. H. and Weinberger, H. F. (1966). On the spectrum of general second order operators, *Bull. Amer. Math. Soc.* **72**, pp. 251–255.

[183] Protter, M. H. and Weinberger, H. F. (1967). *Maximum Principles in Differential Equations* (Prentice-Hall, Englewood Cliffs).

[184] Pucci, P. and Serrin, J. (2007). *The Maximum Principle*, Progress in Nonlinear Differential Equations and Their Applications Vol. 73 (Birkhäuser, Bassel).

[185] Rabinowitz, P. H. (1975). *Théorie de degré topologique et applications à des problémes aux limites non linéaires*, Lecture Notes Lab. Analyse Numerique (Université Paris VI, Paris).

[186] Rayleigh, J. W. S. (1945). *The Theory of Sound*, 2nd edn. (Dover Publications, New York).

[187] Redheffer, R. Personal communication to W. Walter (see p. 295 [224]).

[188] Rellich, F. (1930). Ein satz über mittlere konvergenz, *Math. Nachr.* **31**, pp. 30–35.

[189] Riesz, F. (1910). Untersuchungen über Systeme integrierbarer Funktionen, *Math. Ann.* **LXIX**, pp. 449–497.

[190] Riesz, F. (1918). Über lineare Funktionalgleichungen, *Acta Math.* **41**, pp. 71–98.

[191] Riesz, F. (1934). Zur Theorie des Hilbertschen Raumes, *Acta Sci. Math. Szeged* **7**, pp. 34–38.

[192] Ringrose, J. R. (1959). A note on uniformly convex spaces, *J. London Math. Soc.* **34**, pp. 92.

[193] Rogers, L. J. (1888). An extension of a certain theorem in inequalities, *Messenger of Math.* **17**, pp. 145–150.

[194] Schaefer, H. H. (1967). *Topological Vector Spaces*, Macmillan Series in Advanced Mathematics and Theoretical Physics (Macmillan, New York).

[195] Schauder, J. (1934). Über lineare elliptische Differentialgleichungen zweiter Ordnung, *Math. Z.* **38**, 257–282.

[196] Schauder, J. (1935). Numerische Abschätzungen in elliptischen linearen Differentialgleichungen, *Studia Math.* **5**, pp. 34–42.

[197] Schlag, W. Personal communication to L. C. Evans (see [60]).

[198] Schmidt, E. (1907). Zur Theorie der linearen und nichtlinearen Integralgleichungen, *Math. Ann.* **LXIII**, pp. 433–476.

[199] Schwartz, L. (1950–51). *Théorie des Distributions Vols. I-II* (Hermann, Paris).

[200] Schwartz, L. (1981). *Geometry and Probability in Banach Spaces*, Based on notes taken by P. R. Chernoff, Lecture Notes in Mathematics Vol. 852 (Springer, Berlin and New York).

[201] Serrin, J. (1971). A symmetry problem in potential theory, *Arch. Rat. Mech. Anal.* **43**, pp. 304–318.

[202] Shmulyan, V. L. (1940). Über lineare topologische Räume, *Math. Sbornik, N. S.* **7**, 49, pp. 425–448.

[203] Simon, B. (1983). Semiclassical analysis of low lying eigenvalues I: Nondegenerate minima: asymptotic expansions, *Ann. Inst. H. Poincaré A* **38**, pp. 12–37.

[204] Simon, B. (1984). Semiclassical analysis for low lying eigenvalues II: Tunneling, *Ann. of Math.* **120**, pp. 89–118.

[205] Slobodeckii, L. N. (1958). Generalized Sobolev spaces and their application to boundary problems for partial differential equations, *Leningrad Gos. Ped. Inst. Ucen. Zap.* **197**, pp. 54–112.

[206] Slobodeckii, L. N. (1958). Estimate in L^p of solutions of elliptic systems, *Dokl. Akad. Nauk SSSR* **123**, pp. 616–619.

[207] Smith, H. L. (1995). *Monotone Dynamical Systems. An Introduction to the Theory of Competitive and Cooperative Systems*, Mathematical Surveys and Monographs Vol. 41. (Amer. Math. Soc., Providence).

[208] Smith, H. L. and Thieme, H. R. (1991). Convergence for strongly order-preserving semiflows, *SIAM J. Math. Anal.* **22**, 4, pp. 1081–1101.

[209] Sobolev, S. L. (1937). On a boundary value problem for polyharmonic equations (in russian), *Mat. Sbornik, N.S.* **2**, 44, pp. 465–499.

[210] Sobolev, S. L. (1938). Sur un théorème d'analyse fonctionelle, *Math. Sbornik*, **4**, 46, pp. 471–497.

[211] Sobolev, S. L. (1945). *Certaines Applications de l'Analyse Fonctionelle à la Physique Mathématique* (University of Leningrad, Leningrad).

[212] Stampacchia, G. (1964). Formes bilinéaires coercitives sur les ensembles convexes, *C. R. Acad. Sci. Paris* **258**, pp. 4413–4416.

[213] Stampacchia, G. (1966). *Èquations elliptiques du second ordre à coefficients*

discontinus (Les Presses de l'Université de Montréal, Montreal).

[214] Stampacchia, G. (1965). Le problème de Dirichlet pour les équations elliptiques du second ordre à coefficients discontinus, *Ann. Inst. Fourier* **15**, 1, pp. 189–258.

[215] Stein, E. M. (1970). *Singular Integrals and Differentiability Properties of Functions* (Princeton University Press, Princeton).

[216] Suzuki, T. (1994). *Semilinear Elliptic Equations* (Gakkotosho, Tokyo).

[217] Sweers, G. (1992). Strong positivity in $C(\bar{\Omega})$ for elliptic systems, *Math. Z.* **209**, pp. 251–271.

[218] Takác, P. (1994). A short elementary proof of the Krein–Rutman theorem, *Houston J. of Maths.* **20**, pp. 93–98.

[219] Takác, P. (1996). An abstract form of maximum and anti-maximum principles of Hopf's type, *J. Math. Anal. Appns.* **201**, pp. 339–364.

[220] Triebel, H. (1978). *Interpolation Theory, Function Spaces, Differential Operators* (North-Holland, Amsterdam).

[221] Valentine, F. A. (1964). *Convex Sets* (McGraw-Hill, New York).

[222] Venturino, M. (1978). Primo autovalore di operatore lineari ellittici in forma non-variazionale, *Boll. Un. Math. It.* **5**, pp. 576–591.

[223] Walter, W. (1970). *Differential and Integral Inequalities*, Ergebnisse der Mathematik und ihrer Grenzgebiete Vol. 55 (Springer, Berlin).

[224] Walter, W. (1989). A theorem on elliptic differential inequalities and applications to gradient bounds, *Math. Z.* **200**, pp. 293–299.

[225] Weyl, H. (1940). The method of orthogonal projection in potential theory, *Duke Math. J.* **7**, pp. 414–444.

[226] Wloka, J. (1987). *Partial Differential Equations* (Cambridge University Press, Cambridge).

[227] Yosida, K. (1995). *Functional Analysis*, Classics in Mathematics (Springer,Berlin).

[228] Young, W. H. (1912). On classes of summable functions and their Fourier series, *Proc. Roy. Soc. London Ser. A* **87**, pp. 225–229.

Index

Adams D. R., 128, 152, 268, 319
Adams R. A., 126, 149, 152, 319
adjoint operator, 162
Agmon S., xi, 130, 137, 148, 220, 319
Alexandroff A. D., ix, 319
algebraic ascent, 163
algebraic multiplicity, 163
Alikakos N. D., 184, 319
Amann H., viii, 60, 128, 148–150,
  160, 184, 185, 188, 215, 221–224,
  316, 319
Aronszajn N., 152, 320
Arzela C., 102, 197, 198, 320
Ascoli G., 102, 197, 198, 320

Babuška I., 267, 268, 320
Bakelman I. J., 320
Banach S., x, 2, 64, 78–80, 89, 91, 94,
  96–98, 134, 141, 147, 148, 152, 155,
  157, 159–162, 164, 165, 184, 185,
  195, 198–200, 224, 255, 319–322,
  324, 325, 329
Banach space, 64
Barta J., 224, 320
Beauzamy B., 320
Beltramo A., 267, 320
Berens H., 149, 321
Berestycki H., viii, ix, 220, 221, 223,
  267, 268, 271, 320
Bessel
  generalized kernel, 133
  kernel, 132

potential space, 134
Bessel F. W., 132–134, 152, 320
bilinear form
  coercive, 75
  continuous, 75
Bombal F., 90
Bony J. M., viii, xi, 60, 61, 188, 189,
  221, 320, 326
  minimum principle, 189
boundary lemma
  of E. Hopf, 20
  weak of E. Hopf, 194
Brézis H., x, 80, 81, 89, 128, 152, 159,
  163, 202, 254, 255, 320
Browder F. E., 320
Brown K. J., 311, 312, 320
Butzer P. L., 149, 321

Calderón A. P., xi, 127, 130–134, 136,
  138, 321
Cano-Casanova S., xii, 267–271, 313,
  314, 321
Cantor G., 169
Cantrell R. S., 316, 321
Cauchy A. L., 63, 67, 117, 197, 232,
  239, 243, 256, 262
Cauchy–Schwarz inequality, 63
characteristic function, 58
characterization of the strong
  maximum principle, 216
Chernoff P. R., 329
Chipot M., 326